Equilibrium Structure and Properties of Surfaces and Interfaces

NATO ASI Series

Advanced Science Institutes Series

*A series presenting the results of activities sponsored by the NATO Science Committee,
which aims at the dissemination of advanced scientific and technological knowledge,
with a view to strengthening links between scientific communities.*

The series is published by an international board of publishers in conjunction with the
NATO Scientific Affairs Division

A	Life Sciences	Plenum Publishing Corporation
B	Physics	New York and London
C	Mathematical and Physical Sciences	Kluwer Academic Publishers
D	Behavioral and Social Sciences	Dordrecht, Boston, and London
E	Applied Sciences	
F	Computer and Systems Sciences	Springer-Verlag
G	Ecological Sciences	Berlin, Heidelberg, New York, London,
H	Cell Biology	Paris, Tokyo, Hong Kong, and Barcelona
I	Global Environmental Change	

Recent Volumes in this Series

Series B: Physics

Equilibrium Structure and Properties of Surfaces and Interfaces

Edited by

A. Gonis

Lawrence Livermore National Laboratory
Livermore, California

and

G. M. Stocks

Oak Ridge National Laboratory
Oak Ridge, Tennessee

Springer Science+Business Media, LLC

Proceedings of a NATO Advanced Study Institute on
Surfaces and Interfaces,
held August 18–30, 1991,
in Porto Carras, Greece

NATO-PCO-DATA BASE

The electronic index to the NATO ASI Series provides full bibliographical references (with keywords
and/or abstracts) to more than 30,000 contributions from international scientists published in all
sections of the NATO ASI Series. Access to the NATO-PCO-DATA BASE is possible in two ways:

—via online FILE 128 (NATO-PCO-DATA BASE) hosted by ESRIN, Via Galileo Galilei, I-00044
Frascati, Italy

—via CD-ROM "NATO-PCO-DATA BASE" with user-friendly retrieval software in English, French,
and German (©WTV GmbH and DATAWARE Technologies, Inc. 1989)

The CD-ROM can be ordered through any member of the Board of Publishers or through NATO-PCO,
Overijse, Belgium.

Library of Congress Cataloging-in-Publication Data

Equilibrium structure and properties of surfaces and interfaces /
   edited by A. Gonis and G.M. Stocks.
       p.   cm. -- (NATO ASI series. Series B, Physics ; v. 300)
    Proceedings of a NATO Advanced Study Institute on Surfaces and
 Interfaces, held August 18-30, 1991, in Porto Carras, Greece.
    "Published in cooperation with NATO Scientific Affairs Division."
    Includes bibliographical references and index.
    ISBN 978-1-4613-6499-3   ISBN 978-1-4615-3394-8 (eBook)
    DOI 10.1007/978-1-4615-3394-8
    1. Surfaces (Physics)--Congresses. 2. Electronic structure-
 -Congresses. 3. Surface chemistry--Congresses.   I. Gonis,
 Antonios, 1945-   . II. Stocks, G. M., 1943-   . III. Series.
 QC173.4.S94E68   1992
 530.4'27--dc20   92-26764
                                                          CIP

Additional material to this book can be downloaded from http://extra.springer.com.

ISBN 978-1-4613-6499-3

©1992 Springer Science+Business Media New York
Softcover reprint of the hardcover 1st edition 1992
Originally published by Plenum Press, New York in 1992

This project was sponsored by the

North Atlantic treaty Organization (NATO)
through the Scientific Affairs Division

and was cosponsored by the

U. S. Department of Energy through Lawrence Livermore
National Laboratory
Oak Ridge National Laboratory
University of Kentucky Center for Computational Sciences

PREFACE

It is almost self-evident that surface and interface science, coupled with the electronic structure of bulk materials, play a fundamental role in the understanding of materials properties. If one is to have any hope of understanding such properties as catalysis, microelectronic devices and contacts, wear, lubrication, resistance to corrosion, ductility, creep, intragranular fracture, toughness and strength of steels, adhesion of protective oxide scales, and the mechanical properties of ceramics, one must address a rather complex problem involving a number of fundamental parameters: the atomic and electronic structure, the energy and chemistry of surface and interface regions, diffusion along and across interfaces, and the response of an interface to stress. The intense need to gain an understanding of the properties of surfaces and interfaces is amply attested to by the large number of conferences and workshops held on surface and interface science. Because of this need, the fields of surface and interface science have been established in their own right, although their development presently lags behind that of general materials science associated with bulk, translationally invariant systems.

There are good reasons to expect this situation to change rather dramatically in the next few years. Existing techniques for investigating surfaces and interfaces have reached maturity and are increasingly being applied to systems of practical relevance. New techniques are still being created, which drastically widen the scope of applicability of surface and interface studies.

On the experimental side, new microscopies are bearing fruit. They include, with atomic-scale resolution: scanning tunneling microscopy and its relatives (such as ballistic electron emission microscopy, which can detect buried interfaces), and high-resolution electron microscopy; on the scale of hundreds of Angstroms to microns: high-spatial-resolution secondary ion mass spectroscopy, analytical electron microscopy, low-energy electron microscopy, and photoemission microscopy using synchrotron radiation. These techniques are now providing structural and electronic information about surfaces and interfaces at several complementary length scales which were previously inaccessible.

On the theoretical side, new approaches are generalizing established methods.

Phenomenological theories for total-energy calculations, such as the effective medium theory (EMT, which is based on first principles), the embedded atom method (EAM) and the glue model have succeeded in solving very complex structural problems on the atomic scale, including large surface reconstruction cells, alloy segregation at interfaces and grain-boundary structures. More sophisticated theories have been introduced for the calculation of surface and interface electronic properties, in addition to existing methods based on atomic clusters or on two-dimensionally periodic structures. An example is the real-space multiple-scattering theory (RS-MST), which relaxes the requirements of two-dimensional periodicity without imposing a finite-size cluster geometry.

In 1978, a NATO ASI on Electrons in Metals and at Metallic Surfaces took place in which course some of the issues pertaining to the structure and properties of bulk materials, ordered or disordered, and of surfaces were addressed. As is indicated in the previous paragraphs much has taken place in the ensuing decade. Given the state of affairs described above it was judged both timely and of great importance to convene an international conference, of primarily tutorial nature, as an attempt to unify a number of seemingly diverse aspects of the study of bulk, surface and interface properties. This study should be viewed as a natural continuation of the NATO ASI just mentioned, although the emphasis would now be placed primarily on the surface and interface aspects of metallic systems. At the same time, semiconducting materials were also represented in order to bring about the unifying aspects of this important field of inquiry. In particular, we aimed at a cohesive exposition of:

1. Experimental determination of surface and interface structure and properties. It is in this part that experimental techniques such as LEED and scanning tunneling microscopy along with the underlying physics are presented.

2. The determination of structural and electronic properties of surfaces and interfaces within both semi-phenomenological and first-principles theoretical techniques.

Every attempt was made to assure continuity of presentations between theory and experiment, with ample opportunity for questions and subsequent discussions. It is hoped that the ASI and these proceedings will lead to a serious dialogue between groups of experimental scientists and metallurgists on one hand and theoretical physicists on the other which will hopefully continue long into the future.

In recent years, the meetings which have been held in the general area of surfaces and interfaces have been addressed primarily to senior scientists and have not included a tutorial aspect. One of the unique advantages of a NATO ASI in this area is its prescribed tutorial character which invites participation by younger scientists such as post-doctoral fellows and qualified nonspecialists. In addition, it affords senior scientists the opportunity of presenting material at a level at which differences in concepts and language can be bridged most effectively. Furthermore, this particular ASI would serve to bring about a unified view of the physics and methods of study pertaining to bulk materials and to surfaces and interfaces.

In order to ensure the smooth running and the overall success of the school we have solicited the aid of four scientists with strong international reputation to act in the role of an organizing committee. These scientists are: Dr. Michel Van Hove (US/Belgium), Prof. R. W. Balluffi, (US), Dr. Paul Durham (UK), and Prof. J. F. Van der Veen (Netherlands). They, as well as the lecturers, accepted their respective tasks enthusiastically.

Scientists of eminent standing in their respective fields served as lecturers. Each answered questions from the audience at the end of his/her presentation. A tentative list of invited lecturers and of participants is provided in the last pages of this volume.

As directors of the ASI and as editors of the proceedings, we take this opportunity to thank all those whose efforts contributed to the success of this endeavor. We greatly appreciate the work of the invited speakers to produce lecture materials that were of tutorial character, and of being present during the ASI to answer questions and hold discussions with ASI participants. We also thank the participants for their attention and attendance. Much is owed to the organizational capabilities of Mania Bessieri and her staff of the Congress Center ``Organization Idea" in Athens. They arranged every detail, from meeting arriving participants at the airport, transporting them to the hotel 140 km away, duplicating and distributing material during the conference, and assuring a smooth departure of the participants at the conclusion of the ASI. In addition to funds obtained from NATO, we greatfully aknowledge financial assistance from the US Department of Energy through Lawrence Livermore National Laboratory and Oak Ridge National Laboratory, as well as the US National Science Foundation through the University of Kentucky Center for Computational Sciences. Finally, the staff at Plenum did an excellent job in bringing forth the final product. We hope that these efforts have been well spent .

A. Gonis

G. M. Stocks

May 1992

CONTENTS

INVITED PAPERS

CONTRIBUTED PAPERS

X-RAY SCATTERING FROM SURFACES AND INTERFACES

R.A. Cowley

Oxford Physics
Clarendon Laboratory
Parks Road, Oxford

I.   Introduction

The past 10 years has seen the development of X-ray scattering techniques for the study of the structure of surfaces. Initially this is a surprising development because X-ray scattering has been a well established technique for obtaining bulk crystal structures for many years, while low energy electron scattering techniques, LEED and RHEED, have been very successfully used to give surface structures.   The reason for the development has been three-fold.   Firstly low energy electrons or helium atoms are very strongly scattered by condensed matter, and so the results can only be interpreted in detail if full multiple scattering calculations are also performed. This makes the determination of structures difficult.   In contrast X-rays are only weakly scattered by condensed matter and so the results can then mostly be interpreted in terms of the kinematical theory of X-ray diffraction.   This leads to easier and less unambiguous interpretation of the experimental results.

The second reason for the developments has been improvements in the X-ray scattering techniques.   One obvious improvement is the use of beams from high intensity synchrotron sources which enable high resolution measurements of the scattering to be made.   As important, however, has been the development of suitable diffractometers with sufficient versatility and precision to study surface scattering.   The third reason for the development is that LEED and helium atom scattering are only sensitive to the top one or two atomic layers, and can only be used in high vacuum environments.   The greater penetrability of X-rays enables them to be used in gaseous and even liquid environments, and also to be used to study buried interfaces such as the boundary between two semiconductors or ferroelectric domain walls.

The theory of the X-ray scattering will be described in the next section.   Initially the kinematical theory of the scattering from a simple interface is described.   The simple model is then extended to show the effects on the scattering of a surface overlayer and of the consequential

Equilibrium Structure and Properties of Surfaces and Interfaces
Edited by A. Gonis and G.M. Stocks, Plenum Press, New York, 1992

1

distortion of the top layer of the original material. The model provides a simple way of understanding the type of behaviour which can occur at surfaces. The theory is then further developed to take account of situations in which multiple X-ray scattering is important. This will enable the relationship of the kinematical theory with both the dynamical theory of X-ray diffraction and the theory of X-ray reflectivity to be described.

One of the features of surfaces is that they are only rarely flat. Surface roughness influences the scattering from surfaces, and this in particular gives rise to the possibility of studying the roughness and particularly the roughening transition. In section III, the theory of the scattering from rough, smooth and flat surfaces is described. Section IV describes the experimental techniques used to perform surface scattering measurements, and in section V these techniques and theories are illustrated by some experimental results.

II. Theory of Scattering from Interfaces

1. Kinematical Theory

Initially consider the scattering by a crystal as shown in fig. 1a,

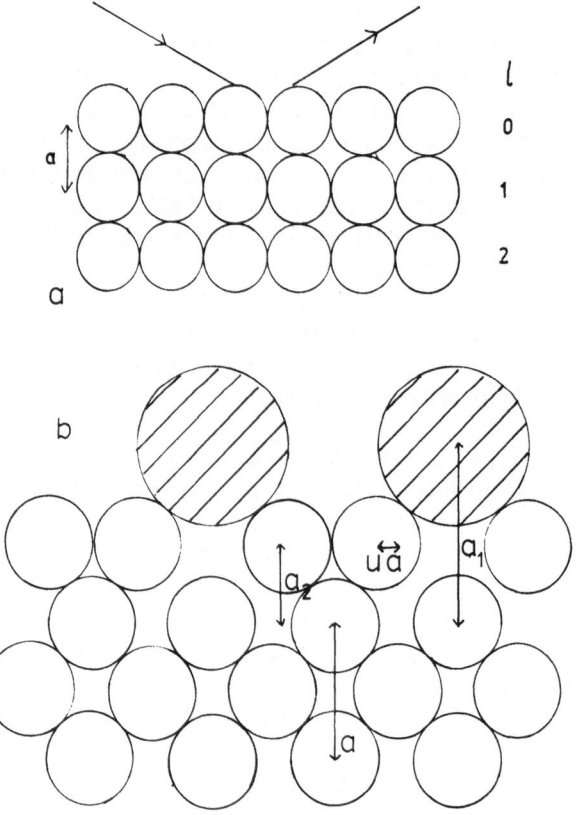

Fig. 1. Part (a) shows a simple square lattice and (b) shows a more complex model with an absorbed layer and consequent distortion of the first layer of the bulk.

where the incident beam with wavevector, k, is incident at an angle θ_1 to the surface and the scattered beam at an angle θ_2. Throughout we shall assume that the scattering is performed in the extended crystal geometry so that the crystal is sufficiently large that it intercepts all the X-ray beam. The changes for other geometries in the formalism are straightforward. If the scattering amplitude from each unit cell of the surface is b the scattered amplitude from the whole crystal shown in fig. 1a is given by

$$A = \sum_{m_1,m_2 = -\infty}^{\infty} \sum_{\ell=1}^{\infty} b \exp(i\, Q_z\, a\, \ell - \mu \ell + i\, Q_x\, a_1\, m_1 + i\, Q_y\, a_2\, m_2) \qquad (2.1)$$

The scattering amplitude b is given by $e^2/mc^2\, F(Q)\, \underline{n}_1.\underline{n}_2$, where $F(Q)$ is the normal structure factor for the scattering and $\underline{n}_1$ and $\underline{n}_2$ are polarisation vectors of the incident and scattered radiation. It therefore depends both on the wavevector transfer, Q, and on the polarisation.

This is then the normal expression for the X-ray scattering except that the attenuation of the beam on passing through each layer is reduced by μ and the summation over ℓ runs from 0 to ∞ instead of $-\infty$ to ∞. The scattering cross-section can then be evaluated as the modulus squared of the amplitude to give

$$S(Q) = 4\pi^2 \left|\frac{b}{a_{xy}}\right|^2 \frac{A_0}{\sin\theta_1} \frac{1}{1 + e^{-2\mu} - 2e^{-\mu}\cos(Q_z a)}\, \Delta(Q_x)\, \Delta(Q_y)\ ,$$

where $\Delta(Q_x)\, \Delta(Q_y)$ is zero unless Q_x and Q_y are reciprocal lattice vectors, τ_x, τ_y, and the integral with respect to Q_x over one reciprocal lattice vector is unity. a_{xy} is the area of the unit cell in the plane, and A_0 the cross-sectional area of the incident beam so that $A_0/\sin\theta_1$, is the illuminated surface area. Usually $\mu \ll 1$ so that

$$S(Q) = 4\pi^2 \left|\frac{b}{a_{xy}}\right|^2 \frac{A_0}{\sin\theta_1} \frac{1}{\mu^2 + 4\sin^2(Q_z a/2)}\, \Delta(Q_x)\, \Delta(Q_y) \qquad (2.2)$$

The cross-section $S(Q) = d\sigma/d\Omega$ is in an experiment often integrated over the detector solid angle. This integral can be performed over the delta functions if the angle subtended by the detector is α_1 in the scattering plane and α_2 perpendicular to that plane, so that $dQ_x = \alpha_1 k \sin\theta_2$ and $dQ_y = \alpha_2 k$. The result is $(k^2 \sin\theta_2)^{-1}$ so that the observed intensity for unit incident intensity becomes

$$I(\tau_x , \tau_y , Q_z) = 4\pi^2 \left|\frac{b}{a_{xy}}\right|^2 \frac{1}{k^2 \sin\theta_1 \sin\theta_2\, (\mu^2 + 4\sin^2(Q_z a/2))} \qquad (2.3)$$

Eqns (2.2) and (2.3) are crucial to the further development because they give rise to scattering streaks whenever Q_x and Q_y are equal to a reciprocal lattice vector, as shown in fig. 2a, and they become large whenever $Q_z = 2\pi L/a = \tau_z$, a reciprocal lattice vector of the bulk material. Writing $Q_z =$

$\tau_z + q$, eqn (2.2) becomes

$$S(q) = 4\pi^2 \left| \frac{b}{v} \right|^2 \frac{1}{\sin\theta_1 \, (q^2 + (\mu/a)^2)} \, \Delta(Q_x) \, \Delta(Q_y) \qquad (2.4)$$

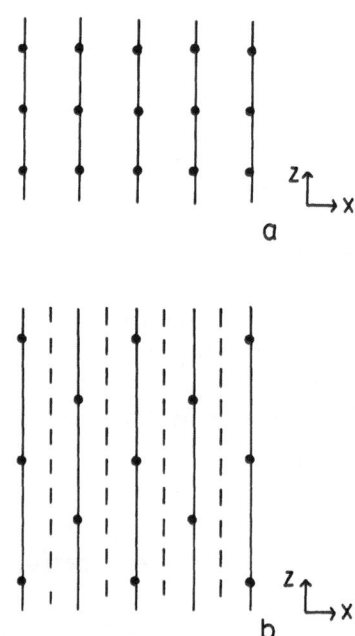

Fig. 2.   Part (a) shows the reciprocal lattice for fig. 1a and the surface streaks are shown by the solid lines.   Part (b) is a similar figure for the surface of fig. 1b showing the bulk streaks and bulk lattice points and the streaks, dotted, arising from the absorbed layer with double periodicity.

where v is the volume of the unit cell.   Integrating this expression over q the intensity becomes proportional to $(q/\mu)(A_0/\sin\theta_1)$, which is the surface area illuminated times the penetration depth and hence the illuminated volume of the crystal.   It is a bulk effect.   In contrast, when $qa > \mu$, the intensity is proportional to $A_0/\sin\theta_1$, and so is proportional to the surface area.   It is therefore by measuring this region of the streaks that we can determine the properties of the surface.   This scattering is usually omitted in the conventional derivation of the scattering because the sum over ℓ in eqn (2.1) is taken from $-\infty$ to ∞.

2.   A Model Surface

A more realistic model of the surface of a crystal is shown in fig. 1b. The square structure of fig. 1a is rotated to give a (110) surface, and on top of the bulk is placed a half-filled monolayer which can simulate either a reconstructed surface or an absorbed monolayer of a large atomic species. The scattering amplitude of the atoms in this layer are taken as 2bB, and the

spacing is a_1 as shown in fig. 1b.   As a result of this overlayer a
reconstruction of the top layer of the bulk is assumed to be distorted by a
lateral displacement of ua, and by a vertical displacement from the next
layer of a_2, fig. 1b.   It is then straightforward to calculate the
cross-section by summing over the successive layers to give in the (x,z)
plane, the result:

$$S(\underline{Q}) = 4\pi^2 \left| \frac{b}{a_{xy}} \right|^2 \frac{A_o}{\sin\theta_1} \; \varDelta(Q_y) \; T(Q_x, Q_z) \; , \qquad\qquad (2.5)$$

where

$$T(Q_x, Q_z) = \left[\frac{1 + C \cos(Q_z a/2) \; \varDelta(Q_x)}{2 \sin^2 (Q_z a/2)} + \left\{ C'(\sin Q_z(a_2 - a/2) \right. \right.$$

$$\left. - C \sin(Q_z a_2)) + B(\sin Q_z(a_1 - a/2) - C \sin(Q_z \; a_1)) \right\}$$

$$\left. \varDelta(Q_x)/\sin(Q_z \; a/2) \right] + (B^2 + (C')^2 + 2B \; C' \cos(Q_z \; (a_2 - a_1))) \; \varDelta(2Q_x) \; (2.6)$$

where $C = \cos(Q_x a/2)$, $C' = \cos Q_x(a/2 + au)$, and the attenuation $\mu = 0$.   The
first term in the expression (2.6) gives the scattering from the undistorted
layers as shown by the solid lines in fig. 2b.   The second and third terms
are the interference terms resulting from the coupling between the
scattering from the bulk and the two distorted surface layers.   They modify
the intensity associated with the surface streaks as shown in fig. 2b.   The
fourth term in eqn (2.3) arises from the two distorted layers.   It gives
rise to streaks between the streaks arising from the bulk because of the
larger, doubled, periodicity of the surface layers.   Whereas the surface
streaks associated with the bulk behave as $1/q^2$ close to the bulk Bragg
reflections, the streaks arising solely from the surface are only slowly
varying in intensity with wavevector.

In order to illustrate these results we have calculated the intensity
$T(Q_x, Q_z)$ for $Q_x = 2\pi/a$ H with H = 0, $^1/_2$, 1, $^3/_2$, and 2 as a function of Q_z and
the results are shown in fig. 3.   The first calculation was of the
scattering from the bulk material; that is with B = 0, $C' = C$ and $a_2 = a/2$.
The result is symmetrical peaks at the Bragg reflections of the bulk
materials (H + L even).   The second model was to add a half layer of
identical atoms on the top, B = $^1/_2$, a_1 = a.   There is then little difference
in the intensity close to the Bragg reflections but in between the Bragg
reflections there is a considerable change in intensity when H is an integer.
Additional streaks also occur for half-integer values of H corresponding to
the new periodicity and the intensity of these is independent of Q_z.   The
next model was to assume the surface layer was a larger atom and hence a
stronger scatterer so that B = 1, and a_1 = 1.2a.   There is no difference in
the streaks with H a half integer but for integral H there is a substantial
difference especially in that there is now a marked asymmetry in the
scattering about each Bragg reflection, and the asymmetry differs for
different Bragg reflections except that the calculated scattering is the
same for H = 0, 2, 4 and the same but different for H = 1, 3, 5.   The origin of
this asymmetry is the interference term in the intensity.   If a_1 = a then the
third term in eqn (2.3) becomes B(1 - 2C $\cos(Q_z a/2)$) which is not singular at

the Bragg reflections. Consequently a marked asymmetry about the Bragg reflections results directly from a different spacing of the atomic layers at the surface. The final model was to allow the top layer of the bulk to relax by choosing $a_2 = 0.6a$ and $u = 0.1$. The scattering is altered and in particular is no longer the same when $H = 0, 2, 4$ and also now varies in intensity when $H = \frac{1}{2}$ and $\frac{3}{2}$.

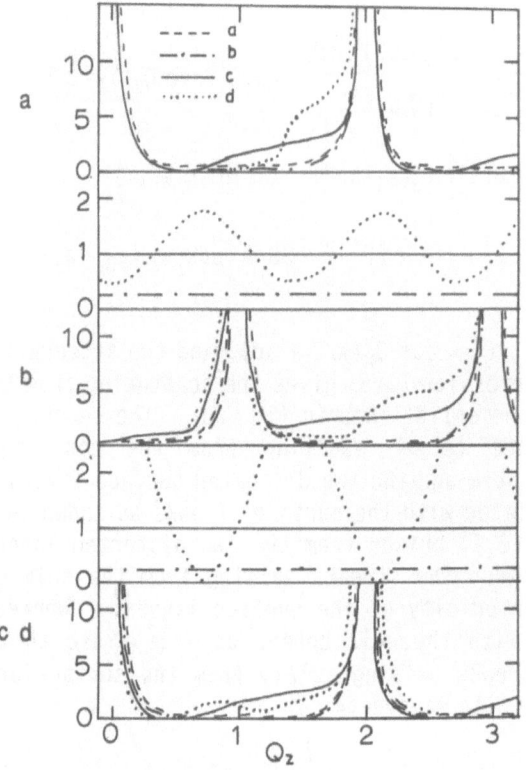

Fig. 3. The intensity of the scattering along the surface streaks for model
(a) the bulk ending without distortion
(b) the bulk with a half-filled top layer at the same atoms as the bulk
(c) the bulk with a half-filled top layer with twice the scattering power of the bulk and $a_1 = 1.3a$
(d) the same as (c) but with the top layer of the bulk relaxed so that $a_2 = 0.6a$ and $u = 0.1a$.

This model then shows how the intensities of the surface streaks depend on the surface structure. The position of new surface streaks gives the periodicity of the surface structure, and the intensity variation of both the bulk and new surface streaks gives information about the surface in-plane and out-of-plane structure. In practice the analysis of the data is more complex than described here. Firstly the scattering amplitudes, b, depend on the form factor and Debye-Waller factors of the atoms. The latter may be different from the bulk. Secondly we have assumed a flat surface and the results can be strongly influenced by surface roughness as described in section III.

3.   Dynamical Theory of X-Ray Scattering

The dynamical theory of X-ray scattering was developed to take account of the multiple scattering of the X-ray beam in perfect crystals. It was developed by Ewald and Darwin and is described in the texts by Zachariasen [1] and by James [2]. The theory is illustrated in the Appendix for a symmetric Bragg reflection as shown in fig. 1a. The result of the theory for extended face geometry of an infinitely thick crystal with negligible attenuation is that there are streaks perpendicular to the crystal surface with maxima around the Bragg reflections as discussed above. The intensity along the streaks is given in terms of the wavevector difference, q, from the Bragg reflections of the kinematic theory as

$$I(q) = F(\frac{q - \delta}{q_c}) \, , \tag{2.7}$$

where

$$F(y) = (y - (y^2 - 1)^{1/2})^2 \qquad |y| \geq 1 \tag{2.8}$$

$$= 1 \qquad |y| < 1$$

and the half-width q_c is given by

$$q_c = \frac{4\pi}{k} \left| \frac{b}{v} \right| (\sin \theta_1 \sin \theta_2)^{-1/2} \, , \tag{2.9}$$

while the displacement of the centre from the centre of the dynamical scattering is given by

$$\delta = \frac{2\pi}{k} \left| \frac{b_0}{v} \right| ((\sin \theta_1)^{-1} + (\sin \theta_2)^{-1}) \, , \tag{2.10}$$

where b_0 is the amplitude, b, evaluated for $Q = 0$. The cross-section can be obtained from these results as

$$S(\underline{Q}) = A_o k^2 \sin \theta_2 F (\frac{q - \delta}{q_c}) \Delta (Q_x) \Delta (Q_y) \, , \tag{2.11}$$

where $Q_z = q + \tau_z$.

For $y > 1$, $F(y) = 1/4y^2$ and hence eqn (2.11) becomes

$$S(\underline{Q}) = A_o k^2 \sin \theta_2 \frac{q_c^2}{4q^2} \Delta (Q_x) \Delta (Q_y) \, .$$

Substituting in the expression (2.9) for q_c this becomes equivalent to eqn (2.4) showing that the $1/q^2$ tails in the dynamical theory are nothing but the kinematical scattering from the surfaces. Ewald was the first person to evaluate X-ray scattering allowing for the surfaces. The result also enables us to understand the difference between the expressions obtained by Ewald and Darwin. Both considered a slab of thickness t and absorption μ but Darwin first took the limit $t \rightarrow \infty$ and then $\mu = 0$. The X-rays were then scattered by only one surface. In contrast Ewald reversed the order of the limits and hence the X-rays were scattered by both surfaces. It is therefore not surprising that his expression for $F(y)$ is, for large y, $1/2y^2$ twice as large as Darwin's result.

A final comment is of order for the symmetrical reflections when $\theta_1 = \theta_2$. For these reflections, q_c can be written as

$$q_c = \frac{8\pi}{\tau} \left| \frac{b}{v} \right| \tag{2.12}$$

which is independent of wavelength. The Darwin width of the Bragg reflection in wavevector space is therefore surprisingly simple, and the displacement of the Bragg reflection in wavevector, δ, is similarly given by

$$\delta = \frac{8\pi}{\tau} \left| \frac{b_o}{v} \right|$$

4. Reflectivity Measurements

Glancing angle reflectivity can be described by the scattering theory developed above. It corresponds to the scattering around the special Bragg reflection at the origin of reciprocal space $\tau_x = \tau_y = \tau_z = 0$. This Bragg reflection is special in that it occurs for all materials whether or not crystalline, and because for X-rays it is usually studied under conditions that θ is small so that $Q_z = q = k(\sin \theta_1 + \sin \theta_2) = k(\theta_1 + \theta_2) = 2k\theta$. The differential scattering cross-section can then be obtained from eqn (2.4) as

$$S(\underline{Q}) = 8\pi^2 \left| \frac{b}{v} \right|^2 \frac{A_o K}{q^3} \Delta(Q_x) \Delta(Q_y) , \tag{2.13}$$

where we have substituted for $\sin \theta$, by q. The reflectivity is then given by integrating over the delta functions in Q_x and Q_y with the detector to give

$$R(q) = \frac{16\pi^2}{q^4} \left| \frac{b}{v} \right|^2 \tag{2.14}$$

which can be written in the form

$$R(q) = q_R^4 / (2q)^4 \tag{2.15}$$

with

$$q_R^2 = 16\pi \left| \frac{b}{v} \right| \tag{2.16}$$

This shows that the reflectivity decreases as q^{-4} unlike the surface scattering near Bragg reflections. The additional factors of q^{-2} arise firstly from the $\sin \theta_1$ from the increasing illuminated area at small θ, and

8

secondly from the $\sin \theta_2$ arising from the integral over the delta Q_x by the solid angle of the detector.

For small wavevectors, q, the result is inadequate because multiple scattering, dynamical effects, are important. We can however use the result of section II.3 with $\theta_1 = \theta_2 = \theta = q/2k$ to give eqns (2.9) and (2.10) as

$$q_c = q_R^2/2q \quad \text{and} \quad \delta = q_R^2/2q \ , \quad \text{when} \quad y = 2\frac{q^2}{q_R^2} - 1 \qquad (2.17)$$

Substituting this expression into eqn (2.1) we obtain

$$R(q) = (2\frac{q^2}{q_R^2} - 1 - 2\frac{q}{q_R}\left(\frac{q^2}{q_R^2} - 1\right)^{1/2})^2 \qquad q \geq q_R \qquad (2.18)$$
$$= 1 \qquad\qquad\qquad q < q_R$$

which is the Fresnel formula [3] for the reflectivity of electromagnetic waves in the limit that the refractive index is slightly less than unity, and that the angles are small. The critical angle θ_c in the Fresnel treatment is given by $q_R = 2k\theta_c$ showing that q_R of eqn (2.16) is the critical reflectivity wavevector.

This derivation shows that it is necessary to be careful whenever either the incident or scattered X-ray beam is at small angles, θ_1 or θ_2 small, because then q_R and δ vary with the wavevector transfer, and this variation can lead to different results as shown by the reflectivity. It is, however, worthwhile briefly discussing the reason for these differences. When the X-ray beam enters the crystal it is refracted so that the angle of propagation in the crystal differs from that outside. This alters the effective momentum transfer, and is the reason for the δ displacement of the Bragg reflections in eqn (2.7). This difference in angle becomes particularly large if θ_1 or θ_2 are small. The second effect of the medium is that the magnitude of the electric field in the crystal is different from that outside, and the difference is given by the Fresnel theory of reflectivity by the amplitude transmission coefficient as

$$T(\theta_1) = \frac{2\theta_1}{\theta_1 + (\theta_1^2 - \theta_c^2)^{1/2}} \ ,$$

while a similar factor connects the amplitude of the scattered wave in and outside the crystal. The scattering theory described in section II.1 can then be corrected for these effects if $S(Q)$ is multiplied by $T(\theta_1)^2 T(\theta_2)^2$. This change in the amplitude with angle is directly accounted for by the dynamical theory and is responsible for the different shape of the reflectivity with q and the Darwin profile. It is worth noting that these factors are not large $T(\theta)$ varies from 1 at large θ to 2 when $\theta = \theta_c$; they are therefore a much smaller surface enhancement effect than the $1/\sin\theta$ factors arising in the intensity from geometrical effects.

9

Finally these enhancement factors cannot be used naively when the angles θ_1 or θ_2 are less than the critical angle. In these conditions the wave in the crystal is a decaying exponential and the development described above is inadequate.

III. Surface Roughening

1. Introduction

In section II.1 the scattering from a flat surface, fig. 1a, was given by eqns (2.1) and (2.2). The summation over the surface m_1 and m_2 in eqn (2.1) led to the delta functions in Q_x and Q_y in eqn (2.2). In reality surfaces are not flat, but the height in the z direction varies with $\underline{R}$ = (x,y) = $(a_1 m_1, a_2 m_2)$ and is given by $h(\underline{R})$. The scattering from such a surface is then given by

$$S(\underline{Q}) = \frac{|b|^2}{\mu^2 + 4 \sin^2(Q_z a/2)} \; F(\underline{Q}) \; , \tag{3.1}$$

where

$$F(\underline{Q}) = \sum_{1,2} \exp\left(i \, Q_z \, (h(\underline{R}_1) - h(\underline{R}_2)) \right) \exp(i \, \underline{Q}_\| \cdot \underline{r}) \tag{3.2}$$

and $\underline{r} = \underline{R}_1 - \underline{R}_2$ and $\underline{Q}_\| = (Q_x, Q_y)$. The expression $F(\underline{Q})$ replaces the factor

$$4\pi^2 \; \frac{A_o}{(a_{xy})^2 \sin\theta_1} \; \varDelta(Q_x) \; \varDelta(Q_y) \; ,$$

and describes the effect of the surface roughness. In this section we shall describe the scattering for various different models of surface roughness but initially remark on the qualitative results. The scattering is given by an ensemble average over the function $F(\underline{Q})$, and the properties are dependent on the behaviour of the heights for very different positions, $\underline{r} \to \infty$. If in these conditions the difference in the heights is bounded the surface is said to be smooth and the scattering has a sharp delta function component. This is similar to the result that a three dimensional crystal can exist provided that the atoms vibrate a finite distance about equilibrium positions when sharp Bragg reflections then occur. As with the thermal motion in three dimensional crystals the surface roughness introduces a Debye-Waller like correction to the intensity of the surface Bragg reflections or streaks. In contrast if the difference in the heights diverge as the distance between the positions increases, then there is no delta function surface scattering, and the scattering is more analogous to that arising from an amorphous or liquid three dimensional material. In this section we shall discuss various models for surface roughness following the work of Berry [4], Andrews and Cowley [5], Wong and Bray [6], Sinha et al [7] and Robinson et al [8].

2. Simple Smooth Surfaces

The evaluation of the scattering requires a knowledge of the ensemble average of the heights, $h(\underline{R})$. The simplest assumption is to assume that the heights at $\underline{R}_1$ and $\underline{R}_2$ in eqn (3.2) are uncorrelated when

$$F(\underline{Q}) = \frac{4\pi^2 A_o}{a_{xy}{}^2 \sin\theta_1} \left| D(\underline{Q}) \right|^2 \varDelta(Q_x)\, \varDelta(Q_y) \qquad (3.3)$$

where

$$D(\underline{Q}) = \big< \exp\left(i\, Q_z\, h(\underline{R}) \right) \big>$$

and the $< \ldots >$ represent the ensemble average. The averaging can be performed if we assume that $h(\underline{R})$ is an integer number of atomic steps an, and that n is a Poisson random variable with variance σ^2. Under these conditionss the Baker-Hausdoff theorem gives,

$$D(\underline{Q}) = \big< \exp(i\, Q_z\, n\, a) \big> = \exp\!\left(-\sigma^2 \sin^2(Q_z a/2) \right) \qquad (3.4)$$

This then gives a correction to the expressions in the previous sections of the form

$$\exp\!\left(-2\, \sigma^2 \sin^2(Q_z a/2) \right)$$

which decreases the intensity of the rod scattering near the Bragg reflections by a factor $D \approx \exp(-\sigma^2 a^2 q^2/2)$, and hence is very similar to the behaviour of the normal Debye-Waller factor except that it depends on $\sin^2(Q_z a/2)$, and so is periodic in reciprocal space. There is also diffuse scattering, but since there is no correlation between the heights in the z direction this is independent of Q_x and Q_y.

It is worthwhile to compare this result with that of a diffuse surface. In section II.1 it was assumed that the interface was abrupt: layers for $\ell =$ 0, 1, 2 . . . were present, and there were no layers above. If the surface is diffuse the scattering amplitude can be written as

$$A = \frac{1}{2\pi} \int_{-\infty}^{\infty} \sum_{\ell=-\infty}^{\infty} b\, \delta\,(z - a\,\ell)\, \phi(z)\, \exp(i\, Q_z\, z)\, d\,z \qquad (3.5)$$

where in comparison with eqn (2.1) the attenuation is neglected and the dependence of the function on Q_x and Q_y is the same as that of eqn 2.1. Eqn (3.5) can be evaluated by writing

$$\sum_{\ell} \delta(z - a\,\ell) = \frac{2\pi}{a} \sum_{\tau_z} \exp(i\,\tau_z\,z),$$

where $\qquad \tau_z = \dfrac{2\pi}{a}\, L.$

Eqn (3.5) can then be evaluated by integrating by parts to give

$$A = \sum_{\tau_z} \frac{b}{i(Q_z + \tau_z) a} \quad \int_{-\infty}^{\infty} \frac{\partial \phi}{\partial z} \quad \exp i \ (\tau_z + Q_z) \ z \ dz \ ),$$

where the definite integral is zero because at $z = -\infty$, $\phi(z)$ is assumed to be zero, no material, and when $z = +\infty$ the attenuation has eliminated the contribution. If the final integral is written $\Phi(\tau_z + Q_z)$ the cross-section becomes

$$S(\underline{Q}) = 4\pi^2 \left| \frac{b}{a_{xy}} \right|^2 \frac{A_o}{\sin\theta_1} \sum_{\tau_z} \frac{\left| \Phi(\tau_z + Q_z) \right|^2}{a^2(\tau_z + Q_z)^2} \ \Delta(Q_x) \ \Delta(Q_y).$$

This shows that a diffuse surface modifies the scattering by the introduction of the form factor $\left| \Phi(\tau_z + Q_z) \right|^2$. If the interface is sharp $\phi(z) = \theta(z)$, all the Fourier coefficients are identical, and the summation over τ_z can be performed by using

$$\sum_{\tau_z} \frac{4}{a^2(Q_z + \tau_z)^2} = \left(\sin^2(aQ_z/2) \right)^{-1}$$

to give eqn 2.2 with $\mu = 0$. In practice $\phi(z)$ may be a slowly varying function, and so $\Phi(\tau_z + Q_z)$ will be non-zero only if $\tau_z + Q_z$ is small, which is the case if $Q_z = -\tau_z + q$. In this case

$$S(\underline{Q}) = 4\pi^2 \left| \frac{b}{a_{xy}} \right|^2 \frac{A_o}{\sin\theta_1} \frac{\left| \Phi(q) \right|^2}{a^2 q^2} \ \Delta(Q_x) \ \Delta(Q_y). \tag{3.6a}$$

One particular form of $\phi(z)$ is $\phi(z) = \frac{1}{2}(1 + \tanh{z/\lambda})$, where λ is the surface width. In this case

$$\Phi = \frac{\pi q \lambda}{2 \sinh(\pi q\lambda/2)} \tag{3.6b}$$

Clearly this type of modulation varies the surface scattering in a way that is difficult to distinguish from the surface roughness given by eqns 3.3 and 3.4.

3. A Correlated Smooth Surface

The model of the roughness of the surface employed in section 2 assumed that there was no correlation between the heights at $\underline{R}_1$ and $\underline{R}_2$ even if the points are very close. This is clearly unreasonable as we would expect the surface height to change only slowly. The probability of the height distribution can be written in terms of $\underline{r} = \underline{R}_1 - \underline{R}_2$ as an isotropic form first introduced by Berry [4],

$$p(h_2\, h_1,\, r) = \frac{1}{2\pi\, S^2(1 - C^2(r))^{1/2}}\, \exp\!\left(-\frac{(h^2_1 + h^2_2 - 2h_1 h_2 C(r)}{2S^2(1 - C(r)^2)}\right)$$

where $S = \langle h^2\rangle^{1/2}$ is the root mean square height of the roughness and the correlation function

$$C(r) = \frac{\langle h(o)\, h(r)\rangle}{\langle h^2(r)\rangle} = \exp(-r^2/d^2)$$

This probability function is determined by two length scales; S, which determines the asymptotic roughness and d which determines the extent in the plane over which the roughness is correlated. Evaluating the cross-section from eqn 3.1 is then performed, following Berry, to give that the cross-section has two components. A surface long range ordered component with $\Delta(Q_x)\, \Delta(Q_y)$ and the same functional form as eqns 3.1, 3.3 and 3.4 but with the exponential decay governed by S^2. The correlation length d does not enter because only the long range properties determine the long range ordered component. The diffuse component is more complex and detailed expressions are given by Berry [4] and by Andrews and Cowley [5]. The wavevector dependence close to a Bragg reflection is given by

$$S(\underline{P},\, q) = \frac{K}{q^4}\left[\, 1 - \exp(-S^2\, q^2)\,\right]^2 \exp\!\left(-p^2 d^2\,(1 - \exp(-S^2 q^2))/(4\, S^2 q^2)\right]$$

where

$$p = (Q_x - \tau_x,\, Q_y - \tau_y)$$

This diffuse scattering is peaked under the surface streak component, p = 0, like thermal diffuse scattering at a Bragg reflection, and the width increases with increasing q. When $q \ll {}^1/S$ then the width in p is 2/L, independent of q, but when $q \gg {}^1/S$ then the width is given by 2Sq/L. The intensity of the diffuse intensity and Bragg-like rod when integrated over p can be shown to be the total scattering from a flat surface.

Berry also introduced another simple model of a step surface with the probability function

$$p(h_1 h_2,\, r) = p_o(h_1)\, \delta(h_1 - h_2)\, C(r) + p_o(h_1)\, p_o(h_2)\,(1 - C(r))$$

with

$$p_o(h) = \left(\frac{1}{2\pi}\right)^{1/2} \frac{1}{S}\, \exp\left(-h^2/(2S^2)\right)$$

and $C(r) = \exp(-r/d)$.

This gives [4,5] the same Bragg-like surface rod as the other model, but the diffuse scattering has a different form

$$S_D(\underline{p}, q) = \frac{K}{d\,q^2} \left(\frac{1}{p^2 + 1/d^2}\right)^{3/2} \qquad \left(1 - \exp(-S^2 q^2)\right) .$$

This is similar in form to the previous result in that as a function of p it is peaked at $p = 0$, but is different in that the width is independent of q. Clearly there is much information in the scattering from smooth interfaces about the correlation in the heights at different distances if it proves possible to measure the diffuse scattering.

4.   Rough Surfaces

A rough surface is one in which the difference in the height of two points on the surface differs on average by

$$< \mid h(o) - h(r) \mid^2 > = A\,r^{2\gamma} \tag{3.7}$$

when eqn (3.2) becomes

$$F(\underline{Q}) = 2\pi \int r \exp(-\frac{A}{2}\,q^2\,r^{2\gamma})\,J_0\,(pr)\,dr ,$$

where we have assumed p and q are small compared with Bragg reflections, and $J_0(pr)$ is the zeroth order Bessel function.

In general this integral cannot be evaluated explicitly but if

$$\gamma = {}^{1}/_2 \qquad \text{then } F(\underline{Q}) = \frac{A\,\pi\,q^2}{\left(p^2 + \left(\frac{A}{2}\,q^2\right)^2\right)^{3/2}}$$

and if

$$\gamma = 1 \qquad \text{then } F(\underline{Q}) = \frac{2\pi}{A\,q_z^{\,2}}\,\exp\left(-\frac{p^2}{2Aq^2}\right)$$

Neither of these show the delta function streaks characteristic of flat surfaces but both give a diffuse peak centred on the position of the rod whose transverse width in p increases with increasing q, but the dependence on q depends on the power, γ.

There is one other important use of a rough surface when

$$< \mid h(o) - h(r) \mid^2 > = \mu \ln r \tag{3.8}$$

which occurs in the theory of surface roughening by Chui and Weeks [9], Den Nijs et al {10} and by Jose et al [11].   The different models then lead to different expressions for the parameter, μ.

Substituting eqn 3.8 into eqn 3.2 for F(Q) we obtain,

14

$$F(\underline{Q}) = 2\pi \frac{2^{1-\eta} \, \Gamma(1 - \eta/2)}{p^{2-\eta} \, \Gamma(\eta/2)} \qquad\qquad (3.9)$$

where the exponent $\eta = \mu q^2/2$. This result shows that there is no surface delta function, but that the scattering is singular as a function of p for p = 0, on the surface rods. The result, eqn 3.9, was obtained by evaluating the Fourier transform in eqn 3.2 as $r \to \infty$. In this limit eqn 3.8 does indeed diverge, but only very slowly. In practice the length, r, is limited to the sample or beam size, and so there may be a significant delta function like component in experiments, and it is difficult to distinguish this from rough smooth surfaces. A summary of the scattering from different types of surfaces is given in fig. 4.

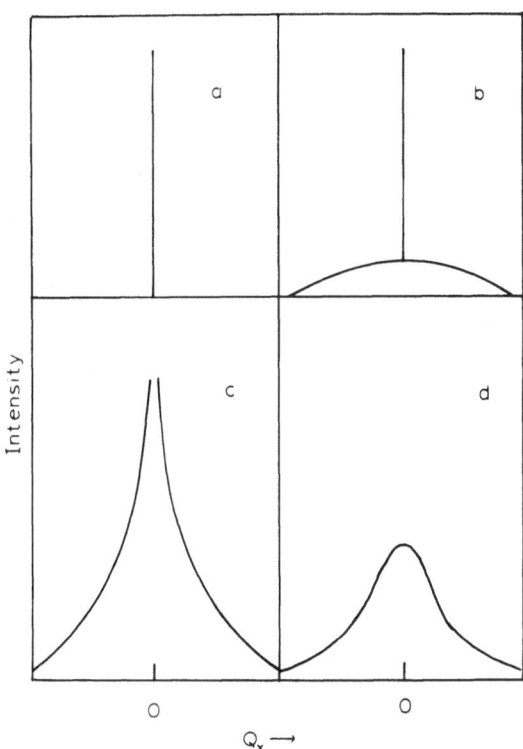

Fig. 4. Schematic scan through the surface streak for fixed q and varying Q_x for (a) a perfectly flat surface, (b) a smooth surface, (c) a surface at the roughening transition, (d) a rough surface.

IV. Experimental Considerations

1. Experimental Equipment
    X-ray studies of the structure of surfaces have been performed with a

few exceptions only over the past 10 years. In part this is because of the development of synchrotron sources of X-ray radiation which provide intense sources for high resolution experiments. It is also, however, because the conventional X-ray crystallographer developed techniques which make it impossible to observe the surface scattering. The crystals used are usually small, much smaller than the beam size, and are ground to give spheres with no well defined faces. The instrumentation usually has relaxed resolution to ensure that the integrated Bragg reflection intensity is correctly obtained. All of these features mitigate against the observation of surface scattering. Most surface measurements are in fact performed with large samples in an extended face geometry, and with high resolution diffractometers.

Most of the experiments have been performed with synchrotron sources which produce a range of X-ray wavelengths in a collimated beam. The range of wavelengths is determined by the energy of the synchrotron and the radius of the bending magnet or the properties of the insertion device. The monochromatic beam is then usually focussed by a mirror onto a monochromator crystal which produces a monochromatic beam at the sample. The X-rays scattered by the sample are then detected either by using a slit system to determine the collimation or by using an analysing crystal before the detector. The latter has the advantage of having higher resolution, ease of alignment and lower background but the disadvantage of lower count rate.

The most difficult part of the experiment arises because in surface science the surface must be prepared clean and then aligned at well defined angles to the X-ray beams. These angles must then be changed to measure the scattered intensity as a function of wavevector whilst preserving the clean surface. This requires the construction of a UHV chamber with thin X-ray windows in which the sample can be accurately aligned. These problems have now been solved and equipment has been constructed by a number of groups [12, 13].

An alternative class of experiments is the study of liquid interfaces. The difficulty with these is that the sample must be kept flat and free from vibration, and so the directions of the incident and scattered X-ray beams must be controlled in a more complicated way. Several groups have overcome these problems.

Some experiments have been performed using laboratory sources rather than synchrotron sources. Laboratory sources produce characteristic Cu or Mo radiation over a wide angular range. Monochromators are then used to collimate the beam, separate $K\alpha_1$ from $K\alpha_2$ and eliminate the white radiation. The advantage of laboratory sources is the possibility of close proximity to other experiments, and, for many, the obvious advantages of working in their own institution. The disadvantage is the lower flux means that the surface rods can often be observed only if $qa \leq 0.2$ whereas with a synchrotron source they can be observed across the whole zone. Clearly more information can be obtained in the latter case.

2. Experimental Resolution

The resolution of an X-ray diffractometer has been discussed in a number of recent papers [15, 16, 17], and it is not our intention to present a derivation of these results again. It is sufficient to comment that at a synchrotron source the resolution of the monochromator system is determined by the angular collimation between the source and the monochromator, and the Darwin width of the monochromator, while in the case of a laboratory source it is determined by the linewidth of the characteristic X-ray and the Darwin width. The resolution of the detector is determined by the slit system or the Darwin width of the analyser crystal. Since the Darwin width is typically ~ 5×10^{-5} and the wavelength of the X-rays is 1 Å, the resolution in wavevector space is around 5×10^{-4} Å^{-1}, and usually much larger ~ 10^{-2} Å^{-1} in the direction perpendicular to the scattering plane at the monochromator.

Surface measurements require good resolution because the surface scattering must be distinguished from the bulk thermal diffuse scattering. The latter usually varies only slowly in reciprocal space while the former lies in the by now familiar streaks. If the resolution in wavevector space is reduced by a factor of say 2, thermal diffuse scattering is reduced by 8, whereas surface streak scattering is reduced only by 2. Clearly surface scattering can more easily be separated from bulk thermal diffuse scattering if good resolution is used.

The other aspect where a knowledge of the resolution is crucial, is for studies close to the roughening transition where the surface delta function scattering changes to a diffuse peak as described in III.4. Before the sharp line disappears there is strong diffuse scattering which is also peaked under the surface line and the extent to which it is included in the scattering depends on the resolution and must be taken account in any analysis of the data, as discussed in detail for liquid surfaces [18] where the surface Rayleigh modes cause the roughening of the surface.

V.   Experimental Studies of Surface Structure

1.   Semiconductors

One of the first X-ray determinations of a surface structure was of the InSb (111) 2 x 2 reconstructed surface [19]. The intensities of the new surface streaks originating from the surface reconstruction were determined. The diffractometer was arranged with a large divergence along the streaks and little intensity change was observed along the streaks. Consequently the results gave information about only the top layer, and the structure was determined from 16 different intensities.

A completely different approach to the study of a clean surface was taken for the Ge(111) 2 x 8 reconstructed surface [20]. The intensity along the bulk surface streaks $^{1}/_{3}(\ell \ell \ell)$ and $^{1}/_{3}(2 + \ell, 2 + \ell, \bar{4} + \ell)$ was measured as shown in figs. 5 and 6. The results were then compared with various models and as shown in figs. 5 and 6, a bulk-like surface layer does not account for the results. The solid line was calculated with a reconstructed surface involving displacement of 19 atoms in the unit cell and gives a good account of the data and is consistent with theoretical considerations.

X-ray techniques have also been used to monitor surface layers on semiconductors. One of the simplest layers occurs when additional layers

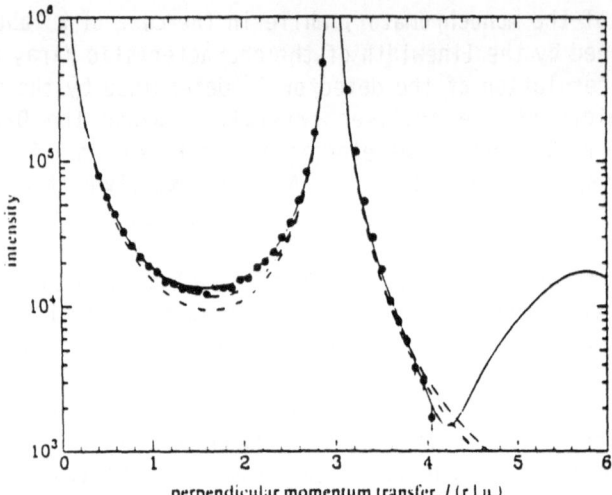

Fig. 5.   The scattered intensity [20] along the $\frac{1}{3}(\ell\,\ell\,\ell)$ streak from a Ge (111) surface.   The dotted line is the intensity calculated from a smooth bulk-like surface, the dot-dash line by a rough bulk-like surface and the solid line by a surface reconstruction.

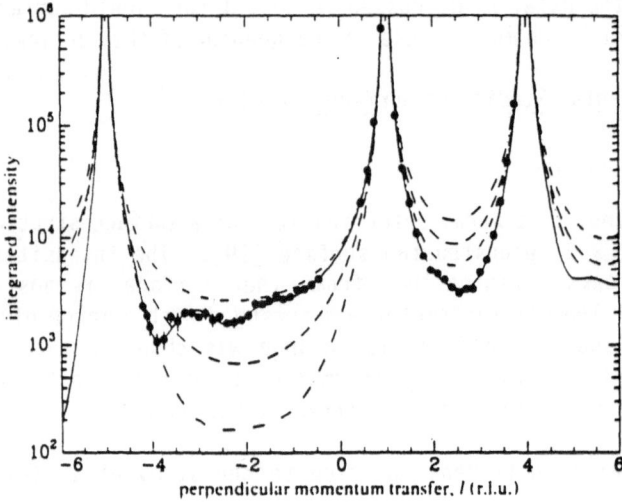

Fig. 6.   The scattered intensity [20] along the $\frac{1}{3}(2 + \ell,\, 2 + \ell,\, 4 + \ell)$ streak from a Ge (111) surface.   The models are as described for figure 5 but the dash-dot lines have bulk-like surfaces with adatoms at different sites.

are only partly grown.   Fig. 7 shows the scattering from a freshly grown (111) Ge surface along the [ℓ ℓ ℓ] direction, and the dramatic difference obtained when an additional half monolayer has been added [21] in a similar way to the simple model described in section II.   Finally we discuss the results obtained for SiO$_2$ layers on top of Si.   This is of interest because the SiO$_2$ is amorphous whereas the Si is crystalline so that both contribute to reflectivity measurements but only the Si crystal lattice to the results near Si Bragg reflections.   Consequently it is possible to study the end of the crystal lattice separately from the amorphous layer.   Initially experiments were performed using a laboratory source [22] and the reflectivity gave information about the thickness of the SiO$_2$ layer and the surface roughness, while the measurements of the streaks near the (111)

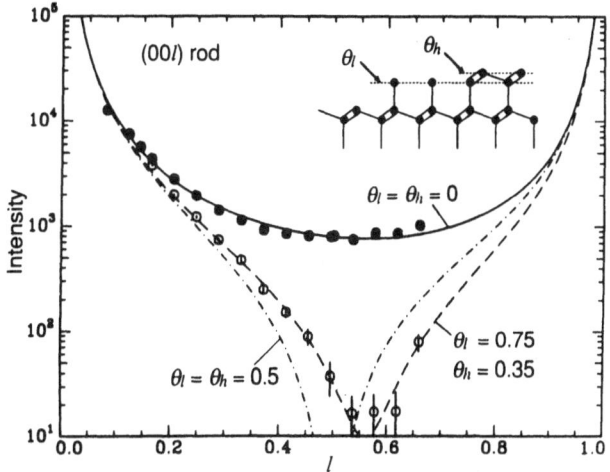

Fig. 7.   The scattered intensity [21] from a Ge (ℓ ℓ ℓ) streak from a freshly sputtered and annealed surface (solid) and after depositing a half layer (open point).

Bragg reflection decreased as $1/q^2$ over the observable range showing that the end of the crystal lattice was flat.   More detailed measurements were made at a synchrotron source with better sample conditions [23].   The results for the $^1/_3(2 + \ell, 2 + \ell, \bar{4} + \ell)$ scan are shown in fig. 8.   The dotted line shows the result of an abrupt termination and the solid line to a model of the termination of the crystal lattice.   These results too led to the conclusion that the Si, Si - 0 interface was flat but the synchrotron results extended further in wavevector and so provided more detailed information about the surface.

    This is certainly not an exhaustive account of X-ray work on semiconductors.   Other experiments have been on amphorous Si on Si (111) [23], NiSi$_2$ on Si (111) [24] and on the √3 x √3 reconstructions induced by

Bi and Ag on Si (111) [25]. There have also been a very large number of studies of semiconductor interfaces to characterise quantum wells and multi-quantum wells but a description of this work is not appropriate here.

2. Metal Surfaces

There have now been a number of studies of the structure of metal surfaces using X-rays: (001) faces of Au [26], Cu [27] and Ag [30], Pb [31], Pt [32] and Cu [33], (111) faces of Pb [31] and Au [34] and (113) faces of Cu [35] and Ni [36]. We shall not review all of these measurements in detail. One of the interesting studies of the structure is that of the Au (001) face [26]. At high temperatures above 1170K the results are consistent with a disordered surface, but below 1170K the surface layers distorts to a hexagonal structure superimposed on the cubic bulk structure. This is shown by fig. 9 which shows a scan perpendicular to the surface streaks showing

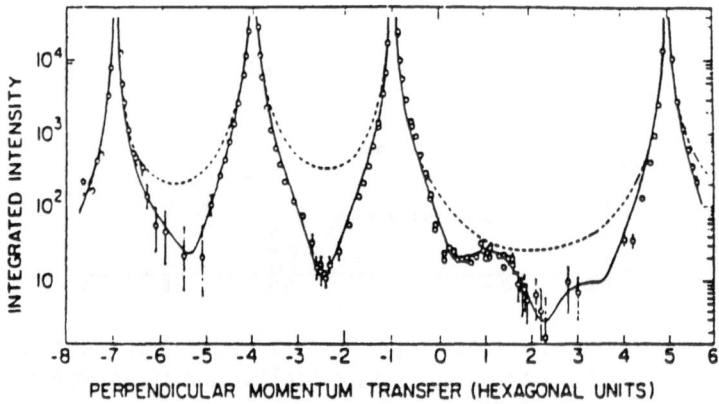

Fig. 8. The scattered intensity [23] from a Si(111) interface with Si O_2. The intensity is observed for Q = $^{1}/_{3}(\ell \ell \ell)$.

additional peaks below 1120K which lie on a hexagonal lattice. On cooling the hexagonal lattice is distorted by the underlying cubic structure and then below 970K the hexagonal structure rotates slightly by 0.81° to a distorted hexagonal structure as shown in fig. 10. The surface layer is therefore reconstructed to be similar to a close packed (111) bulk plane and it is perhaps surprising that the nearest neighbour lattice constant in the layer is 2.77 Å slightly smaller than the bulk 2.885 Å. This experiment illustrates well the power of the X-ray technique and also the complexity of the behaviour which can occur at surfaces.

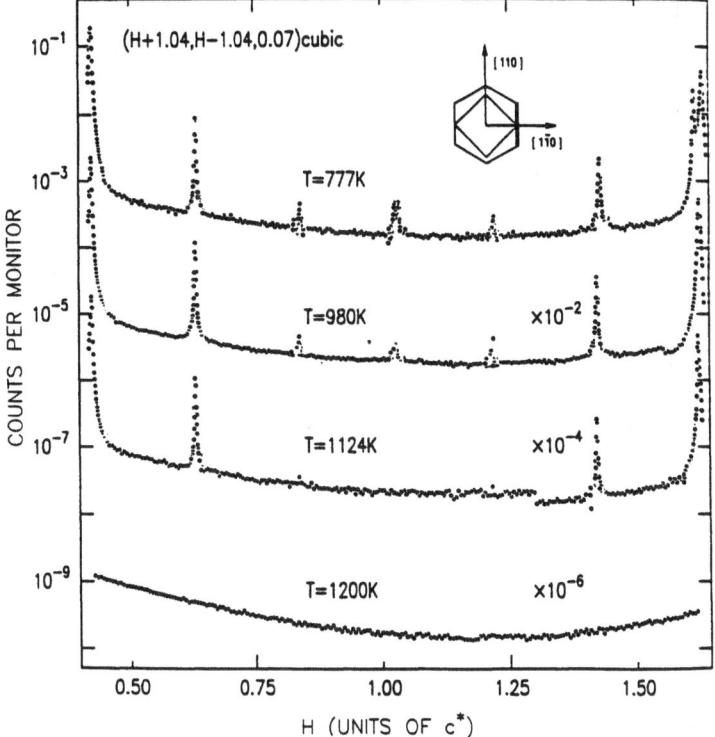

Fig. 9. The intensity observed [26] for the Au (001) face as a function of temperature. Above 1170K there are no surface peaks but below the peaks correspond to a hexagonally ordered layer distorted by the square substrate. The scan in reciprocal space is shown in the insert.

The behaviour of the (111) surface of Au is also interesting. Since the (111) surface is close packed it might be expected that the surface would be a simple termination of the bulk. Measurements [34] show however that the behaviour is far more complex. Above 865K the surface layer does have hexagonal symmetry with a lattice parameter similar to that of the hexagonal structure on the Au (001) face. Below a temperature of 865K this structure distorts to give a series of discommensurations. In the high temperature phase the surface streaks suggest that the structure is only partially ordered presumably because the discommensurations are then disordered.

Much of the work on clean metal surfaces has been directed towards the study of the roughening transition which is expected to occur at sufficiently high temperatures for simple surfaces [9]. Probably the most successful study was that on the Ni (113) surface [36], which below the roughening transition fitted a Gaussian long range ordered component and a Lorentzian diffuse component to simulate the diffuse scattering, while for temperatures above the roughening transition the theoretical form of eqn 3.9 was fitted. The results are consistent with theoretical expectations but the authors admit that it was not possible to test the form of the divergence

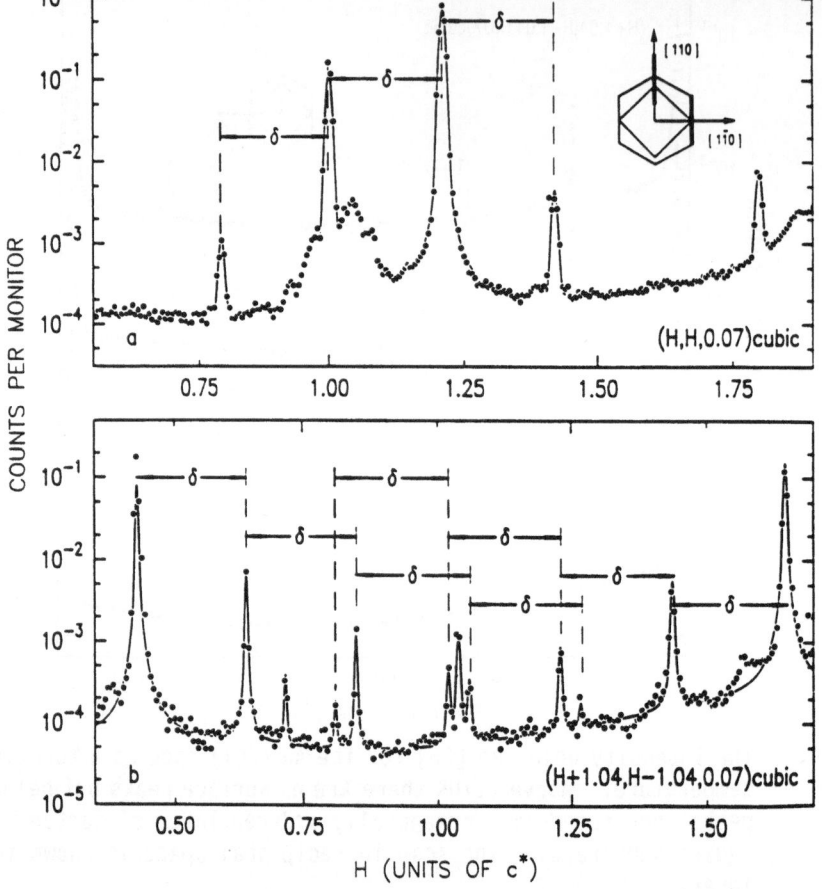

Fig. 10. The intensity scattered [26] at room temperature from the Au(001) face. The results show the peaks arising from the rotated and distorted hexagonal structure. The scans are shown by the inset in the figure.

in eqn 3.9 uniquely. Other surfaces have shown different behaviour on heating. In the case of the Ag (110) face [30] the results suggest that on heating the surface separates into flat (110) faces and rough slightly inclined faces, and that the fraction of the area covered by the flat (110) faces decreases on heating. Similar behaviour is observed for Cu (110) faces.

A final example is that of the Pt (110) surface which has a (1 x 2) reconstruction corresponding to alternate missing rows in the surface layer. Initially it was expected that this structure would disorder by randomly occupying both layers and then at a higher temperature would roughen [37]. The experiments showed [32] that both the roughening and disordering occurred simultaneously although the exponents were consistent with those of a two-dimensional ordering or Ising model.

3. Liquids

There have been extensive studies of the properties of liquid surfaces because they are difficult to study using electron techniques. As discussed in section III, they are also of interest because the capillary surface waves roughen the surface so that the deviation in the heights diverges as given by eqn 3.8.   This is because the capillary waves on the surfaces have frequencies cq, where c is a velocity and so the mean displacement is divergent for two dimensional systems.   In liquids this divergence is eliminated by the effect of gravity but this only influences the fluctuations for wavevectors q much smaller than the resolution effects in the experiments.   Consequently the scattering is given by eqn 3.9 convoluted with the resolution function of the instrument.

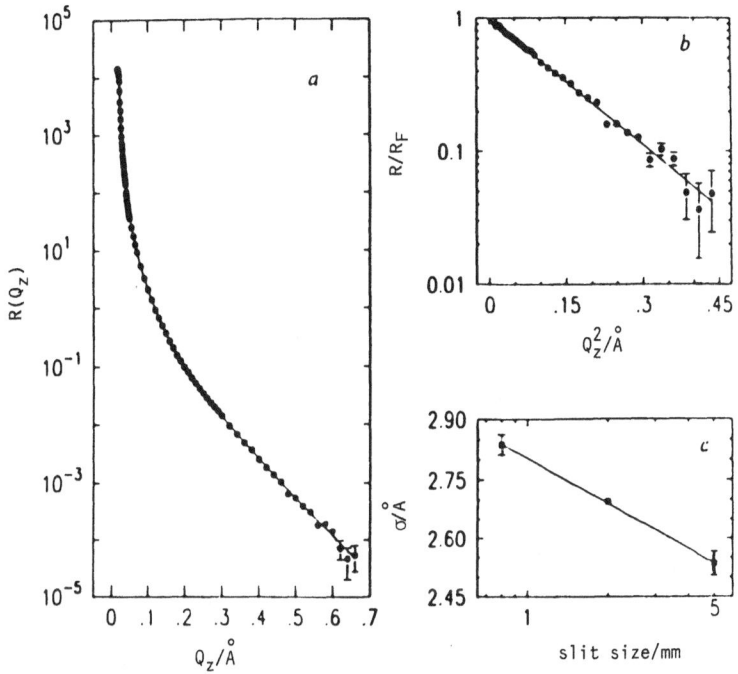

Fig. 11.   The measured reflectivity [38] from water (a) as a function of Q_z. Part (b) shows the fall-off due to the diffuse surface, and (c) the parameter of that fall-lff compared with theory for different experimental resolutions.

Experimentally [38] much of the work has concentrated on studying the water-vapour interface, and typical results are shown for the reflectivity in fig. 11.   The scattering shows a sharp peak when $Q_y = Q_y = 0$ which decreases in intensity with increasing Q_z.   The results are well described by the scattering expected from a sharp surface multiplied by a Gaussian as given in eqn 3.4.   This is surprising initially in view of the theoretical results for a rough interface, but there is the difficulty of distinguishing between long range order and diffuse scattering as discussed in section III.4.   The experiments have now been repeated with different resolutions

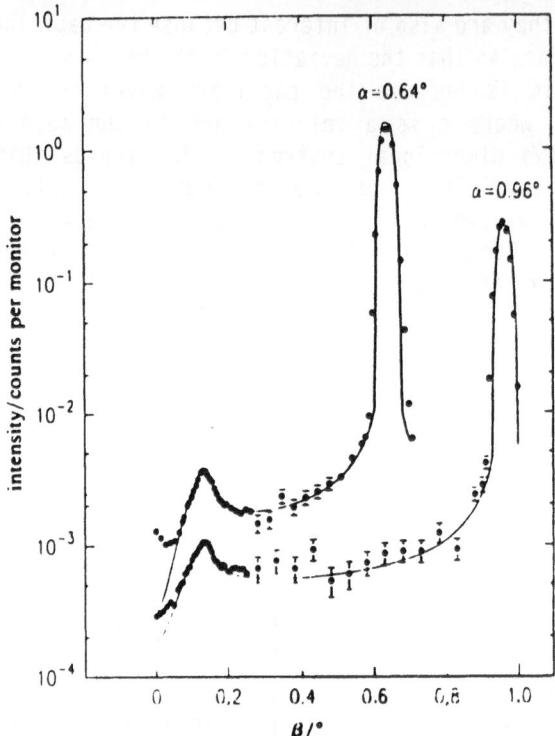

Fig. 12. The scattered intensity from H_2O as a function of the scattered angle, β, for fixed incidence angles, α, = 0.64 and 0.96°. The solid lines are the theoretical predictions based on the capillary wave theory and show the specular peak and the peak when β is the critical angle.

and the results are similar except that the Gaussian parameter S increases as the resolution volume decreases. This is because the diffractometer then integrates over less of the diffuse scattering. These results are then in agreement with a parameter free theory of the scattering not only for the reflectivity but also for the diffuse scattering as shown in fig. 12. It is nevertheless disappointing that the experiments have not yet managed to confirm the detailed theoretical form for the scattering.

X-ray scattering and neutron scattering are now being extensively applied to study films on liquids including soaps and surfactants. The techniques enable the detailed structure of the films to be determined which is proving of great importance for developing paints and detergents, for example. X-ray scattering has by far the best resolution but neutron scattering has the advantage of having a higher cross-section for hydrogen and of being able to modify the scattering by isotopic substitution.

4. Ferroelectric Domain Walls

A different application of these techniques is to the study of buried interfaces and a great deal of work has been done to study semiconductor interfaces. A quite different application is to ferroelectric domain walls and we shall illustrate this by explaining the results obtained for KH_2PO_4,

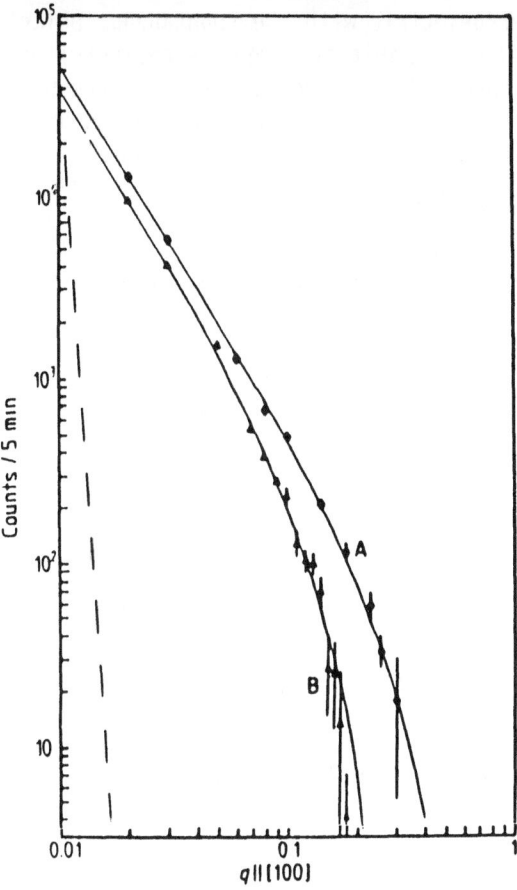

Fig. 13. The wave-vector dependence of the scattering from domain walls in
KDP at 30K and 119.5K compared with fits to obtain the domain wall
widths [40].

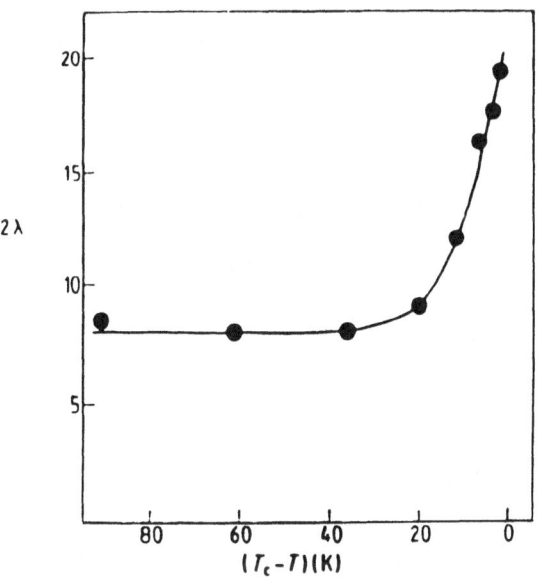

Fig. 14. The temperature dependence of the domain wall width in KDP [40].

KDP. KDP is a ferroelectric with the spontaneous polarisation along the c-axis. Below T_C it breaks into different ferroelectric domains to minimize the electrostatic energy and the motion of the corresponding domain walls determines the electrical properties. In the ferroelectric phase the crystal structure also distorts from its tetragonal structure to give an orthorhombic structure and this distortion occurs differently in different unit cells. Each of the distortions then gives rise to a different Bragg reflection of the low temperature phase. Simple geometric considerations then show that the domain walls separating two different domains must be aligned perpendicular to the line joining the Bragg reflections and will give rise to the by now familiar surface scattering along these directions in reciprocal space [39]. Measurements were made on KDP [40] and a scan along one of these streaks is shown in fig. 13. The intensity decreases with q as given by eqns. 3.6 showing that the domain wall width, about one unit cell, increases with temperature, fig. 14. The measurements were also repeated around several different Bragg reflections from which data a detailed atomic picture of the domain wall was constructed.

References

1. Zachariasen, Theory of X-ray Diffraction in Crystals, Wiley New York (1944)
2. James
3. J.D. Jackson, Classical Electrodynamics, Wiley New York (1975)
4. M.V. Berry, Phil. Trans. A276, 611 (1973)
5. S.R. Andrews and R.A. Cowley, J. Phys. C: Solid State Phys. 18, 6427 (1985)
6. P. Wong and A.J. Bray, Phys. Rev. B37, 7751 (1988)
7. S.K. Sinha, E.B. Sirota, S. Garoff and H.B. Stanley, Phys. Rev. B38, 2297 (1988)
8. I.K. Robinson, E.H. Conrad and D.S. Reed, J. de Physique 51, 103 (1990)
9. S.T. Chui and J.D. Weeks, Phys. Rev. B14, 4978 (1976)
10. M.Den Nijs, E.K. Reidel, E.H. Conrad and T. Engel, Phys. Rev. Lett. B16, 1217 (1985)
11. J.V. Jose, L.P. Kadanoff, S. Kirkpatrick and D.R. Nelson, Phys. Rev. B16, 1217 (1969)
12. P.H. Fuoss and I.K. Robinson, Nucl. Inst. & Methods A222, 171 (1984)
13. E. Vlieg, A. van't Ent, A.P. de Jongh, H. Neerings and J.F. van der Veen, Nucl. Inst. & Methods A262, 522 (1987)
14. J. Als Nielsen and P.S. Pershan, Nucl. Inst. & Methods 206, 545 (1983)
15. R. Pynn, Y. Fujii and G. Shirane, Acta Cryst. A39, 38 (1983)
16. R.A. Cowley, Acta Cryst. A43, 826 (1987)
17. C.A. Lucas, E. Gartstein & R.A. Cowley, Acta Cryst. A45, 416 (1989); Ibid A46, 576 (1990)
18. A. Braslau, P.S. Pershan, G. Swislow, B.M. Ocko and J. Als Neilsen, Phys. Rev. A38, 2457 (1988)
19. J. Bohr, R. Feidenhansl, M. Nielsen, M. Toney, R.L. Johnson and I.K. Robinson, Phys. Rev. Lett. 54, 1275 (1985)
20. R.G. van Silfhout, J.F. van der Veen, C. Norris and J.E. Macdonald, Faraday Discuss. 89, 169 (1990)
21. E. Vlieg, A.W. Denier van der Gon, J.F. van der Veen, J.E. Macdonald and C. Norris, Phys. Rev. Lett. 61, 2241 (1988)

22. R.A. Cowley and T.W. Ryan, J. Phys. D. 20, 61 (1987); R.A. Cowley and C. Lucas, J. de Physique C7 50, 142 (1989)
23. I.K. Robinson, Phys.Rev. Lett. 57, 2714 (1986)
24. I.K. Robinson, R.T. Tung and R./ Feidenhansl, Phys. Rev. B 38, 3632 (1988)
25. T. Takahasi, S. Nakatani, T. Ishikawa and S. Kikuta, Surf. Sci. 191, L825 (1987); Jap. J. Appl. Phys. 27, L753 (1988)
26. D. Gibbs, B.M. Ocko, D.M. Zehaer and S.G.J. Mochrie, Phys. Rev. B42, 7330 (1990)
27. I.K. Robinson, E. Vlieg and S. Ferrer, Phys. Rev. B42, 6954 (1990)
28. I.K. Robinson, A.A. MacDowell, M.S. Altman, P.J. Estrup, K. Evans-Lutterodt, J.D. Brock & R.J. Birgeneau, Phys. Rev. Lett. 62, 1275 (1989)
29. I.K. Robinson, Phys. Rev. Lett. 50, 1145 (1983)
30. I K. Robinson, E. Vlieg, H. Harris & E.H. Conrad (to be published)
31. P. Fuoss, L. Norten and S. Brennan, Phys. Rev. Lett. 60, 2049 (1989)
32. I.K. Robinson, E. Vlieg and K. Kerr, Phys. Rev. Lett. 63, 2578 (1989)
33. B.N. Ocker and S.G.J. Mochrie, Phys. Rev. B38, 7378 (1988)
34. A.R. Sandy, S.G.J. Mochrie, D.M. Zehaer, K.G. Huang and D. Gibbs, Phys. Rev. B43, 4667 (1991)
35. K.S. Liang, E.B. Sirota, K.L. D'Amico, G.J. Hughes and S.K. Sinha, Phys. Rev. Lett. 59, 2447 (1987)
36. I.K. Robinson, E.H. Conrad and D.S. Reed, J. de Physique 51, 103 (1990)
37. J. Villain and I. Vilfan, Surf. Sci. 199, 165 (1988)
38. for a review see P.S. Pershan, Faraday Discuss. Chem. Soc. 89, 231 (1990)
39. D.A. Bruce, J. Phys. C: Solid State Physics 14, 5195 (1981)
40. S.R. Andrews and R.A. Cowley, J. Phys. C: Solid State Physics 19, 615 (1986)

APPENDIX

Dynamical Theory of X-ray Scattering

Consider a crystal as shown in fig. A with a beam incident of amplitude A_ℓ, and a beam of amplitude B_ℓ. Then if the planes are parallel to the

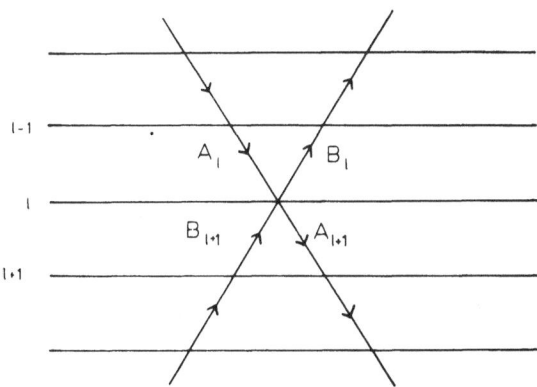

Fig. A. The scattering from the different planes in the dynamical theory of diffraction.

surface of the crystal the scattering at a plane gives the relations

$$B_\ell = - i g A_\ell + (1 - i g_0) e^{-i\psi} B_{\ell+1} \qquad\qquad \text{A1}$$

$$A_{\ell+1} = (1 - i g_0) e^{i\psi} A_\ell - i g e^{i2\psi} B_{\ell+1} \qquad\qquad \text{A2}$$

where $- i g A_\ell$ is the amplitude of the incident wave scattered into the diffracted wave at the ℓth plane, and $- i g_0 A_\ell$ is the amplitude scattered in the same direction. The scattering is $\pi/2$ out of phase with the incident wave and the magnitudes of g and g_0 can be obtained from the scattered intensity, eqn. 2.3, by omitting the term arising from the sum over ℓ to give

$$g = \frac{2\pi b}{k a_{xy} \sin\theta} \ , \qquad\qquad \text{A3}$$

and ψ is the phase difference between the planes

$$\psi = Q_z a/2$$

B_ℓ can be eliminated from eqn A1, to give a three term recurrence relationship for A_ℓ. This can be solved by noting that the scattering from each plane is small so that $A_{\ell+1} = x A_\ell$ with $|x|$ close to one. The result is

$$(1 - i g_0) (x^2 + 1) = x(g^2 e^{i\psi} + (1 - i g_0)^2 e^{i\psi} + e^{-i\psi})$$

Now assuming the scattering is close to a Bragg reflection we can write $x = (1-\beta)e^{im\pi}$, and $\psi = m\pi + \Delta$ where $\Delta = qa/2$ with β and Δ small to give

$$\beta^2 = g^2 - (g_0 - \Delta)^2, \qquad\qquad \text{A4}$$

and the ratio of the incident to scattered amplitudes at the surface of the crystal from eqn A2 becomes

$$\frac{B_0}{A_0} = \frac{- g}{(g_0 - \Delta) + (g_0 - \Delta)^2 - g^2)^{1/2}} \qquad\qquad \text{A5}$$

Now introducing

$$y = \frac{\Delta - g_0}{g} = \frac{q - \delta}{q_c}$$

the intensity scattered by the crystal is

$$I(q) = y - (y^2 - 1)^{1/2} \qquad\qquad |y| \geq 1 \qquad\qquad \text{A6}$$

$$= 1 \qquad\qquad\qquad\quad |y| < 1$$

as given in eqn 2.8.

SCANNING TUNNELING MICROSCOPY

Horst Niehus

Institut für Grenzflächenforschung und Vakuumphysik
Forschungszentrum Jülich
Postfach 1913, D-5170 Jülich, Germany

1. Introduction

Probably the most fascinating aspects of microscopy originate from the suggestive power of images offering the direct reproduction of the 'real world' without the need of complex transformations. The interest in looking for smaller and smaller details stimulated the development originally of optical microscopes with high magnification and ended for the present by the build-up of completely new microscopic techniques based on electronic effects. For long time the possibility of imaging features on an atomic scale seemed to be unrealistic. After the invention of the field ion microscope, at least atomic resolution at specifically shaped thin tips could be obtained. It was Binnig and Rohrer who showed in 1981 that indeed such tips could be also used as local probes to image the atomic structure at flat surfaces under utilization of the well known tunneling effect. In the following ten years after birth of the scanning tunneling microscopy (STM /1/), the technique has become very popular and is nowadays in fact already widely used in a variety of different groups, among others in physics, chemistry, biology and medicine. The main attraction of STM is still its ability to show directly images with atomic resolution implying that we can literally see the atoms at the surface. The resolution lateral to the surface comes as close as about 0.5Å. The resolution in normal direction is often limited by the mechanical properties (as in particular the vibrational noise and thermal drift) and has been improved within the last years down to better than 0.05Å. Because the field of STM is rapidly growing, we also recommend to the reader some of the recent comprehensive review articles /1-3/.

Equilibrium Structure and Properties of Surfaces and Interfaces
Edited by A. Gonis and G.M. Stocks, Plenum Press, New York, 1992

During scanning tunneling microscopy a sharp metal tip as the local probe is laterally scanned at a close distance to the surface. After application of a small voltage to the tip and sample the induced tunneling current gives a measure for the separation of tip and surface. This current flows either by tunneling of electrons to or from the surface, depending on the voltage polarity. The electronic states at the surface (or tip) are probed within a few electron volts on either side of the Fermi level, in other words the STM images show electronic features which are closely related to the position of individual atoms but must not represent in any case the geometric position of the nuclei itself, e.g. dangling bonds may well point away from the actual atom position and skew the topographic part in the image. STM data usually contain a mixture of information from topographic and electronic effects, a separation is not always simple and straightforward.

Following this introductory part, the basic principles underlying STM will be recapitulated. The influence of local electronic effects and its application for scanning tunneling spectroscopy (STS) will be discussed before a brief description of recent STM equipment is given. The application of STM and STS is illustrated for semiconductor surfaces, i.e. for clean silicon, carbide formation at Si(111) and the investigation of growth of V and Fe silicides. Pure topographic effects can be followed easiest at clean metals, e.g. at Cu(110), whereas the influence of adsorbates (e.g. oxygen or nitrogen) already complicates the interpretation. Methods revealing direct information on nuclei positions like ion scattering techniques seem to be a natural supplement to STM. A fruitful combination of STM with low energy ion scattering techniques (in particular NICISS /4/) will also be briefly discussed. Finally the usefulness of STM even without achieving atomic resolution in the images will be shown in case the metal alloy NiAl(111).

2. Basic Principles

In general in all STM investigations a tip with a sharp curvature at its end is positioned in Z direction (normal to the surface) by piezoceramic elements close (5 - 10 Ångstroms) to the probed surface. The scanning motion X, Y of the tip with respect to the surface (or vice versa) is actuated by scanning voltages U_x and U_y applied to piezo elements. A small voltage U_{Tip} is given to the tip and the corresponding tunneling current I_{Tip} appears to be extremely sensitive to the separation between tip and sample. The tunneling current is determined by the overlap of wave functions of the tip and the sample as illustrated in fig.1 and can be described on the basis of the transfer Hamiltonien concept in the Bardeen tunneling formalism which neglects the interaction of the two electrodes and can be written /5/ for the limit of small voltages:

$$I = 2\pi e/\hbar \ \sum \ f(E_\mu) \ [\ 1 - f(E_\nu - eU)] \ \delta(E_\mu - E_\nu) \ | \ M_{\mu\nu} \ |^{\ 2} \tag{1}$$

$$M_{\mu\nu} = \hbar^2/2m \ \int \ dS \bullet (\ \psi_\mu^* \ \nabla \ \psi_\nu - \psi_\nu \ \nabla \ \psi_\mu^*) \tag{2}$$

with the tunneling matrix element $M_{\mu\nu}$ between the states ψ_ν and ψ_μ at the tip and surface, respectively. $f(E_\mu)$ and $f(E_\nu - eU)$ are the Fermi-Dirac occupation factors at the energies E_μ and $E_\nu - eU$ with a small Voltage U applied across the electrodes (fig. 1). The matrix elements are than evaluated over the entire vacuum region lying between the tip and the surface.

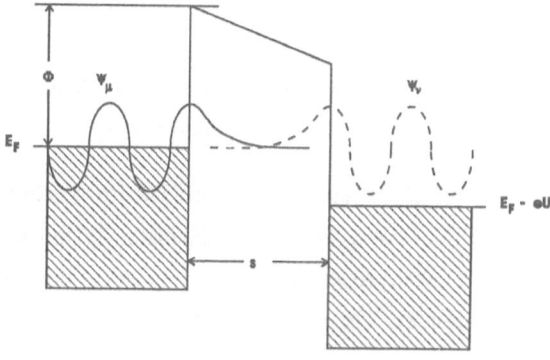

Fig.1.   Schematic energy diagram for tunneling. From ref./6/.

In its simplest form consider a one dimensional potential barrier in order to get an impression for the functional dependency of the electron current on electrode separation. If we neglect any electrical field in the junction and consider only an average work function ϕ it is easy to describe the wave function damping into the simple potential barrier /6/:

$$\psi_\mu = \psi_\mu{}^0 \exp(-\kappa\, z) \tag{3}$$

and

$$\psi_\nu = \psi_\nu{}^0 \exp[-\kappa\,(s - z)] \tag{4}$$

Here the inverse decay length κ of the wave functions is defined as

$$\kappa = \sqrt{2m\phi/\hbar^2} \tag{5}$$

Inserting the two wave functions in equn. (2) it is possible to rewrite the matrix element:

$$M_{\mu\nu} = \hbar^2/2m \int dS\ 2\,\kappa\ (\psi_\mu{}^0)^* \psi_\nu{}^0 \exp(-\kappa\, s) \tag{6}$$

and the tunneling current becomes:

$$I_{Tip} \propto \Sigma\ |\psi_\mu{}^0|^2\ |\psi_\nu{}^0|^2\ \exp(-\kappa\, s) \tag{7}$$

31

where it has to be summed over all relevant states which have appropriate energies. It can be noted that the dependency on Z has dropped out and the current can be evaluated for any surface separating the two electrodes.

This equation for the one dimensional tunneling configuration /7/ yields already the exponential dependence of the current on the distance variation. The same relation has been earlier obtained by Duke /8/ before the exploration of tunneling microscopes. In order to get an impression of the variation of the current I_{Tip} we may estimate the magnitude for a typical work function of $\phi = 4.5$ eV. From equn. (5) a decay length of 1.1 Å $^{-1}$ can be calculated which gives then rise to a variation of I_{Tip} typically by one order of magnitude per Å change in the tip to sample distance.

In order to evaluate the spatial resolution of an STM it is necessary to calculate the tunneling current in three dimensions. Tersoff and Hamann /9/ introduced the most simple model for describing the local probe (tip) as a mathematical point source with s-wave symmetry located at a position r_t. The motivation other than mathematic simplicity was the intention to extract the information of the probed surface properties neglecting the tip characteristics and also to achieve the best lateral resolution. Tersoff and Hamann /9/ obtained for the electronic current I_{Tip}:

$$I_{Tip} \propto \sum |\psi_\nu (r_t)|^2 \delta(E_\mu - E_F) \equiv \rho(r_t, E_F) \tag{8}$$

For this idealized STM with a pure point tip and no varying electronic structure at the tip the tunneling current would be a direct measure for the local density of states (DOS) $\rho(r_t, E_F)$ at the Fermi energy E_F, resulting exclusively from the investigated sample surface properties.

So far, the influence of the tip voltage U_{Tip} between tip and sample was left out of consideration. One might hope that the effects of nonzero voltage can be incorporated by rewriting the tunneling current formula as follows:

$$I_{Tip} \propto \int_{E_F}^{E_F + eU_{Tip}} \rho(r_t, E) \, dE \tag{9}$$

In such an approximation the voltage dependency of matrix elements and the density of states of the tip have been neglected. Also the effect of a finite voltage on the wave functions in the vacuum region is ignored in equn. (9).

By including the voltage dependence and indirect transitions, the inverse decay length κ depends not only on the workfunktion but also on the applied voltage U and the parallel wave vector $k_{||}$ involved in the electronic transition:

$$\kappa = \sqrt{(2m\phi - \tfrac{1}{2}U_{Tip}e)/\hbar^2 + |\mathbf{k}_{||}|^2} \qquad (10)$$

Selloni et al. /10/ suggested to generalize the results for a transition with $\mathbf{k}_{||} = 0$ by considering the influence of the voltage outside of the electrodes by introducing the barrier transmission factor $T(E,U)$:

$$I_{Tip} \propto \int_0^{eU_{Tip}} \rho(\mathbf{r}_t, E)\ \rho_{Tip}(\mathbf{r}_t, E - eU_{Tip})\ T(E,U)\ dE \qquad (11)$$

or neglecting the state density effects resulting from the local probe ρ_{Tip} itself:

$$I_{Tip} \propto \int_0^{eU_{Tip}} \rho(\mathbf{r}_t, E)\ T(E,U)\ dE \qquad (11a)$$

Feenstra /6/ showed, that several general features of the current I_{Tip} as function of the voltage can be evaluated, although, in order to obtain this equation, the already mentioned approximations have to be incorporated. In the WKB approximation the transmission term can be expressed as follows:

$$T(E,U) = \exp\{ -2s[2m/\hbar^2 (\phi - E + \tfrac{1}{2} eU_{Tip})]^{1/2}\} \qquad (12)$$

From this equation we can immediately see that mostly the highest lying states will contribute to the tunneling current. The situation is schematically shown in fig.2.

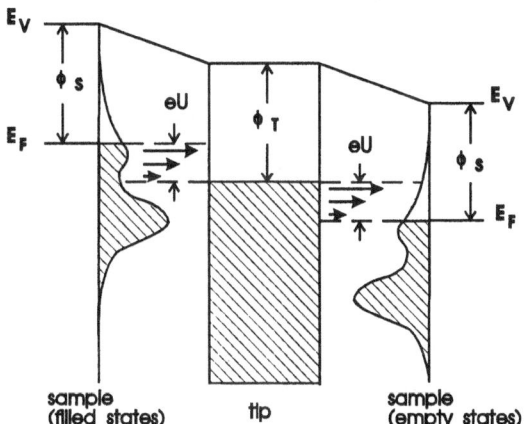

Fig.2.   Schematic description of the electronic effects in the tunneling current energy diagram. Left side: positive bias at the tip; tunneling from the sample to the tip; - probing occupied states. Right side: negative bias at the tip; tunneling from the tip into the empty states at the sample. From J.Behm in ref./3/.

In case of a positive bias U_{Tip} the maximum of tunneling current is expected at E= eU (left side in fig.2); electron current from the sample to the tip, i.e. probing occupied (filled) states of a semiconductor. The occupied states contribute most effectively at the upper side of $\Delta E = eU$ because here they feel the smallest tunnel barrier. With decreasing energy levels, the density of states is superimposed by the exponential damping from the transmission factor (equn. 12) towards lower E values. Hence probing occupied states with STS be scanning eU (the potential barrier height) is complicated because of the overlap with this damping effect. On the other hand for the negative bias, the maximum current is obtained at $\Delta E = 0$ eV, in this situation one can effectively scan the density of states for the non occupied (empty) states by varying the tip voltage U. These sprectra might be compared with results from inverse photoelectron spectroscopy. It has to be remembered, that associated features of the density of states from the tip are still neglected, schematically shown in fig.2 by the structureless region in the middle part representing the tip properties. For quantitative interpretations, the superposition of both, tip and surface DOS features have to be considered.

2.1 Surface Topography

In most STM investigations the microscope is operated in the constant current mode, i.e. I_{Tip} is held constant by appropriate variation of the tip to sample distance with the help of a feedback circuit. Thus by scanning the tip laterally across the sample the measured contours would image the local state density in front of the surface. The STM pictures essentially reflect the lowest Fourier components of the spatial distribution of the electronic state density since the higher components decay faster into the vacuum gap. Since these states at the surface are probed within a few electron volts on either side of the Fermi level, the connection between these electronic states and the atomic structure depends on the investigated system: in case of metals these states generally follow the atoms in a uniform manner and thus topographic information of probed metal surfaces can be expected.

In the Tersoff Hamann formalism described above, the tip influence was confined to spherical geometry with s-wave symmetry. In the constant current mode, the tip would follow the constant charge distribution at E_F. The charge density contours above the surface in case of clean metals can be approximated by the linear superposition of radial symmetric charges positioned at the nuclei sites. As already shown above, the Tersoff Hamann calculations lead to an exponential decay of the tunneling current on the tip to sample separation, which allows a rough estimate on the measured corrugation to be expected in an STM experiment at metal surfaces. The estimated corrugation should be considerably weaker compared with e.g. thermal energy atom scattering (TEAS /11/)

because the distance of closest approach in TEAS experiments is closer ($\sim 3\text{Å}$) to the surface corresponding to the typical STM gap distance of $\sim 10\text{Å}$. It has been shown that in the s-wave approximation the STM corrugation even of highly structured (rough) surfaces as the Au(110)-(1x3) reconstructed phase should be negligible. In the same mainframe the lateral resolution has been determined to $\sqrt{2(R_t + s)}$ with R_t being the tip radius and s the tip to surface separation, which leads to about 5Å lateral resolution with rather sharp tips at small separations s.

Atomic resolution for metal surfaces would not be expected on the basis of these estimations. Fortunately, the experimental situation is better than proposed by the Tersoff Hamann theory! In fact, atomic resolution on Au(111) has been first reported by Hallmark et al. in 1987 /12/, shortly after also on Al(111) by Wintterlin et al. /13/. In the meantime also on other metals such as Ru, Pt, Ni and Cu atomic resolution has been measured /14-20/. In fig.3 a high resolution STM image of the clean Cu(110)-(1x1) surface is shown.

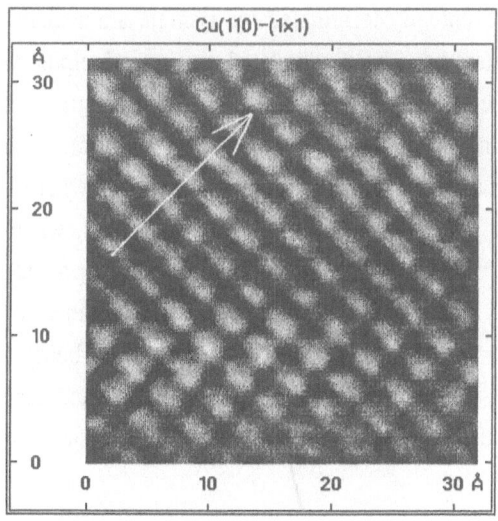

Fig.3. High resolution STM image of the Cu(110)-(1x1) surface in a gray scale plot. The area of 32 x 32 Å^2 is shown as measured in the constant current mode. The schematically inserted arrow points in a <110> direction. Gray scale: 0Å (dark) to 0.3Å (white); $T_{Tip} = -40\text{mV}$, $I_{Tip} = 5\text{nA}$.

The gray scale image for a 32 x 32 Å^2 area is acquired in the constant current mode ($i_{Tip}=5\text{nA}$) at a tip to sample bias of -40mV. The gray scale in fig.3 ranges from 0 to 0.3Å (black to white) and the Cu atoms in the unit cell (model in fig.20a) are well resolved (2.56Å apart in the close packed $<110>$ rows along direction of the schematic arrow in fig.3, and 3.6Å apart in $<001>$ direction).

The puzzle of 'why do we observe atomic resolution at metal surfaces' is still not finally solved. Originally, Hallmark et al./12/ attributed the effect at Au(111) to the presence of localized surface states near E_F at the probed surface, an effect which is very unlikely to appear in general at all other surfaces showing atomic resolution. It has been also argued that some elastic tip to surface interactions might be responsible. Recently, Chen /21/ introduced a theory for localized surface states at the tip as opposed to the s-wave model of Tersoff and Hamann. From experimental observations showing that the corrugation and resolution measured with STM often changes abruptly during the data acquiring process, we may conclude that basically the degree of measured corrugation and resolution depends strongly on the features of the tip itself. During these sudden 'switches', the surface properties have not changed, instead the tip properties itself influenced the image quality. Thus probably many described tip sharpening procedures /12,13,22,23/ in fact result in a migration of tip atoms under the influence of a strong electric field until, finally, if a good (sharp) tip builds up, a (tungsten) cluster is formed at the end of the tip eventually terminated by a single atom in its apex. The electronic states of tungsten clusters have been calculated on first principle methods /24/ and it has been found that in case of W_4 or W_5 clusters, localized dangling bonds can be identified with a d-state protruding from the apex atom. Such configurations are not supposed to be very stable. This is in agreement with experimental studies where it is shown that stable conditions by scanning with a blunt tip might stay for hours, whereas good resolution is often only achieved for very limited time and ends usually by a 'tip switch' to loose good tip properties. In this mainframe, the localized surface states at the tip may enhance measured corrugation amplitudes in an STM experiment being magnitudes larger as expected on the basis of the s-wave model from Tersoff and Hamann. The calculated amplitudes in Chen's theory agree in fact well with the measured data from Wintterlin et al. /13/ at Al(111).

In case of semiconducting surfaces the effect of corrugation and atomic resolution is inherently less important compared with metals. That is, because localized bonds protrude already from the surface into the vacuum gap. These dangling bonds give rise to strong corrugational effects, and much higher amplitudes are measured compared with metal surfaces. A classical example for microscopy at semiconducting surfaces is the Si(111)-(7x7) structure, a high resolution STM image is shown in fig.4. This image displays an area of 190x190 $Å^2$ and the ad-atoms in the 7x7 structure are clearly visible. From the tip to surface voltage polarity $U_{Tip} = -1.87V$ it is obvious that electrons from the tip are tunneling to the surface, i.e. into empty states at the Si surface atoms. As will be shown below, probing empty or filled states gives different (spectroscopic) information, whereas the topographic and spectroscopic parts cannot be clearly separated but are intermixed. Here, and also in case of adsorbates at metal surfaces the problem of extracting topographic information in STM starts to play an important role.

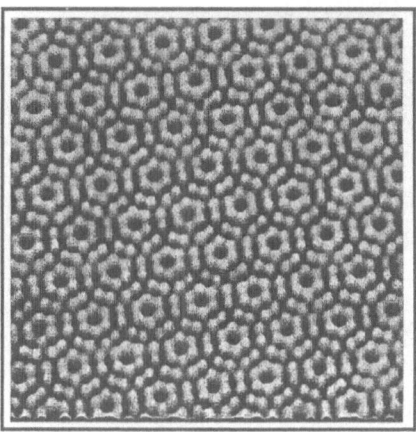

Fig.4. High resolution STM image of Si(111)-(7x7) surface. Gray scale: from dark to white: 2.5Å; U_{Tip} = -1.87V, I_{Tip} = 1nA.

The measured corrugation in this example is high, and with 1.6Å across ad-atoms to corner holes in fact much larger than measured corrugation amplitudes at clean metals which are less than 0.1Å. Indeed, Si(111)-(7x7) with its directional bonds and large interatomic distances of the ad-atoms is the ideal candidate for an STM experiment.

2.2 Local Spectroscopy

In the deduction of the tunneling current it has been shown (equn. 11) that I_{Tip} is proportional to the local density of states $\rho(\mathbf{r}_t , E_F + \Delta E)$. Different from metal surfaces, semiconductors show a strong variation of ρ with U_{Tip}. Already the choice of the tunneling voltage in a constant current STM experiment determines wether electrons are tunneling into the valence band or are coming out of the conduction band. Hence the simplest way of getting crude spectroscopic information would be to take two constant current images from the same sample area but measured with opposite tip voltages. As a result, the two STM images represent the local distribution of filled and empty states, respectively. This strategy has been first applied for GaAs samples /26/. It has been proposed that the empty states should be preferentially localized at the As and the filled states at the Ga atoms /9/. In fact by imaging the GaAs(110) surface with both polarities 'simultaneously', Feenstra et al. /26/ showed the appearance of alternating zig-zag rows from Ga and As atoms in the (110) surface.

The voltage dependence of STM images can be also used for getting the information of 'obscured' surface structure, like the uncovering of the stacking fault in the Si(111)-(7x7) unit cell in the light of the filled states /27-29/. Also adsorbates and compounds have been investigated with the same technique and an example of Si-carbide formation will be given in chapter 4.2.

The method of voltage dependent STM in the constant current mode obviously has to fail for investigations with close to or at zero tip voltage. The same situation occurs for semiconductors when states are probed which lie very close to, or even in the band gap. Thus in a next step the instrumentation for STS measurements have been changed in such a way that the stabilization of the tip to surface distance is still performed in the constant current mode, however has been separated in time from the determination of the I-U dependence. Basically two different approaches have been undertaken in the recent years. First, by just stopping the X-Y scanning for a certain time at a desired point of interest in an STM measurement, consequently fix the actual tip distance for that time interval and measure the tunneling current I(U) at the fixed tip position while the tip voltage is swept over the voltage interval ΔU. In order to advance the dynamical range for the measurement the tip to surface distance can be varied in addition in a controlled manner (exponential variation of the current with applied voltage!). Several tricks have been applied, the most fruitful approach seem to be the 'variable separation spectroscopy' invented by Feenstra /6,30/. Another possibility is to acquire the I-U data not only for one or a few points in an STM image but instead for all pixels in the STM image. The acquisition of the 'current imaging tunneling spectroscopy' (CITS/31/) data is performed by moving the tip for each pixel (X_j, Y_k) in the image, controlling the distance in a constant current measurement at a specific tip voltage U_{Top}, and then maintaining this distance while scanning the tip voltage U_i for storing each corresponding current value I_i in the corresponding image($_i$) for pixel (X_j, Y_k). In order to reduce the flood of data, usually the number of voltage points and hence the number of stored images($_i$) is reduced in the CITS measurement with respect to STS at few points.

Following the rapid development of small efficient computers the degree of the experimental techniques to measure STS features has reached a rather high standard. However, although data acquisition seems nowadays not to be the problem, the data explanation still is! Besides the attracting aspect to get direct information for the local density of states spatially on an atomic scale, it has to be realized that a number of approximations enter the interpretation of the acquired data. It has been already emphasized, that equation (11) cannot be deduced from first principle theories but gives just a generalization of the small voltage s-wave results from Tesoff and Hamann where the effect of finite voltage across the tunnel junction is placed into the transmission term T(U,E). Lang /32/ showed, that even under the assumption that this generalization is valid, it will not lead straightforwardly to the interpretation of the tunneling I-U curve to obtain the energy spectrum. It can be hoped although that the derivation dI/dU can be expanded for not to large voltages to extract the relationship with the state density $\rho(r_t$, $E_F + \Delta E)$. Nevertheless, the strong voltage dependence of the transmission term $T(E,U)$ gives rise to an additional complication. In order to evaluate the spectrum in a first approximation, it has been proposed by Feenstra et al. /33-35/ to normalize dI/dU on the conductance I/U which leads to the logarithmic derivation $dlnI$ / $dlnU$. As a result the exponential dependence in $T(E,U)$ cancels out. Difficulties arise of course at zero voltage or close to the band gap which leads to singularities in the logarithmic derivation. A slight smoothing of the I/U term may suppress the singularities /36/. A comprehensive description of the field of STS can be found in ref./6,37/. To recapitulate: a proper treatment of the STS problem is not yet available, data interpretation depends at present on the validity of a number of not fully understood approximations.

Another general problem in STM origins from the difficulty to separate clearly between topographic and electronic features. In practice, e.g. adsorbed gas molecules or atoms often disturb a smooth electronic structure of a metal substrate so that topographic and electronic (spectroscopic) effects cannot be strictly resolved. This is also true for semiconductor surfaces and in both cases the distinction between electronic states and surface topography is complex and the data interpretation in terms of geometric structure of the surface becomes non trivial; e.g. the interpretation of features in CITS images obtained at different separation defining voltages U_{Top} can lead to quite contradictory results unless the complete spectrum of surface states has been examined /53/. In conclusion STM images usually contain a mixture of information from topographic and electronic effects, a separation is often simple for clean metals, not as simple for semiconductors and even more complex for adsorbate covered materials. Adsorbates sometimes show up as discrete objects by monitoring the difference in the STM images for variations of the tip to sample voltage U_{Tip}. They may or may not yield both, protrusions or depressions in the STM images. On the other hand, the lack of clear mass specificity in STM leaves room for the combination with other surface analytical tools!

3. Equipment

Since the advent of STM with the pioneering work of Binnig and Rohrer the number of STM users has expanded tremendously (e.g. the number of participants of the recent STM conferences /2/ went up from initially about hundred to more than thousand). Many companies offer in the meantime the equipment for all different kinds of investigations as for in-air, in-vacuum and in-liquid applications. The widespreading of the field would not be possible without the rapid development of STM probes within the last years. Whereas at the beginning the important credit of Binnig and Rohrer was the belief in the idea, nowadays as it is well established that the STM concept works, it is much easier to develop new STM layouts and a variety of small pocket sized microscopes have appeared. The way of simplifying the instruments led to new (sometimes ingenious) pieces of equipment. The quality of all modern instruments have to be tested by showing atomic resolution for the 'standards': graphite or Si(111)-(7x7).

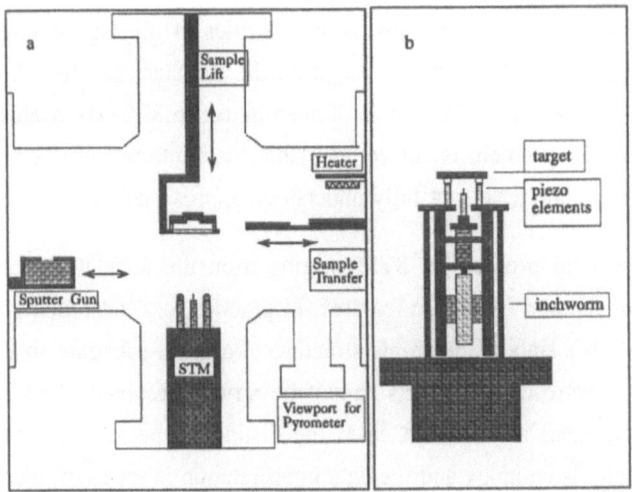

Fig.5. Schematic of a simple UHV STM equipment in a small 4.5" O.D. cross including a) the 'beetle' STM /40/. Alternatively the 'beetle' STM or an inchworm STM on a 2 3/4 " O.D. UHV flange b) can be mounted to the small chamber.

The basic components of an STM are the X-Y and Z scanner elements (piezo ceramics), the coarse approach and the sample holder. Technically, one has to bridge the gap from the 'cm-world' down to the atomic resolution of some Å. In such a microscopic world any unintentional X-Y or Z movement of fractions of an Å either from the tip or the target immediately reduces the quality of the instrument. Among others, the most important sources of the difficulties are the temperature influenced length variation of piezo ceramics (temperature drift) and the coupling to any vibrations from the outside world to the microscope. Vibration isolation in the original Binnig and Rohrer design /1/ was accomplished by suspending the whole microscope at springs and damp the resonant motion of the springs by eddy current damping in connection with permanent magnets. This type of damping is still in use for modern microscopes, however the main effort in many designs is put in a small and ridged construction of the instrument in order to achieve a high resonance frequency of the microscope itself and thus the effort for damping circuits can be minimized. Similarly the influence of temperature drift has been reduced in several STM's by the construction of self-compensating arrangements. The coarse approach, originally made by the famous 'louse' of Binnig and Rohrer /38/ has been replaced either by tricky mechanical screw driven constructions, linear piezo ceramic motors (inchworm, cf. fig.5b) or other computer controllable approaches. Also the rather bulky tripot X-Y an Z scanner is in most cases replaced by a single piezoceramic tube scanner.

Most of the STM measurements presented in this article were performed in a small ultra high vacuum (UHV) chamber with a commercial 'beetle' type microscope /39/. The whole microscope fits on a 2 3/4 " O.D. UHV flange and can be attached directly to the UHV system. Vibration damping of this microscope is achieved by three small pieces of viton below the base plate. Three outer piezo tubes which are terminated by small spherical balls act as support for the sample holder. The piezo ceramic tubes are covered by segmented metallic films in order to actuate movements of the three tubes by application of electric voltages. Consequently, the sample can be manipulated by shifting the sample holder laterally or by rotating it. Rotation of the sample holder leads to an up - or down movement of the target by the particular construction of the holder /40/ and servers as a controlled coarse approach. The tip itself is located at a fourth piezo tube in the middle of the array. Because the three tube elements holding the sample and the middle Z - piezo holder are of the same material, a temperature compensation for the whole equipment occurs. All four piezo tubes are completely fixed in the base plate making the instrument very rigid and leads to high mechanical resonance frequencies. In addition we developed a small (to fit on a 2 3/4 " O.D. UHV flange) inchworm STM (fig.5b) which is independent of specially shaped target holders.

In its simplest configuration (fig.5.) the instrument was operated in a 4 1/2 " O.D. cross as UHV chamber. The sample could be moved in the STM chamber without breaking the vacuum from the microscope into a position where it could be sputter cleaned or loaded with low energetic gas ions (e.g. N^+), it could be also transferred to another position for annealing up to temperatures of $< 1500K$, in the same position also metal evaporation onto the sample could be performed. The piezoceramic elements have been calibrated in normal direction by a Sloan Dectac instrument /41/, and the lateral X, Y magnification factors are usually obtained by imaging the known Si(111)-(7x7) structure. Most of the STM images shown in the following sections have been acquired in the constant current mode within 20sec to 60sec per image consisting of 512 x 512 pixels.

Finally it should be noted that STM investigations of surface properties made in UHV of course require well prepared surfaces which are well defined and extremely clean. Even smallest amounts of contamination hardly or not visible with normal surface analytical tools as e.g. Auger electron spectroscopy immediately show up in STM to disturb the images and often ruins the atomic resolution.

41

Also the tip properties are influenced by adsorbates not only by adsorption directly at the tip but also during scanning in the STM measurements itself just by atom-hopping from the surface to the tip. In fact the tunneling tip remains the most uncharacterized part in the experiment. Only in a few cases the shape has been controlled in situ by field ion microscopy /42,43/. However, even after the preparation of a sharp tip, a monoatomic apex at the tip will be probably destroyed during the first approach to the surface and the actual tip configuration responsible for the measured STM image is formed during the scanning across the sample. Thus in general the tip shape looks different before and after the measurement. Several recipes are known in order to achieve sharp tips which remain stable over larger periods and give atomic resolution /12,13,22,23/. In fact, the not well defined properties of tips in combination with surface contamination also hinders strongly the investigation of non crystalline samples, an example might be amorphous H:Si or organic materials. Whereas at crystalline surfaces the condition of the system tip-sample can be checked by looking into the acquired STM images which should express at least some crystallographic properties of the sample (e.g. appearance of straight lines, crystallographic determined angles, etc.), a similar check is not possible for completely unknown 'arbitrary' structures as e.g for amorphous surfaces. Consequently, the decision wether or not a measured STM feature result from the real structure of non crystalline surfaces or just expresses the poorly defined tip to sample properties remains difficult in these examples.

4. Semiconductors

4.1 Clean Si(100) and Si(111)

The first STM images with atomic resolution have been obtained at Si surfaces /44/. The electronic features, i.e. dangling bonds sticking out of the surface give rise to large corrugation amplitudes and facilitates the distinctions between surface and noise properties in the data sets. The reflection of directional bonds in STM on the other hand alters also the imaged surface topography in the experiment. In case of the Si(001)-(2x1) surface the spatial shift of empty and filled states becomes most obvious. In order to minimize the free surface energy, the initial two dangling bonds per surface atom at Si(001) are reduced to only one by shifting the atoms by about 0.2Å to form pairs of atoms (dimers). As a consequence, filled states are located between the two dimer forming atoms with the two dangling bonds left for the empty states which are spatially separated, pointing away from the dimer complex. In such a reconstruction long dimer rows are formed, there direction changes on every terrace separated by a monoatomic step with 90°. An STM image from the filled states of a clean Si(001)-(2x1) surface is presented in fig.6. Three terraces, separated by monoatomic steps of 1.4Å height are visible. The straight lines are formed by dimer rows, the individual dimers are not resolved in the rows, nor the nuclei position of the two atoms forming one symmetric dimer can be seen in this image. Locked at some defects or step edges zig-zag lines are also visible which reflect the occurrence of non symmetric 'buckled' dimers /45-47/.

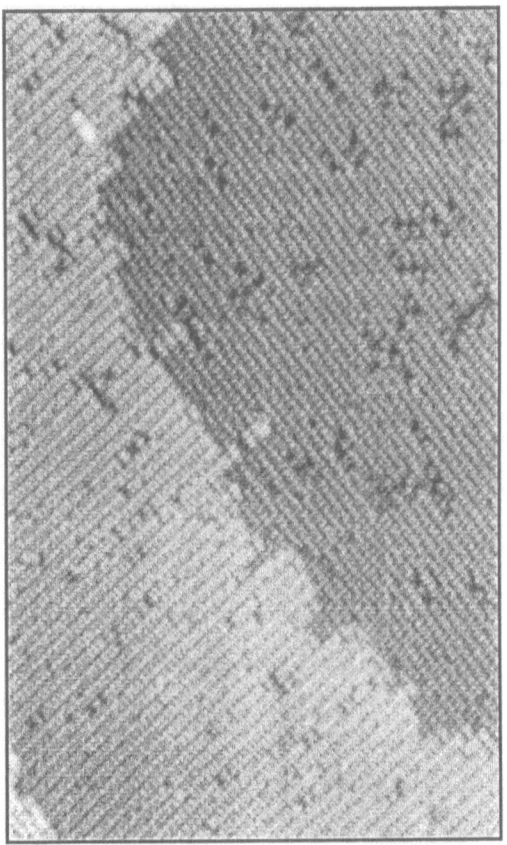

Fig.6. STM topograph of an area about 150Å x 250Å of Si(001)-(2x1). Tunneling from the surface to the tip, i.e. probing filled states. Three terraces separated by single atomic steps of 1.4Å height are shown. From ref./48/.

Reversing the voltage polarity between tip and sample gives the image of the empty states in fig.7. The STM image of an Si(001)-(2x8) reconstructed surface is shown as will be obtained after small metal (e.g.Ni) contamination /48/. The dimer rows although still present on this surface and just separated by some (dark appearing) vacancy channels are hardly visible. These vacancy channels organize themselves after annealing to long rows resulting on average in a (..x8) superstructure. Dimer rows on the large terrace in the lower left of fig.7 are running from the upper left to the lower right. The reason of hiding the dimers in the image results from the measurement of the dangling bonds (empty states) pointing away from the nuclei positions of the Si-Si dimer pair. This is in contrast to filled states imaged in fig.6 being confined between the two dimer atoms.

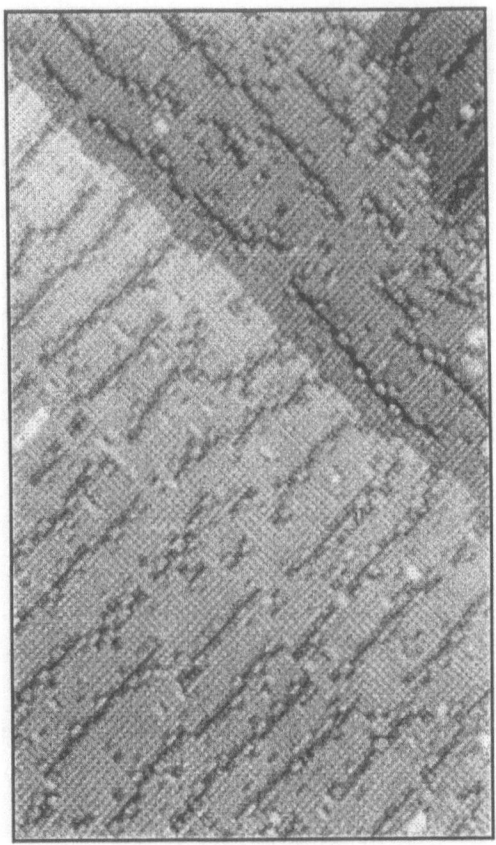

Fig.7. STM topograph of an area about 150Å x 250Å of Si(001)-(2x8). Tunneling from the tip to the surface, i.e. probing empty states. Three terraces separated by single atomic steps of 1.4Å height are shown. The (2x8) superstructure is formed by self organization of vacancy channels. From ref./48/.

The (2x8) superstructure has been explained by the appearance of vacancies and 'split off' dimers occurring along the dimer rows at a distance of approximately every eights of the dimers /48,49/, the cause of which is probably surface stress which will be relieved by this vacancy channel arrangement /50/.

Similar electronic effects appear at other semiconductor surfaces as e.g.. GaAs/26/ or Si(111)/51/. Even the modification in the electronic structure at the surface by the appearance of a stacking fault below the surface atoms which does not influence the position of the topmost layer can be seen with STM. A well known example is the Si(111)-(7x7) surface imaged for filled and empty states, respectively. A sequence of 63 CITS images is shown in fig.8.

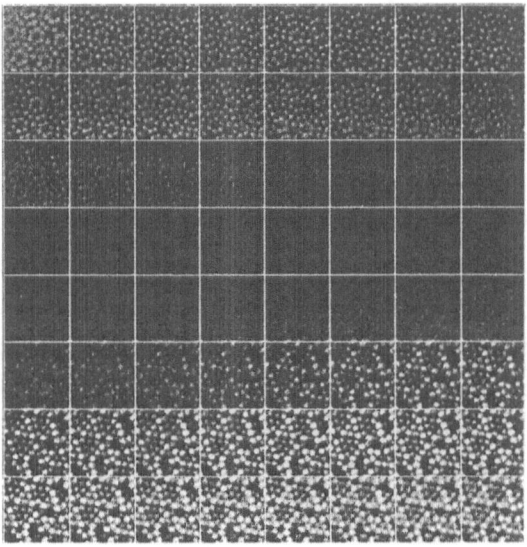

Fig.8.   Spectroscopy at Si(111)-(7x7): First image in the upper left shows the topography of the sampled area of about 60x60Å². A sequence of 63 CITS images is measured from U_{Tip} +1V (upper left) to U_{Tip} -1V (lower right).

The first image ('topography') in the upper left has been obtained with a tip to sample voltage of U_{Top}= +1V at a constant current of 1nA. All other images are measured in the CITS configuration scanning the voltage pixel for pixel /31/, i.e. without activation of the feedback and thus showing the local current distribution for various tip voltages; the second image in the first row starts at U_{Tip}= +1V and for the following images the voltage is decremented by about -0.03V steps down to the last image in the lower right with U_{Tip}= -1V. The images in the upper row show the spatial distribution of empty states and are very much the same as shown in fig.4. The ad-atoms and the corner holes are clearly visible. The middle part appears structureless because no tunneling current can flow in the band gap region of the Si semiconductor. In the lowest row the filled states of the same area as before are imaged. Here again ad-atoms and corner holes are clearly resolved. In addition the six ad-atoms in the upper half of the unit cell clearly look brighter than the remaining six in the lower half. This asymmetry in the image of the unit cell has been shown to be a result of the stacking fault below one half of the unit cell /52-54/.

4.2 Carbide Formation at Si(111)-(7x7)

Imaging of electronic bond states gives also the opportunity to add a limited element specificity to the STM data. We can take advantage of such an effect for monitoring adsorbates or compound formation. When a Si(111) surface is prepared in situ to result in a smooth and clean surface, different procedures have ben applied so far, either Ar^+ sputter cleaning and subsequent annealing at about 1300K or simply annealing at higher temperatures without sputtering. Annealing a Si(111) crystal in good vacuum ($< 10^{-9}$ Torr) at 1500K leads to a surface of good quality consisting of large flat terraces covered by the (7x7) superstructure. However annealing at the same temperature with the pressure rising to about 10^{-8} to 10^{-7} Torr by outgassing walls of the UHV chamber in the neighborhood (basically CO partial pressure) induces some decoration at the surface which can be monitored after cooling down to room temperature as 'Yeti footprints' in the STM images.

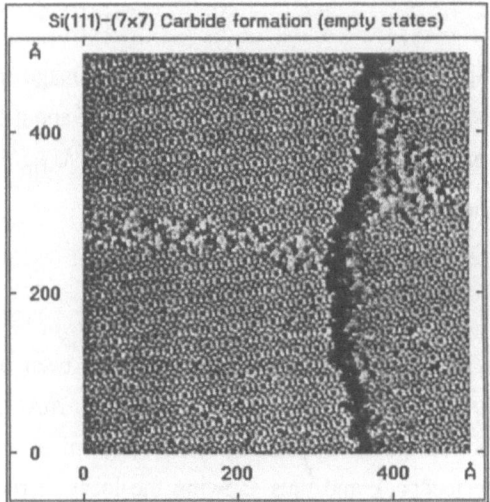

Fig.9. STM image of the empty states from a Si(111)-(7x7) surface showing the onset of carbide formation. $U_{Tip}=-1.0V$, $I_{Tip}=$ 1nA.

Fig.9 shows an STM image of an area of 500x500Å² for negative tip bias, i.e. imaging empty states. Two terraces of Si(111)-(7x7) are visible separated by the monoatomic step running from the top to the bottom. In the lower part of fig.9 some point defects seem to appear on both terraces. The most remarkable features are the defective looking areas which can be followed at the step edge and also occurs as a small stripe -'Yeti footprints'- running almost horizontally across the middle of the upper left terrace. On the basis of this empty state image it is hard or even impossible to decide whether or not all of the features are just due to crystalline defects in the (7x7) superstructure. It can be noted though, that the 'Yeti footprints' separate two (7x7) superstructure domains which will not match if their structures are extrapolated to either sides. The effect can be seen if we follow the trace given by the corner holes from the lower left to the upper right.

Imaging the same area but for a positive tip bias (filled states) immediately shows the difference (fig.10): as 'real defects' remain the point defects on some parts on the terraces (e.g. at the lower part), however the step decoration and the 'Yeti footprints' appear now as protrusions instead of depressions for the empty states. Also the phase boundary at the upper left terrace is now better visible.

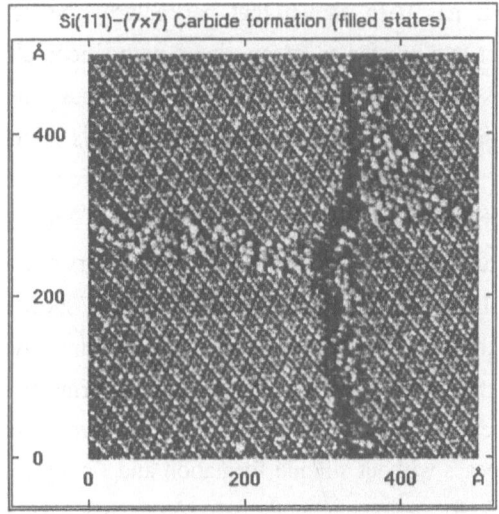

Fig.10. STM image of the filled states from exactly the same Si(111) - (7x7) surface area shown in fig.9. $U_{Tip} = +1.2V$, $I_{Tip} = 1nA$.

From such voltage dependent imaging a clear distinction between defects and ad- or compound- atoms can be drawn. Nevertheless, the species of the 'Yeti footprint' forming atoms cannot be extracted from these measurements, at least not in a straight forward manner. If larger fraction of the surface are covered by this species, which can be obtained by long time annealing at 1500K at poor vacuum, in situ Auger electron spectroscopy showed spurious amounts of carbon. In the STM images large triangular shaped areas with straight contour lines along the <110> directions can be seen, this appears similar to the LEEM /55/ images obtained for the phase transition of (7x7) to (1x1) at higher temperatures. In contrast to such a reversible phase transition the above shown effect is clearly impurity stabilized, probably by Si-carbide formation.

4.3 Growth of V and Fe silicide

High interest in the understanding and description of the metal silicon contact is based on its importance for the semiconductor device technology /56/. Si as a single component semiconductor is used for the majority of electronic devices and the interface from the semiconductor to the outside world is usually performed by metal contacts. Both, (rectifying-) Schottky and simple Ohmic contacts have been observed. In a first example for silicide formation we will look for a silicide of metallic conductance.

Transition metal silicides are of great technological importance and thus many surface analytical studies are already known /56-58/ using techniques as Auger electron spectroscopy (AES), low energy electron diffraction (LEED), ultraviolet photoemission spectroscopy (UPS). It has been shown that particularly the near and noble metals (Ni,Pd,Pt,Ag,Au) react to a greater or lesser extend with silicon already at room temperature to form silicides /59/. The question of room temperature reaction is still controversial for the refractory metal/silicon systems. Some metals including chromium, niobium and titanium are supposed to undergo intermixing at room temperature instead of forming directly the silicide phases. Even no, or only small intermixing has been reported for molybdenum, tungsten and vanadium /60/ metal layers at room temperature. In a comprehensive AES,LEED and UPS study Clabes et al. /61/ showed some basic features of V/Si which makes the system also interesting for a structure investigation with STM. Namely it was found that room temperature deposited V forms an abrupt interface with Si and no reaction occurs, upon annealing, in a temperature regime of 500K to 650K strong intermixing appears without silicide formation and finally at annealing temperatures of 800K the silicide formation starts, the completely reacted silicide interface shows metallic behavior and is composed of vanadiumdisilicide (VSi_2) with no other V silicide components.

Initially V sticks at the crystal sites where it hit the Si(111)-(7x7) substrate coming from the vanadium evaporator. This can be seen in fig.11, which shows a 0.7 monolayer (ML) V coverage on Si(111)-(7x7) as obtained at room temperature deposition. Vanadium atoms or small vanadium clusters are visible as bright spots (protrusions) ontop of the (7x7) superstructure. In contrast to room temperature silicide forming metals as Pd /62/ there is no preferred adsorption site on the (7x7) unit cell.

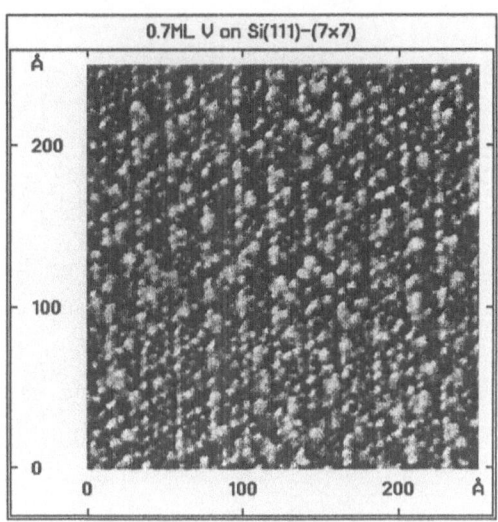

Fig.11. STM image of the filled states from a Vanadium covered Si(111)-(7x7) surface. Coverage: 0.7ML V, evaporated at 300K substrate temperature, no annealing. Small Vanadium clusters are visible on top of the (7x7) superstructure cells. U_{Tip}= +0.85V, I_{Tip}= 1nA.

Annealing at 600K results in vanadiumdisilicide formation, for thin films up to a coverage of approximately 10ML small three dimensional silicide islands are formed. On a large scale STM it can be seen that the silicide islands are spread over the terraces and seem to have slightly lager size at step edges. In a close-up view (fig.12) the reappearance of Si (7x7) structure elements from the initially fully iron covered Si surface can be seen. One may also note in fig.12, that at several areas the (7x7) superstructure is incomplete and (5x5) or (9x9) superstructures do also appear close to the silicide islands. Unfortunately with this solid phase epitaxy there is no success to form smooth continuos V-silicide films. For technological aplications other silicide formation procedures (e.g. co-evaporation of Si and V) have to be introduced.

In addition to metallic like silicides, also narrow band gap semiconductor silicides can be produced. These semiconductors are expected to open a way towards integration of opto- and microelectronic devices on single Si-chips. Namely ß-FeSi$_2$ has been proposed recently for device application since it is known to form an orthorombic crystal lattice and exposes a direct band gap of 0.87eV /63,64/. The investigation of ß-FeSi$_2$ grown by solid phase epitaxi on Si(111)-(7x7) with STM shows a similar growth mechanism as obtained for vanadiumsilicide. Room temperature deposition of Fe on the Si substrate leads also not to silicide formation or Fe-Si intermixing.

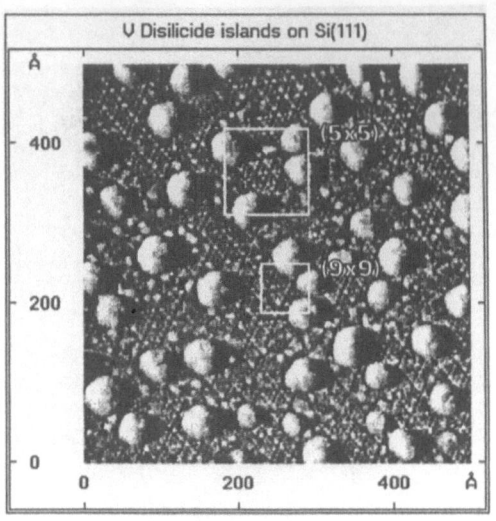

Fig.12. STM image of Vanadiumdicilicide islands formed after five ML Vanadium deposition subsequently annealed at 600K. Note the imperfections in the reappeared Si(111) surface. Besides larger areas of (7x7) also (5x5) and sometimes (9x9) structures are visible (see marked areas). $U_{Tip} = +1.2V$, $I_{Tip} = 1nA$.

From the phase diagram it can be seen that annealing results first in Fe mono-silicide formation which upon heating to higher temperatures (900K) changes into ß-FeSi$_2$. On the basis of an STM investigation we found that annealing of ultra-thin Fe layers (less than 10ML) at 900K rips up also in this case the metal layer and leads to a roughening of the surface (fig.13). Surprisingly enough, the STS measurement clearly shows that this type of iron-disilicide has no band gap and appears to be metallic. On the left side in fig.13 a hexagonal superstructure is visible resulting from the iron silicide, where we also notice the azimuthal alignment of this epitaxial layer with respect to the Si(111)-(7x7) substrate seen on the right side of fig.13. The (2x2) superstructure in combination with the metallic conductance suggests a direct comparison with Ni or Co disilicides /65,66/, which are much better investigated and are known to grow as a cubic phase in a CaF$_2$ lattice type. Thus a new Fe disilicide phase has been found which can be described by a (stressed) cubic lattice. This phase occurs at small coverages of less than 10ML.

Higher initial coverages of Fe lead in fact to the expected semiconducting Fe silicide phase which however shows a rough and faceted surface. Initially, after deposition of thick metal films (~100ML) in both cases, V or Fe, a smooth metal surface with a corrugation of only a few Ångstroms high will be found.

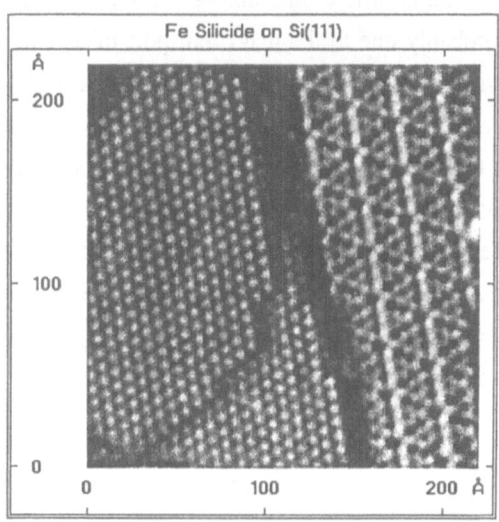

Fig.13. STM image of Ironsilicide on Si(111)-(7x7). After 10Ml Fe deposition at room temperature, the sample has been annealed at 900K to chemically react from iron to the silicide. Left side: silicide, right side Si(111)-(7x7) substrate. $U_{Tip} = +1.8V$, $I_{Tip} = 1nA$.

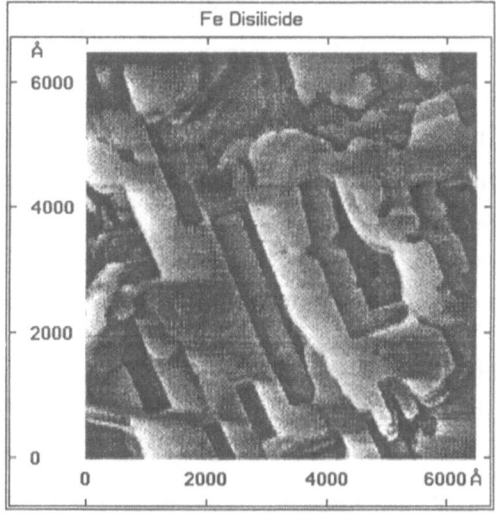

Fig.14. Large area STM image of thick (100ML) FeSi$_2$. Gray scale ranges from 0Å (dark) to a maximum of 90Å (white). Note also the ghost shadows in the image occurring from double tip imaging due to the steep sides of the FeSi$_2$ islands. $U_{Tip} = -1.8V$, $I_{Tip} = 1nA$.

Annealing at temperatures where intermixing and reaction does occur immediately roughens the surface, probably due to the mass transport of Si and the formation of the disilicide compounds. In the large scale survey in fig.14 of such a surface the development of several flat Fe disilicide islands can be seen. As a consequence of the large corrugation ranging up to 80Å in height one also may notice several 'ghost structures' in the image. Such 'ghost structures' appear quite frequently at sharp and steep surface structures and are the result of imaging the very same surface area with different parts of the tip during scanning. Care has to be taken, when in the STM images several equally shaped surface structures appear, normally an effect of multiply imaging the same object with a 'double' or 'multiple' tip.

It has been shown that scanning tunneling microscopy gives a new instrumentation to uncover the surface structure of semiconductors on an atomic scale. The scanning tunneling spectroscopy helps to identify the local electronic structure. Some element specificity can also be achieved in special cases by voltage dependent microscopy. In general the experimental STM data of semiconductors are characterized by there high quality of the signal to noise ratio.

5. Metals

STM investigations of metal surfaces require considerable lower noise background in order to maintain atomic resolution as compared with semiconductor surfaces. The cause has already been discussed in chapter 2 and is related to the different bond configurations and the resulting small corrugation amplitudes of less than 0.1Å for metals. Hence, recent STM investigations of metal surfaces showing atomic resolution are rare compared with semiconductors. It can be accentuated, that a variety of surface analysis problems can be solved also without explicit atomic resolution. An example will be given below (chapter 5.4) in case of the binary alloy surface of NiAl.

Although the situation for topography discovered by the STM is better for metals, in case of adsorbates the overlap with electronic information remains still a problem. A combination with other structure sensitive techniques provides more reliable structure analysis. In particular the ion scattering techniques show a direct measure for nuclei positions and are therefore well suited for such combined investigations. A brief description of one of the methods shall be included here.

In low energy noble gas impact collision ion scattering spectroscopy with neutral detection (NICISS /67/) the formation of the shadow cone fig.15a) and blocking cone fig.15b) can be directly used in order to determine the backscattered flux of particles.

The method has been described already in detail /4,68/ and can be summarized as follows: a low intensity noble gas ion beam is focussed onto the surface and the backscattered fraction of the beam is measured in a certain energy window which makes the method also mass discriminative as a function of the incidence angle ψ_{in}. With increasing ψ_{in} the shadow cone is swept over the neighbor atoms of the first layer. At a critical incidence angle ψ_c a steep increase of the backscattering intensity is measured.

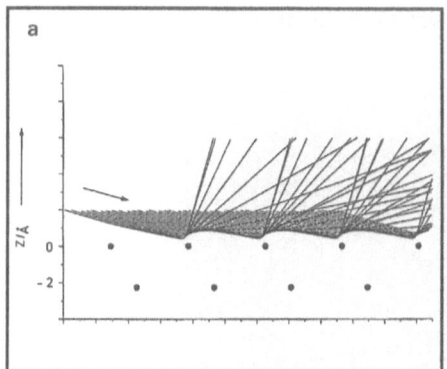 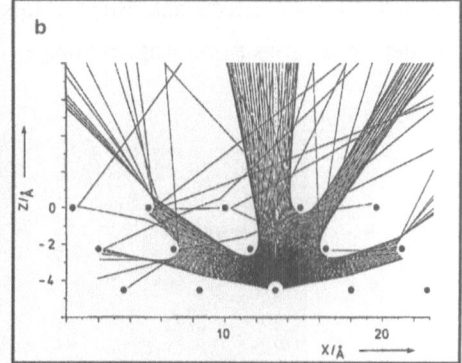

Fig.15. Ion scattering: a) Calculated ion trajectories for grazing angle of incidence. Two dimensional plot for 1000 eV Ne$^+$ ions impinging in a parallel beam from the left side. Scattering at the Pt atoms in the first layer leads to the shadow cone formation. b) Blocking cone formation by ion trajectories starting from a point source (scattering center in an atom in the third layer). From ref./4/.

The shape of the intensity variation with ψ_{in} is basically determined by the strong trajectory focusing at the shadow cone (fig.15a) and blocking cone (fig.15a) edge of involved atoms. Up to grazing angles of incidence of roughly $\psi_{in} < 30°$ the NICISS patterns contain basically information owing to the first layer. At larger angles ψ_{in} the deeper layers start to contribute to the spectra (shadow cone focusing onto second and third layers). The advantage of combination of STM with NICISS will be demonstrated below.

5.1 Cu(110)

The STM image of clean Cu(110) obtained with atomic resolution has been already shown in fig.3. The in situ surface preparation of such a surface when it is finally mounted into the UHV system after careful ex situ orientation and polishing /69/ involves Ar$^+$ sputtering and consequent annealing. The fcc (110) surfaces of a variety of metals seem to undergo easily surface reconstruction often of the missing row type which involves quite much of mass transport for moving surface atoms.

Surprisingly, the mass transport of Cu atoms needed for the reconstruction can be provided even at room temperature. In fig.16 the time evolution of step edges on a clean Cu(110) surface is shown by imaging subsequently the same 500 x 500 Å^2 area in 15sec time intervals. Although the average position of the steps is not changed (in fact the steps are pinned at certain defects not visible in this section of the image), the individual shape of the step edge varies from image to image.

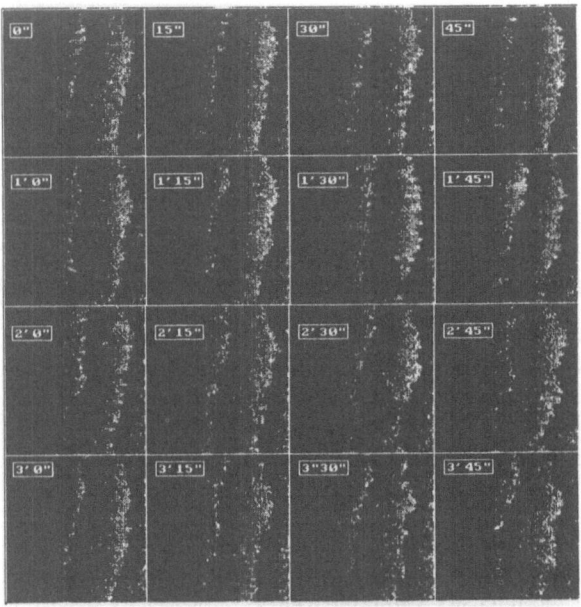

Fig.16.   Mobility of Cu atoms at step edges on room temperature clean Cu(110): STM images of the same 450 x 450 Å^2 area taken sequentially every 15sec. Note that the individual shape of each of the imaged two single step edges varies from image to image.

The movement of the step edges has been found to be independent of the STM X-Y scan direction and can be seen for Cu only on the clean surface. Similar results for Cu are reported by Jensen et al./70/. Also at Ag surfaces the high mobility of surface atoms at room temperature has recently been demonstrated in an STM experiment /81/.

Once an adsorbate covered domain is formed e.g. Cu(2x1)O or Cu(2x3)N, steps are immediately locked, usually exposing low-indexed step directions and their shape remains constant, a step configuration locked by oxygen is seen in fig.17.

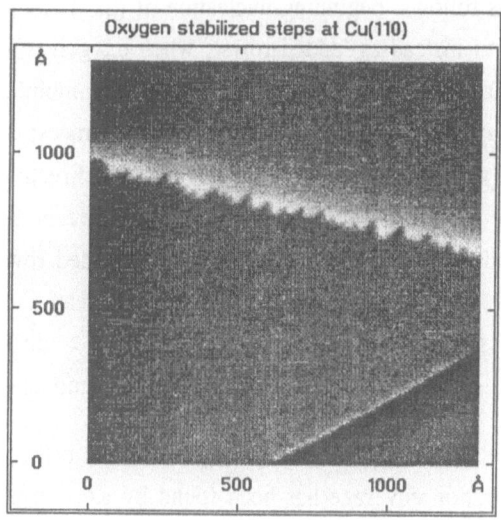

Fig.17. STM images at Cu(110) with oxygen locked single step edges. U_{Tip}= 200mV, I_{Tip}= 1nA.

Here the possibility of following dynamic evolutions are a real advantage of STM. Although nowadays in many STM investigations controlled temperature variation of the sample still remains problematic, which in fact confines the possibility of this type of measurements to a limited number. Sometimes also a possible interaction of the tip and surface cannot be totally excluded, so that changes induced by the STM measurement itself might also occur. But in general the possibility of monitoring surface dynamics, individual step shapes and defects are positive credits of STM.

5.2 O/Cu(110)

The oxygen copper system has been studied extensively beginning with the pioneering work of Ertl /71/, who already suggested a missing row (MR) model from the occurrence of strong half order LEED reflexes in the Cu(110)-(2x1)O system. However it took more than twenty years to get a complete agreement on the development of the MR structure in case of the 0.5 monolayer saturation coverage for the (2x1) superstructure /70,72-74/. The growth mechanism of the oxygen induced (2x1) phase has recently been uncovered by STM investigations of Coulman et al. /73/. Since the first proposal of the oxygen induced missing row reconstruction the question has arisen how the necessary transport of matter during the development of the reconstruction happens. Thanks to the STM investigations /70,73/ this problem has been resolved in the last years and the process can be viewed as follows: beginning nucleation of (2x1) elements does not occur as single MR elements but instead as 'added rows'; when oxygen is adsorbed on terraces it may be locked into O-Cu- components by aggregation with mobile Cu atoms on the terraces, which can evaporate from step edges onto the terraces; additional oxygen and copper then starts to form long O-Cu- chains in <001> direction. These long one dimensional strings are then also aggregating in <110> direction leaving always a one row space between adjacent O-Cu strings. Neighboring added rows are 5.1Å apart, i.e. the lattice spacing in <110> is twice as in the case of clean Cu(110) and can be also described by the 'missing' of every other Cu row in <001> direction. Finally, in case of the saturated (2x1) structure missing row and added row model are equivalent.

Recently a novel phenomenon of long range self organization of the (2x1) domains occurring below saturation coverage has been found by Kern et al. /75/ in a TEAS experiment. Upon annealing of an oxygen covered Cu(110) surface with e.g. $\Theta_0 \approx 0.25$ ML at 640K and subsequent cooling to room temperature they found a periodic supergrating with a spacing from 140Å to 60Å (!), depending on the surface preparation conditions. A real space investigation can be made by STM to substantiate this astonishing finding. After annealing the clean Cu(110) crystal at 900K, the crystal was exposed subsequently to about $2x10^{-6}$ Torr s oxygen while cooling the sample down to room temperature. A constant current image of a 1400 x 1400 Å² area of the partially oxygen covered Cu(110) surface is shown in Fig.18. Indeed, long range ordering can be seen in such a 'piano keyboard' like pattern!

It is interesting to realize, that in spite of the fact that the O-Cu added rows are located above the remaining clean Cu areas, owing to the different local electronic structures of clean and oxygen covered Cu, the Cu areas appear as bright, and the (2x1) Cu-O areas as dark stripes. A similar effect for imaging oxygen on copper has been found by Coulman et al. /73/ who showed that at low bias voltages (also used in this investigation) (2x1) Cu-O islands are imaged as concavities instead of protrusions on clean Cu metal areas. Thus it has to kept in mind, that the apparent corrugation ontop of the large terraces between Cu (2x1)O and Cu (1x1) areas is not a topographic but rather an electronic effect.

The 'piano keyboard' is typical for a 0.25ML oxygen coverage (half saturation) of the annealed oxygen copper surface and depicts clearly the long range spatial self organization of the (2x1) Cu-O islands. In the high coverage STM image of fig.19 the aggregation of individual Cu-O rows (along <001>) into stripes separated by Cu(1x1) islands can be clearly recognized. In this image the two stripes consist of 11 and 14 rows, respectively. Also at low oxygen coverages a tendency of saturation in the aggregation of the Cu-O added rows has been found.

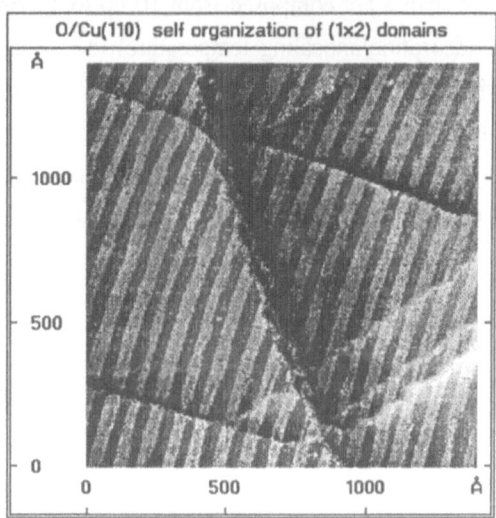

Fig.18. Long range self organization of Cu(110)-(2x1)O islands at oxygen coverages below saturation: 'piano keyboard' image of a 1400 x 1400 Å² area covered about half with Cu(2x1)O domains. Stripes of Cu(1x1) (bright) and Cu(2x1)O (dark) domains are running in <001> direction, Note: stripes are crossing the monoatomic steps but not the multiple step array at the middle. U_{Tip} = -780mV, I_{Tip} = 1nA.

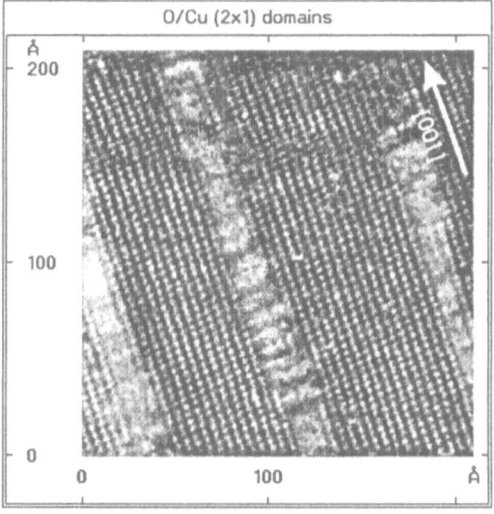

Fig.19. STM image for high coverage Cu(2x1)O islands at Cu(110). Basically two Cu(2x1)O stripes consisting of 11 and 14 O-Cu- rows are visible. The bright islands separating the O-Cu rows are uncovered Cu areas. The <001> direction is indicated. U_{Tip} = 60mV, I_{Tip} = 1nA.

Typically, after the stripes are composed from 10 up to 15 rows (50 to 80Å wide, depending also on the adsorption/annealing temperature) they stop to grow in <110> direction and upon additional oxygen exposure new islands at different parts on the Cu(1x1) surface start to grow instead. Such a distinct saturation behavior might be explained either by a stress induced confinement (elastic interaction) of the width (in <110> direction) for individual arrays (stripes) of the long O-Cu added rows (in <001> direction) or alternatively by electrostatic interaction.

5.3 N/Cu(110)

The synergic effectiveness of combining STM with other techniques can be also seen for the nitrogen induced (2x3) reconstruction at Cu(110). Heskett et al. /76/ have recently reported in a LEED, TDS and HREELS study of the adsorption of atomic nitrogen on Cu(110) that upon annealing the N-loaded surface to 600 - 700K a very sharp and stable (2x3) LEED pattern is developed. These results were interpreted as a nitrogen-induced reconstruction of the Cu(110) surface. In addition, based on the HREELS data, the nitrogen atoms were assigned to be bonded in the long bridge position sites and similar to the bulk copper nitride, Cu_3N complexes may be formed.

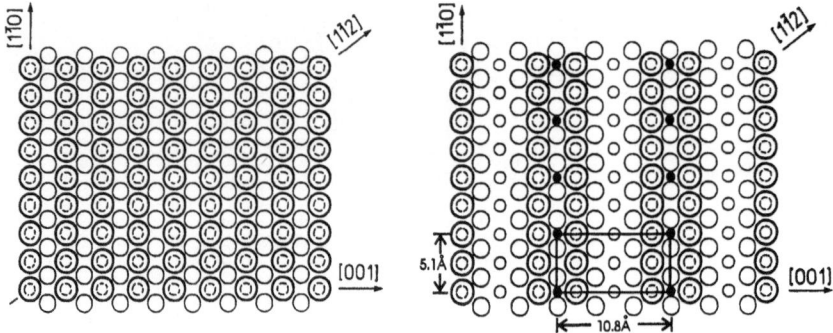

Fig.20. Surface structure model of the clean (1x1) (left) and nitrogen induced (2x3) reconstructed (right) Cu(110) surface. On top view of the first four monolayers. Bold circles: first layer Cu atoms; filled dots: nitrogen atoms in the long bridge position. The unit cell of the (2x3) structure is indicated. From ref. /77/.

The model of the clean and reconstructed surface as deduced from a combined structure analysis is shown in fig.20 First the NICISS measurements for the <112> scattering plane shall be discussed /77/. Here obviously only one type of array consisting of Cu atoms normal to the surface can contribute to the in-plane scattering shown in the inset in fig.21. The intensity of 180° backscattered He projectiles as a function of the angle of incidence is plotted.

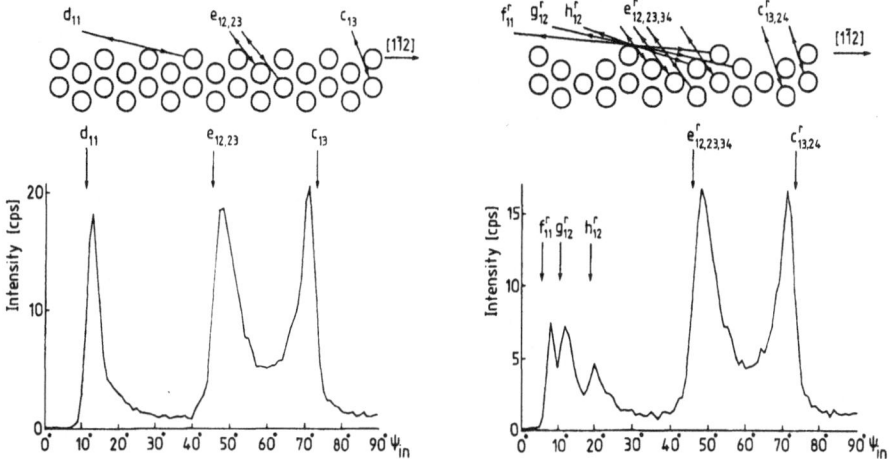

Fig.21. Intensity of He backscattered at 180° by head-on collision with Cu atoms versus the angle of incidence ψ_{in} in the <112> scattering geometry. insets show the scattering plane normal to the surface and the critical trajectories. a): (left side) clean Cu(110)-(1x1). b): (right side) N/Cu(110)-(2x3). From Ref. /77/.

In case of the head on collision geometry (sketched in the insets) a strong intensity peak is expected. On clean well ordered Cu(110)-(1x1) the NICISS pattern shown in fig.21a is obtained: at low angles of incidence ψ_{in} the intensity of projectiles backscattered from Cu atoms is very small, indicating a smooth and almost defect free surface; at $8° < \psi_{in} < 20°$ the head-on backscattering from surface atoms leads to a sharp 'surface peak' (indicated as slope d_{11}) and finally, at larger ψ_{in} the backscattering from second and third layer Cu atoms leads to two additional peaks (slopes $e_{12,23}$ and c_{13}).The occurrence and angular position of the peaks can be easily understood on the basis of the two dimensional in-plane atom arrangement normal to the surface (see also inset in fig.21a).

The corresponding NICISS pattern of the Cu(110)-(2x3)N surface is shown in fig.21b. The appearance of five edge slopes in spite of only three for the non reconstructed surface (fig.21a) immediately shows that nitrogen induces in fact a reconstruction of the Cu surface. The origin of the five slopes is schematically shown in the inset in fig.21b. From the peak position and peak shape near $\psi_{in} \approx 15°$ (slope g^r_{12} and d^r_{11}) a modified missing row structure can be deduced as shown in fig.20. It is a missing row structure with every third <110> row is missing. This leads to a threefold periodicity in the <001> direction. Additional NICISS measurements in the <110> scattering plane show practically no difference between the patterns of the Cu(110)-(1x1) and Cu(1110)-(2x3)N surface, hence a Cu reconstruction in the <110> direction can be excluded. Consequently, the twofold periodicity in <110> has been ascribed to the arrangement of the nitrogen atoms, which are probably located in the long bridge positions as already proposed from the HREELS data /76/ shown in fig.20. Such a structure results in a chainlike structure in <110> direction.

The STM investigation /20/ shows directly the chainlike structure in the high resolution image shown in fig.22 (52Å x 52Å). The distance between equidistant chains is determined from the STM image to 11Å in the <001> direction. The distance between two minima within each chainlike structure in the <110> direction is measured to 5Å. This is very close to the dimensions of the (2x3) unit cell of 10.8Å x 5.1Å as deduced from the schematic structure shown in fig.20. Each chain link consists of an oval black center hole enclosed between four bright spots. The distance between the spots is approximately 3.5Å and 5Å in the <001> and <110> directions, respectively.

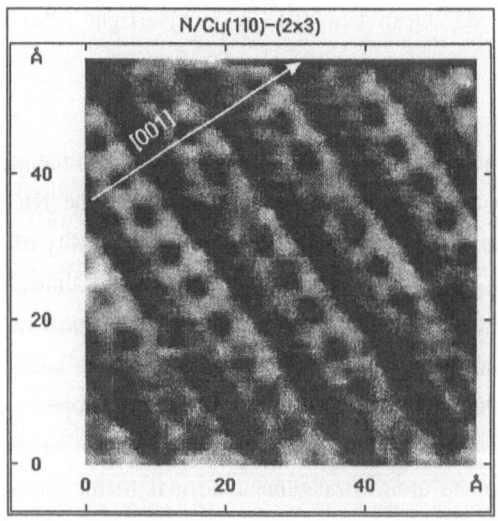

Fig.22. High resolution STM image of the Cu(110)-(2x3)N surface. Surface area 52x 52Å². Chains are running from the upper left to the lower right in <110> direction. Gray scale: 0Å (dark) to 0.5Å (white); $U_{Tip} = -120mV$, $I_{Tip} = 1nA$. From ref. /20/.

In view of the structure model in fig.20, where the relative positions of the Cu atoms are accurately determined from the NICISS patterns, the bright spots in the STM image can be faithfully associated with Cu atoms. Note, however, that unlike the non reconstructed Cu surface shown in fig.3, not all Cu atoms present in the close packed <110> rows (mutual Cu-Cu distance 2.55Å) are imaged in fig.22 but only every other Cu atom instead. Knowing positively from the NICISS measurements, that the mutual positions of the Cu atoms within the <110> rows on the (2x3) reconstructed surface are practically the same as on the non reconstructed (1x1) surface, the absence in the STM image of every other Cu atom in the <110> rows cannot result from a topographic effect, but has to be ascribed to the influence of the N atoms. It appears that similar as for oxygen on Cu, the change in the electronic state of a Cu atom due to the bonding to the N atom changes its 'visibility' in the STM pattern.

Similar to the depressions imaged by STM for the position of oxygen we here are inclined to associate the holes in the chains with the location of the N atoms. In any case, from the fact, that the visibility of every other Cu atom is changed we can clearly discard the fourfold site for nitrogen and conclude that the N atoms are located in the long bridge position in agreement with ref.76 (cf. fig.20). On the basis of the STM and NICISS data, the nitrogen coverage cannot be determined exactly. Also AES and XPS give contradictory results /77,80/. Hence additional nitrogen compared with the N atoms shown in fig.20 might be located in <110> troughs and cannot be excluded at present. These additional N atoms will not alter the general features of the reconstruction model shown in fig.20 and are supposed to yield in the same NICISS patterns, as well as the same STM images.

In summary, the combination of NICISS and STM resulted in the finding of a new structure model for the nitrogen induced (2x3) reconstruction of Cu(110). The modified missing row model for the Cu atoms was first based on NICISS measurements. STM directly shows and fully confirms the model. It excludes the fourfold hollow site position for the N atoms and confirms the bridge location as also deduced from HREELS measurements. The double periodicity in <110> directions is directly visible in the STM images and in accordance with the arrangement of the N atoms along <001> lines separated by 5.1Å.

5.4 NiAl(111)-(1x1)

The NiAl(111) bcc lattice results in a rather open surface. The truncated bulk positions would yield either a Ni or an Al terminated surface. NICISS is expected to uncover the surface type easily: at grazing incidence the surface peak should be indicative for He backscattering from Ni or Al surface atoms. From NICISS data obtained at slightly oxygen contaminated surfaces a domain mixing of Ni and Al terminated surfaces has been found /78/. As steps are always present at surfaces, a surface configuration with pure Ni and pure Al domains separated by mono-atomic steps might explain the data. Domain mixing for similarly prepared surfaces has been independently deduced from a detailed LEED analysis of NiAl(111) /79/.

In order to prove the step height distribution (mono- vs. double- step height) immediately, the NiAl(111) surface was imaged by scanning tunneling microscopy /25/.

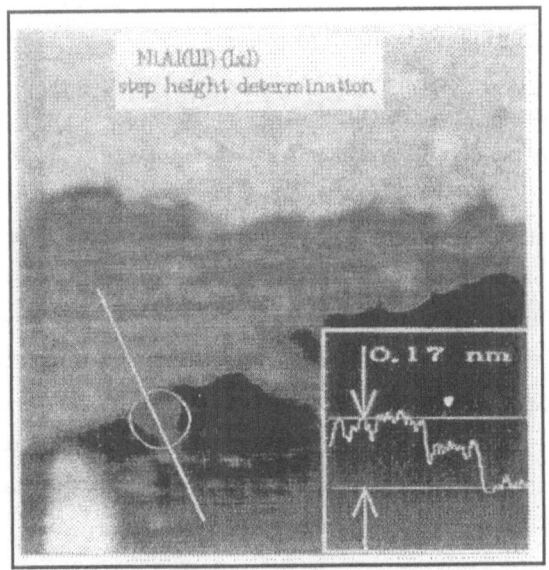

Fig.23. STM image of a 200 x 200Å NiAl(111) surface area. Three terraces separated by steps are shown. One triangular shaped Al island can be seen adjacent to one of the step edges. For the step height determination a line scan crosses this triangular domain going from the upper to the lower terrace. The step height of the terraces is measured to 1.7Å and the domain height to about 0.8Å, i.e. double- and single-atomic steps, respectively. U_{Tip} = -600mV, I_{Tip} = 1nA. From ref./25/.

It has been found that terraces were covered by small (on average 25Å wide) triangular islands. In a few cases it was possible to locate one of those islands adjacent to a step separating two terraces. Such a configuration directly shows a comparison of the heights of steps and islands. An example is given in fig.23 for an area of 200 x 200 Å.

The line scan shown has been positioned in such a way that it crosses the triangular domain near the step edge when going from the upper (gray) terrace down to the lower (dark) terrace. At the right side of fig.23 the measured vertical height along the line scan is plotted and the mean terrace step height of 1.7 Å is also indicated. All over the surface measured at many areas, we never found terrace step heights considerably less than 1.7Å. Surprisingly enough, this step height shows that only double steps occur. It can be also easily recognized that the height of one triangular island is roughly half of the double step height, i.e. about 0.8 Å.

Summarizing the STM data, we found large flat terraces separated by double atomic steps. The terraces are covered with small triangular islands a single step high azimuthally ordered with respect to the substrate. These findings strongly suggest that the terraces consist of one atom species (Ni as has been identified by NICISS, see fig.24 below) covered with small domains of the other alloy component (Al) which may have segregated over the top layer under the influence of oxygen.

On the basis of in situ AES measurements, small amounts of oxygen were detected, which could only be removed by annealing at temperatures above 1300K. Consequently, these clean surfaces where reinvestigated with NICISS. The 180° backscattering patterns for the Ni and Al signals are shown in fig.24.

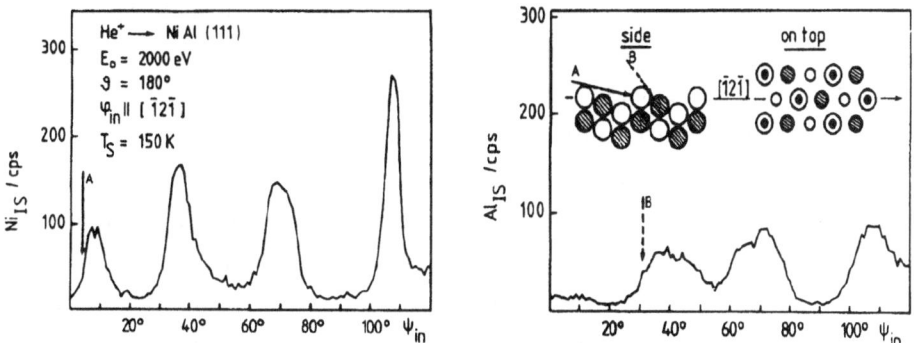

Fig.24. NICISS pattern for 180° backscattering at clean NiAl(111). He backscattered at a) Ni- and b) Al- atoms. The inset shows a schematic side and on top view of the scattering geometry. Open circles: Ni; hatched circles: Al atoms. Head-on backscattering situation from first layer atoms (A) and second layer atoms (B) is indicated by arrows. From ref./25/.

The most striking difference with the oxygen contaminated surface /78/ can be recognized from the Al signal in fig.24b: there is no Al first layer peak! The onset of the Al_{IS} signal occurs at $\psi_{in} > 25°$ with peak B, the second layer signal. The NICISS pattern of the Al line (fig.24b) shows also that the second layer indeed contains Al atoms (peak B) and that there is not an Al depletion region due to the annealing treatment. The Ni_{IS} signal in fig.24a obviously exhibits the first layer peak A at grazing angle of incidence. Thus for oxygen free NiAl(111) the first layer consists exclusively of Ni atoms. Small amounts of oxygen have a large effect on the surface conditions and lead in particular to Al surface segregation resulting in small, about 25Å wide Al islands (probably with oxygen in subsurface location) on large Ni terminated terraces. Such an oxygen contaminated surface appears in LEED or ion scattering investigations as a surface with Ni and Al terminated domains (domain mixture). Finally it should be noted, that also in this'investigation just the combination of the two techniques - NICISS and STM - lead to a consistent surface structure model. Basically the step height has been determined by STM whereas the atomic species was discovered with NICISS.

6. Conclusions

STM investigations at crystalline well defined surfaces in ultra high vacuum improved significantly the understanding of mechanistic processes occurring on the nanometer scale and opened a new way for the characterization of the surfaces with respect to their local electronic and topographic structures. For semiconductors the STM images are dominated by the electronic features (dangling bonds) leading to large measured corrugations which makes experiments relatively easy. However, the transformation of measured local I-U curves into the local state density function ρ is at present only partially successful, in particular, effects of the tunneling matrix elements on the deconvolution are not completely solved yet. By acquiring a large body of local I-U curves including voltages close to the work function (larger tip-to-sample separations) the tunneling matrix can be approximated for these higher voltages (here ρ is only a small perturbation element to the measured tunneling current). The hope is to extrapolate the matrix elements towards small voltages when ρ becomes in fact the dominant factor in the tunneling current. Clean metal surfaces are supposed to express directly the topographic features; here the interpretation of the images becomes easy, however, the experiments are more delicate because of the measured small corrugation of less than 0.1Å. The effects of surface dynamics (step evolution), adsorbate induced restructuring of surfaces and detailed defect analysis are only a few of the promising application of STM. Mass specificity remains still a problem for STM and furthermore the mixture of topographic and electronic information in STM points to the fruitfulness of a combination with other surface structure sensitive techniques. In this context the complimentary character of STM and ion scattering has been demonstrated.

Acknowledgement

It is a pleasure to acknowledge the fruitful discussions with W.Raunau, who also allowed to preprint some STM images for silicide formation from his thesis work.

References

1. G.Binnig, H.Rohrer, Ch.Gerber, E.Weibel, Phys.Rev.Lett.**49**, 57 (1982)
2. Proceedings of the Second Intern. Conf. on STM, J.Vac.Sci.Technol.**A6** (1988)
   also, Proceedings of the Third Intern. Conf. on STM, J.Microscopy **152** (1988)
   also, Proceedings of the Fourth Intern. Conf. on STM, J.Vac.Sci.Technol.**A8** (1990)
   also, Proceedings of the Fifth Intern. Conf. on STM, J.Vac.Sci.Technol.**B9** (1991)
3. Scanning Tunneling Microscopy and related Methods, Eds. R.J.Behm, N.Garcia, H.Rohrer, NATO ASI SERIES E: Applied Science Vol.**184** Kluwer Academic Publishers, Dordrecht NL, 1990
4. H.Niehus, R.Spitzl, Surf. Interface Anal. **17**, 287 (1991)
5. J.Bardeen, Phys. Rev. Lett. **6**, 57 (1961)
6. R.M.Feenstra, in :Scanning Tunneling Microscopy and related Methods, Eds. R.J.Behm, N.Garcia, H.Rohrer, NATO ASI SERIES E: Applied Science Vol.**184** , p.211 ff Kluwer Academic Publishers, Dordrecht NL, 1990
7. R.M.Feenstra, W.A.Thompson, A.P.Fein, Phys. Rev. Lett. **56**, 608 (1986)
8. C.B.Duke, Tunneling in Solids (Academic Press, New York, 1969), p.253
9. J.Tersoff, D.R.Hamann, Phys. Rev. Lett. **50**, 1998 (1983), Phys. Rev. **B31**, 805 (1985)
10. A.Selloni, P.Carnevali, E.Tosatti, C.D.Chen, Phys. Rev. **B31**, 2602 (1985)
11. B.Poelsema, G.Comsa, 'Thermal Energy Scattering from Disordered Surfaces', Springer Tract. in Modern Physics 115 (Springer, Berlin Heidelberg 1989)
12. V.M.Hallmark, S.Chiang, J.F.Rabolt, J.D.Swalen, R.J.Wilson, Phys. Rev. Lett. **59**, 2879 (1987)
13. J.Wintterlin, J.Wiechers, H.Brune, T.Gritsch, H.Höfer, R.J.Behm, Phys. Rev. Lett. **62**, 59 (1989)
14. J.V.Barth, H.Brune, G.Ertl, R.J.Behm, Phys. Rev. B42, 9307 (1990)
15. E.Kopatzki, R.J.Behm, Surf. Sci. **245**, 225 (1991)
16. F.Jensen, F.Besenbacher, E.Laesgarard, I.Stensgaard, Phys. Rev. **B41**, 10233 (1990)

17. R.Feidenhans'l, F.Gray, M.Nielsen, F.Besenbacher, F.Jensen, E.Laesgarard, I.Stensgaard, K.W.Jacobsen, J.K.Norskov, R.L.Johnson, Phys. Rev. Lett. **65**, 2027 (1990)

18. A.Samsavar, E.S.Hirschorn, T.Miller, F.M.Leibsle, J.A.Eades, T.C.Chiang, Phys. Rev. Lett. **65**, 1607 (1990)

19. D.Coulman, J.Wintterlin, J.V.Barth, G.Ertl, Surf. Sci. **240**, 151 (1990)

20. H.Niehus, R.Spitzl, K.Besocke, G.Comsa, Phys. Rev. **B43**, 12619 (1991)

21. C.J.Chen, J. Vac. Sci. Technol. **A9**, 44 (1990)

22. Ch.Wöll, S.Chang, R.J.Wilson,P.H.Lippel, Phys. Rev. **B39**, 7988 (1989)

23. J.E.Demuth, U.Köhler, R.J.Hamers, J.Microsc. **151**, 299 (1988)

24. S.Ohnishi, M.Tsukuda, Solid State Commun. **71**, 391 (1989)

25. H.Niehus, W.Raunau, K.Besocke, R.Spitzl, G.Comsa, Surf. Sci. **225**, L8, (1990)

26. R.M.Feenstra, J.A.Stroscio, J.Tersoff, A.P.Fein, Phys. Rev. Lett. **58**, 1192 (1987)

27. R.S.Becker, J.A.Golovchenko, D.R.Hamann, B.S.Schwartzentruber, Phys. Rev. Lett. **55**, 2032 (1985)

28. G.Binnig, H.Rohrer, F.Salvan, Ch.Gerber, A.Baro, Surf. Sci. **157**, L373 (1985)

29. R.M.Tromp, R.J.Hamers, J.E.Demuth, Phys. Rev. **B34**, 1388 (1986)

30. C.K.Shih, R.M.Feenstra, G.V. Chandrashekhar, Phys. Rev. **B43**, 7913 (1991)

31. R.J.Hamers, R.M.Tromp, J.E.Demuth, Phys. Rev. Lett. **56**, 1972 (1986)

32. N.D.Lang, Phys. Rev. **B34**, 5947 (1986)

33. J.A.Stroscio, R.M.Feenstra, A.P.Fein, Phys. Rev. Lett. **57**, 2579 (1986)

34. R.M.Feenstra, P.Martensson, Phys. Rev. **B39**, 7744 (1989)

35. J.A.Stroscio, R.M.Feenstra, J.Vac.Sci.Technol. **B6**, 1472 (1988)

36. R.M.Feenstra, P.Martensson, Phys. Rev. Lett. **61**, 447 (1988)

37. J.Tersoff, in: Scanning Tunneling Microscopy and related Methods, Eds. R.J.Behm, N.Garcia, H.Rohrer, NATO ASI SERIES E: Applied Science Vol.**184** , p.77 ff Kluwer Academic Publishers, Dordrecht NL, 1990

38. G.Binnig, H.Rohrer, Ch.Gerber, E.Stoll, Surf.Sci. **144**, 321 (1983)

39. K.H.Besocke, Surf.Sci.**181**, 145 (1987)

40. K.H.Besocke, Delta Phi Electronic, Postfach 2243, D-5170 Jülich, Germany

41. Sloan Dektak II, Sloan Technol. Corp., Santa Barbara, CA, USA

42. Th. Michely, K.H.Besocke, M.Teske, J.Microsc. **152**, 77 (1988)

43. Y.Kuk, P.J.Silverman, H.Q.Nguyen, J.Vac.Sci.Technol. **A6**, 524, (1988)

44. G.Binnig, H.Rohrer, Ch.Gerber, E.Weibel, Phys. Rev. Lett. **50**, 120 (1983)

45. R.J.Hamers, R.M.Tromp, J.E.Demuth, Phys. Rev. **B34**, 5343 (1986)

46. A.J.Hoeven, D.Dijkamp, J.M.Lenssinck, E.J.van Loenen, J.Vac.Sci.Technol. **A8**, 3657 (1990)

47. R.J.Hamers, U.K.Köhler, J.E.Demuth, J.Vac.Sci.Technol. **A8**, 195 (1990)

48. H.Niehus, U.K.Köhler, M.Gopel, J.E.Demuth, J.Microsc. **152**, 735 (1988)

49. A.Tamura, J.Vac.Sci.Technol. **A8**, 192, (1990)

50. J.A.Martin, D.A.Savage, W.Moritz, M.G.Lagally, Phys. Rev. Lett. **56** 1936 (1986)

51. R.S.Becker, B.S.Swartzentruber, J.S.Vickers, T.Klitsner, Phys. Rev. **B39**, 1633 (1989)

52. H.Neddermeyer, S.Tosch, Festkörperprobleme **29**, 133 (1989)

53. Th.Berghaus, A.Brodde, H.Neddermeyer, Surf.Sci.**193**, 235 (1988)

54. J.E.Demuth, U.Köhler, R.J.Hamers, J.Microsc. **152**, 299 (1988)

55. W.Telips, E.Bauer, Ultramicroscopy **17**, 57 (1985)

56. L.J.Chen, K.T.Tu, Mat.Sci.Rep. (1991)

57. R.Tung, Phys. Rev. Lett. **52**, 461, (1984)

58. W.Mönch, Rep.Prog.Phys. **53**, 221 (1990)

59. G.Rubloff, P.S.Ho, Thin Solid Films **93**, 21 (1982)

60. C.Achete, H.Niehus, W.Losch, J.Vac.Sci.Technol. **B3**, 1327 (1985)

61. J.G.Clabes, G.W.Rubloff, K.Y.Tan, Phys.Rev. **B29**, 1540 (1984)

62. U.K.Köhler, J.E.Demuth, R.J.Hamers, Phys. Rev. Lett. **60**, 2499 (1988)

63. M.C.Bost, J.E.Mahan, J.Appl.Phys. **58**, 2696 (1985)

64. N.Cherief, C.D'Anterroches, R.C.Cinti, T.A.Nguyen Tan, J.Derrien, Appl.Phys.Lett. **55**, 1671 (1989)

65. S.P.Murarka, Silicides for VLSI Applications (Academic, Orlando 1983)

66. H.von Hänel, J.Henz, M.Ospelt, J.Hugi, E.Müller, N.Onda, Thin Solid Films, **194**, 259 (1990)

67. H.Niehus, Surf.Sci. **166**, L107 (1986)

68. H.Niehus, Appl.Phys. **A53**, 338 (1991)

69. U.Linke, B.Poelsema, J.Phys. **E18**, 2627 (1985)

70. F.Jensen, F.Besenbacher, E.Laensgaard, I.Stensgaard, Phys.Rev.**B41**, 10233 (1990)

71. G.Ertl, Surf.Sci.**6**, 208 (1967)

72. H.Niehus, G.Comsa, Surf.Sci. **140**, 18 (1984)

73. D.J.Coulman, J.Wintterlin, R.J.Behm, G.Ertl, Phys.Rev.Lett. **64**, 1761 (1990)

74. E.van de Riet, J.B.J.Sneets, J.M.Fluit, A.Niehaus, Surf.Sci. **214**, 211 (1989)

75. K.Kern, H.Niehus, A.Schatz, P.Zeppenfeld, J.Goerge, G.Comsa, Phys.Rev.Lett. **67**, 855 (1991)

76. D.Heskett, A.Baddorf, E.W.Plummer, Surf.Sci. **195**, 94 (1988)

77. R.Spitzl, H.Niehus, G.Comsa, Surf.Sci. 250, **L335** (1991)

78.   H.Niehus, Nucl.Instr.Meth. **B33**, 876 (1988)

79.   J.R.Noonan, H.L.Davis, J.Vac.Sci.Technol. **A6**, 722 (1988)

80.   A.P.Baddorf, D.M.Zehner, Surf.Sci. **238**, 255 (1990)

81.   J.Wolf, B.Viczenci, H.Ibach, Surf.Sci. **249**, 233 (1991)

ATOMISTIC STRUCTURE OF INTERNAL INTERFACES BY HIGH RESOLUTION ELECTRON MICROSCOPY

Manfred Rühle

Max-Planck-Institut für Metallforschung,
Inst. für Werkstoffwissenschaft,
Seestraße 92, 7000-Stuttgart 1, Germany

Abstract

With the advent of the newest generation of high resolution electron microscopes (HREM) it is now possible to analyse structures and defects in materials. However, a naive (direct) interpretation of the high resolution images is not possible since lens aberrations of the instrument may dramatically influence the observable contrast. A quantitative evaluation of the HREM images requires simulation of images. The best possible configuration can be extracted by a trial-and-error-method.

In this paper the HREM technique will be described and applied for the characterization of the interface between a Nb film (grown by molecular beam epitaxy) and a sapphire substrate.

1. Introduction

Recently, the resolution of commercially available electron microscopes has been significantly improved. The point resolution, which is defined more accurately in Sect. 3.1, is now better than 0.17 nm for instruments with an acceleration voltage of 400 kV. A new generation of high-voltage, ultra-high-resolution instruments pushes the resolution limit to about 0.1 nm. Considerable advances have also been made in the analysis of micrographs obtained by high-resolution electron microscopy (= HREM) [1-3]. Methods and programs have been developed [3,4] which allow the simulation of HREM images of any given atom arrangement. Those simulated images are used to interpret experimentally obtained micrographs. The recent developments enable us to use HREM as an important method for solving problems in materials science. Atomic structures of different lattice defects, such as phase boundaries, grain boundaries and dislocations, can be determined by HREM. The difficulties in the interpretation of HREM are similar to those in conventional transmission electron microscopy of lattice defects in crystalline materials: the experimentally obtained micrographs do not usually present a direct image of the object.

The extraction of information on the structure of lattice defects from HREM micrographs is complicated [1,2]. Despite this, very useful information has been obtained on the structure of lattice defects using HREM, particularly for semiconductors. The new generation of instruments allows also characterization of defects in ceramics and metals by HREM. This paper summarizes the possiblities of HREM and then discusses applications of HREM in materials science. The various steps needed for

Equilibrium Structure and Properties of Surfaces and Interfaces
Edited by A. Gonis and G.M. Stocks, Plenum Press, New York, 1992

69

successful working with HREM are described. Specifically, results for a Nb/Al$_2$O$_3$ interface are described in detail.

2. Different Transmission Electron Microscopy Techniques

The different possibilities offered by modern transmission electron microscopy is summarized schematically in Fig. 1. Conventional TEM techniques (CTEM) involve bright field (BF) and dark field (DF) imaging and selected area diffraction (SAD). These are used for morphological analysis, phase and defect identification studies. Spectroscopy can be performed with high spatial resolution. The probe size in scanning transmission electron microscopy ranges from ~ 1.0 nm to 50 nm. Energy dispersive spectroscopy (EDS) uses X-rays emitted from the specimen for a chemical characterization, while characteristic energy losses can be used for identifying qualitatively and quantitatively the light element by energy loss spectroscopy [5-7]. The surrounding of atoms can be probed by extended energy-loss fine structure studies (EXELFS) whereas energy-loss near-edge structure (ELNES) investigations result in information on the excitation state of atoms or ions. High resolution electron Microscopy (HREM) studies result in structure images. This paper focuses on the latter studies.

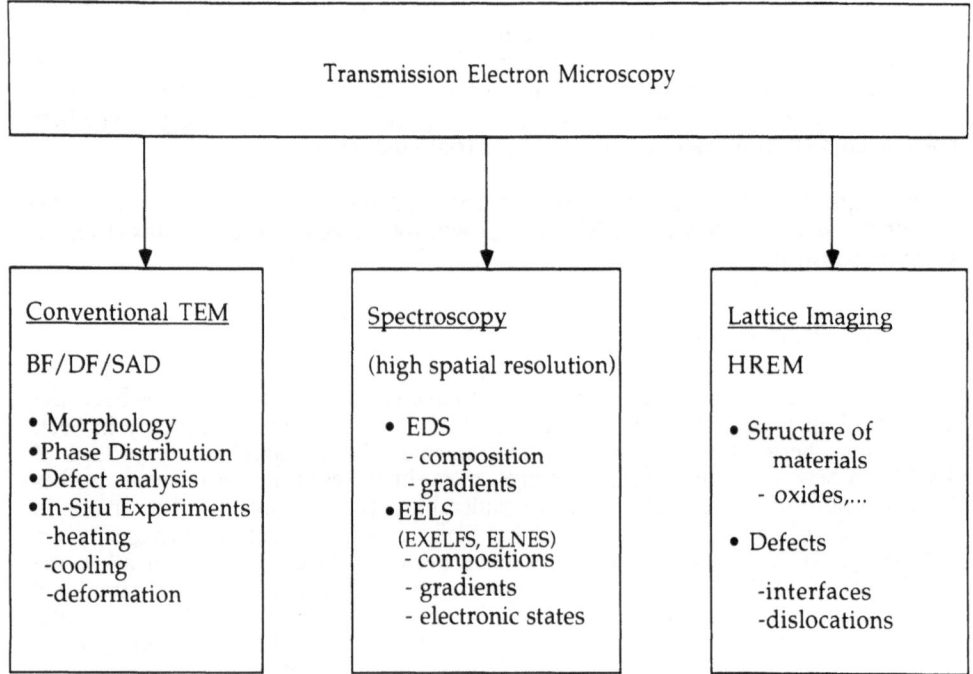

Fig. 1.   Schematic illustrations of different TEM techniques
(BF = bright field imaging; DF = dark field imaging; SAD = selected area diffraction; EDS = energy dispersive spectroscopy; EELS = electron energy-loss spectroscopy; EXELFS = extended energy-loss fine structure; ELNES = electron energy loss near-edge structure; HREM = high resolution electron microscopy).

3. Comments on Direct Lattice Imaging of Distorted Materials Using HREM

The geometric beam path through the objective lens of a TEM is shown in Fig. 2a. Beams from the lower side of the object travel both in the direction of the incoming and diffracted beam. All beams are focused by the objective lens in the back focal plane to form the diffraction pattern. In the image plane, the image of the object is produced by interference of the transmitted and diffracted beams. Fig. 2b uses wave optics to describe physical processes which contribute to the image formation. From the lower side of the foil, a wave field emerges that can be described by a transmission function q(x,y). For an undistorted lattice, q(x,y) represents a simple periodic amplitude and intensity distribution. The transmission function of a complex lattice with a large periodicity (e.g., a periodic grain boundary) is very complicated and q(x,y) is a non-periodic function for the distorted region of a crystal.

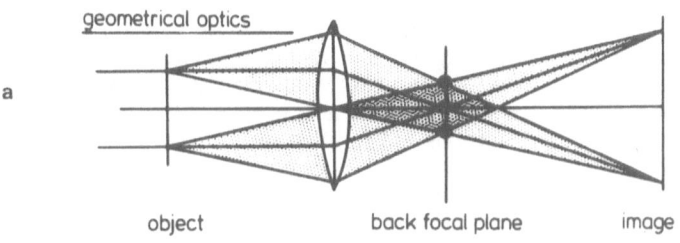

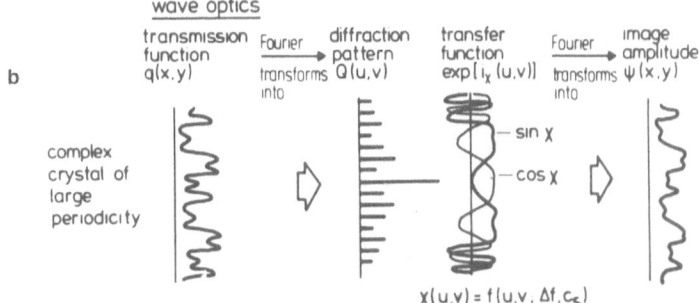

Fig. 2.   Image formation by the objective lens of a transmission electron microscope
a) Geometrical optical path diagram
b) Wave optical description (see text for explanation).

The intensity distribution in the diffraction pattern is given by the Fourier transformation Q(u,v) of the transmission function q(x,y) where u, v are the coordinates in the diffraction plane (reciprocal space). Since spherical aberration cannot be avoided with rotationally symmetrical electromagnetic lenses [1,2], the beams emerging from an object at a certain angle (Fig. 2a) undergo a phase shift relative to the direct beam. Imaging under a slightly defocusing mode, Δf, leads also to a phase shift which depends on the sign and magnitude of the defocus value, Δf. The influence of the lens errors and the defocus on the amplitude of the diffraction patterns can be described by the contrast transfer function $\chi(u,v)$:

$$\chi(u,v) = \frac{\pi}{\lambda}\left[\Delta f(u^2 + v^2)\,\lambda^2 - \frac{1}{2}\,C_s\,(u^2 + v^2)^2\,\lambda^4\right],$$

λ = wave length of electrons, C_s-constant of spherical aberration. The dependence of the contrast transfer function as a function of reciprocal distance ("space frequency") is shown in Fig. 3 for a modern 400 kV instrument and for the atomic resolution microscope at the National Center for Electron Microscopy in Berkeley [8].

The image is formed by a second Fourier transformation of the amplitude distribution in the diffraction pattern multiplied by the contrast transfer function. The amplitude in the image plane, $\Psi(x,y)$, is therefore not identical to the wave field in the object plane (transmission function $q(x,y)$). The image is severely modified if scattering to large angle occurs, since the influence of the spherical aberration increases strongly with increasing scattering angle. The modification is most severe if the components in the diffraction pattern coincide with the oscillating part of the contrast transfer function (large values of u,v). If, however, the wave vectors lie within the first wide maximum of the contrast transfer function $\chi(u,v)$, it can be assumed that characteristic features and properties of the object can be directly recognized in the image [1,2]. For good HREM imaging, the first zero value of the contrast transfer function under the optimum defocusing condition (Scherzer focus) must be at sufficiently large reciprocal lattice spacings so that lattices of many materials can directly be imaged. Good imaging conditions are fulfilled for lattices with large lattice parameters [9-12] If, however, deviations in the periodicity exist, components of the diffraction pattern appear at large diffraction vectors. It is then most likely that certain Fourier components possess reciprocal lattice distances larger than the Scherzer focus. Lens aberrations and defocusing cannot be neglected.

3.1   Resolution

The image contrast obtained in a high-resolution electron microscope (Fig. 2) is produced by interference of the diffracted beams with the transmitted beam. The interference can also be represented as an impulse response in the form of a delta function, the location and shape of which is highly dependent on the focusing distance [1,13]. From this description, it can easily be concluded that the resolution is not uniquely defined and that various distances can be adopted to define the resolution: d_1 is defined as the Scherzer resolution, which is determined from the focusing distance Δf at which the influence of negative defocusing compensates as far as possible for the influence of the spherical aberration [1,2]. The information limit, d_2, represents the smallest resolvable distance on an image at an optimum defocusing distance.

For a HREM analysis of crystalline materials, the first step is to establish the number and location of the positions of atom rows. It is assumed that the lattice to be investigated is periodic in the direction of the incoming beam so that the projected potential on the lower surface of the object is formed by the superposition of identical atoms that lie upon each other in the z direction (Fig. 4). To determine the atomic structure of distorted regions (or regions of large lattice periodicity), the resolution is defined below. The evaluation of the number of atomic rows and their exact positions are discussed in the next section. Different zone axes of crystalline materials can be used for the evaluation of crystal structures and of defects if the lattice plane distances are larger than the Scherzer resolution d_1 of the instrument.

If lattice defects must be analyzed by HREM, one must first establish the number of atomic rows that exist in the surroundings of a lattice defect. Naturally, no information can be obtained in any HREM investigation on defects which are not parallel to the direction of the incoming electron beam. HREM images always represent a two-dimensional projection of electron distribution on the exit surface of the transmitted foil. For the image evaluation, it is implied that the atomic rows are exactly parallel to the incoming electron beam.

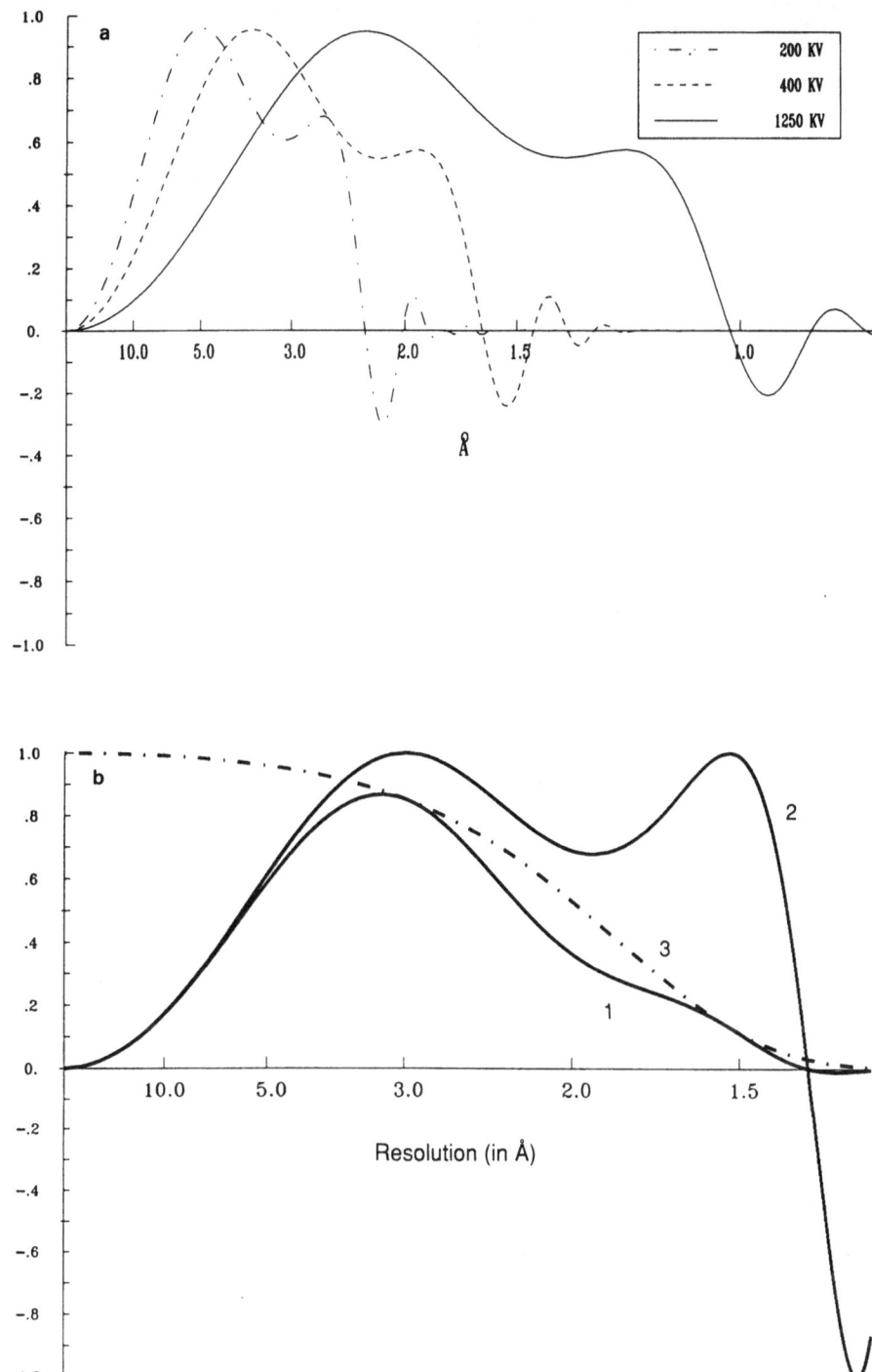

Fig. 3.   Contrast Transfer Function of High Resolution Electron Microscopy
a) 400 kV instrument
b) Atomic Resolution Microscopy. Actual calculated Contrast Transfer
Function (CTF) (1), undamped CTF (2), and damping envelope (3) of
the Atomic Resolution Electron Microscope (ARM) at the Center for Electron Microscopy, Berkeley (Hetherington et al. [8]). Operating conditions:
voltage 800kV, defocus - 55 nm.

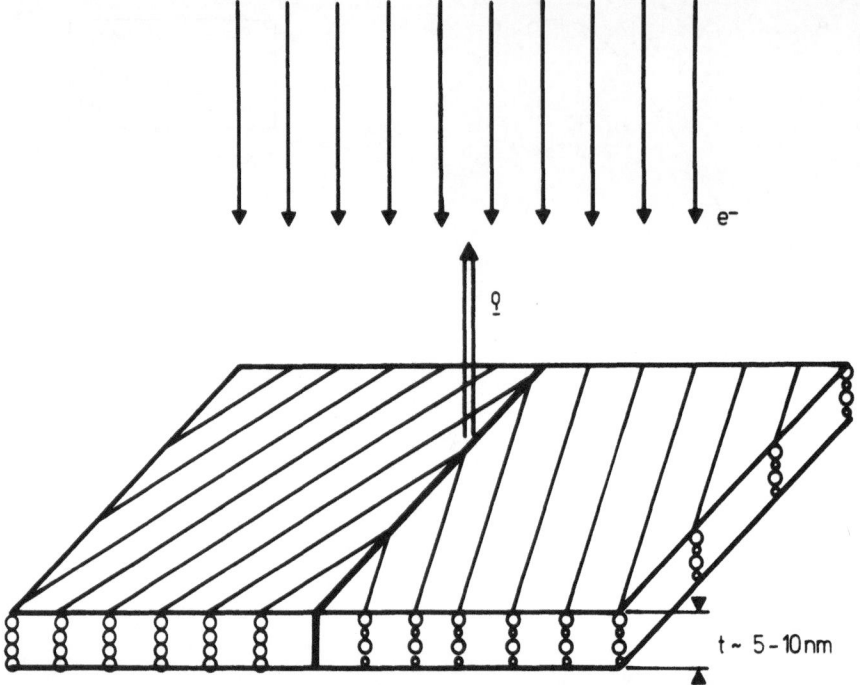

Fig. 4.   Direct lattice imaging by HREM. The crystalline specimen must be adjusted so that the direction of the incoming electron beam coincides exactly with the orientation of atomic rows. The schematic drawing includes a grain boundary. HREM can be successfully performed for pure tilt boundaries with tilt axis parallel to the direction of the incoming electrons.

In order to discuss the possibilities of determining the atomic structure, it is necessary to introduce a parameter d_0 representing the minimum distance of two neighboring atom columns in the region of the lattice defect. Comparing d_0 with d_1 and d_2 enables us to decide whether the distorted structure (or structural unit) in the lattice defects can be resolved in all details or not. Three cases can be distinguished:

a) $d_0 > d_1$. In this case, all atom columns can be observed; artefacts are not introduced into the Scherzer image. For thin foils (typical thicknesses 5 to 10 nm), channels in the lattice structure appear as white spots on the image. If certain a *priori* information on the atom configuration is available, the observed HREM micrographs can be directly compared with the atomic structure [1,2].

b) $d_1 > d_0 > d_2$. The determination of the number of atom columns of a distorted structure requires micrographs taken under a series of defocusing distances which differs typically by small values, ~ 5nm. A quantitative evaluation requires the comparison of micrographs taken under specific focusing conditions with computer simulated images in order to differentiate interferences produced by artefacts and real lattice distortions such as additional atom columns, for example.

c) $d_0 < d_2$. The existing atom rows can no longer be resolved as such, and thus only an absolute measurement of the intensity distribution would give the exact number of atom rows which produce a light (or dark) contrast spot on an HREM image. This evaluation would be very difficult, especially since any surface defects introduced during specimen preparation may cause phase shifts which modify drastically the intensity distribution on the HREM images.

From this brief summary of the optimum conditions for observing structures in the vicinity of lattice distortions (grain boundary, dislocations), the following conclusions may be drawn:

a) d_1, the point resolution in the Scherzer focus, must be as small as possible.

b) Analyses of structures surrounding lattice defects are confined to those cases in which strict symmetry exists in the direction of the incoming electron beam.

c) The specimen thickness must be as small as possible and less than an effective extinction length of the selected zone axes. In most cases, the foil thickness should be between 5 and 15 nm. The exact values of the foil thickness t and the defocusing value Δf are very important parameters which must be precisely determined. Computer simulations for which the experimentally determined parameters t and Δf are required are thus indispensable if details are to be resolved which are of magnitude d_1 or below d_1.

d) Extremely careful adjustment of the electron microscope has to be carried out in order to obtain a good correlation between the image and the projected structure.

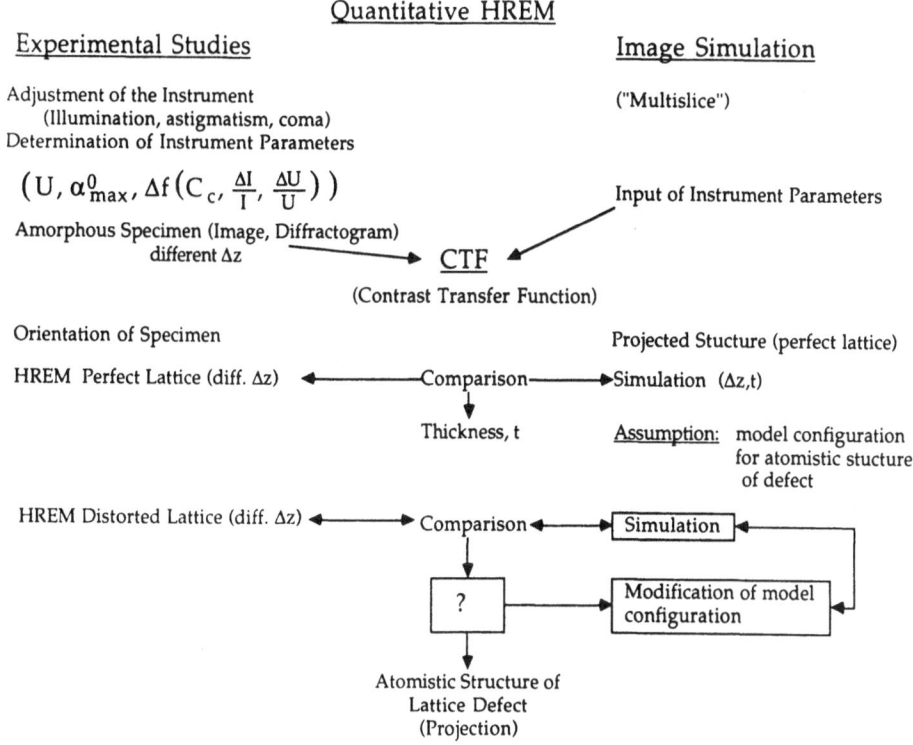

Fig. 5.   Quantitative HREM – a flow chart.

3.2   Determination of the Atom Column Locations

After determining the number of atom columns in the vicinity of the lattice defect, the coordinates of the individual atom columns must be determined with the best possible accuracy. The evaluation demands a comparison between experimentally obtained micrographs and computer–simulated images. A "trial and error" method is usually applied. The comparison is made by superimposing the two images and visually comparing the location of the individual contrast features. It is important that the location of all contrast maxima and minima coincide. Very reliable data of atomic coalescence coordinates are obtained for $d_0 < d_1$: the accuracy is $\pm\,0.01\,d_1$

[9,13]. The accuracy of the atom coordinate determination is thus substantially better than the actual resolution. This can easily be explained by the fact that very small changes of the phases at different waves may result in an observable change in the intensity distribution and in the location of the intensity maxima during formation of a phase contrast image [1,2]. The relative displacement of perfect lattices against each other (e.g., in the neighborhood of phase boundaries or even at grain boundaries) can be analyzed from HREM micrographs. The values obtained by HREM are often sufficiently accurate for a differentiation between different grain boundary models. The resolution $d_1 = 0.16$ nm (400 kV HREM) is not sufficient to determine small displacements in the vicinity of ordered grain boundaries (e.g., twins) in metals. Figure 5 shows a flow chart for quantitative HREM.

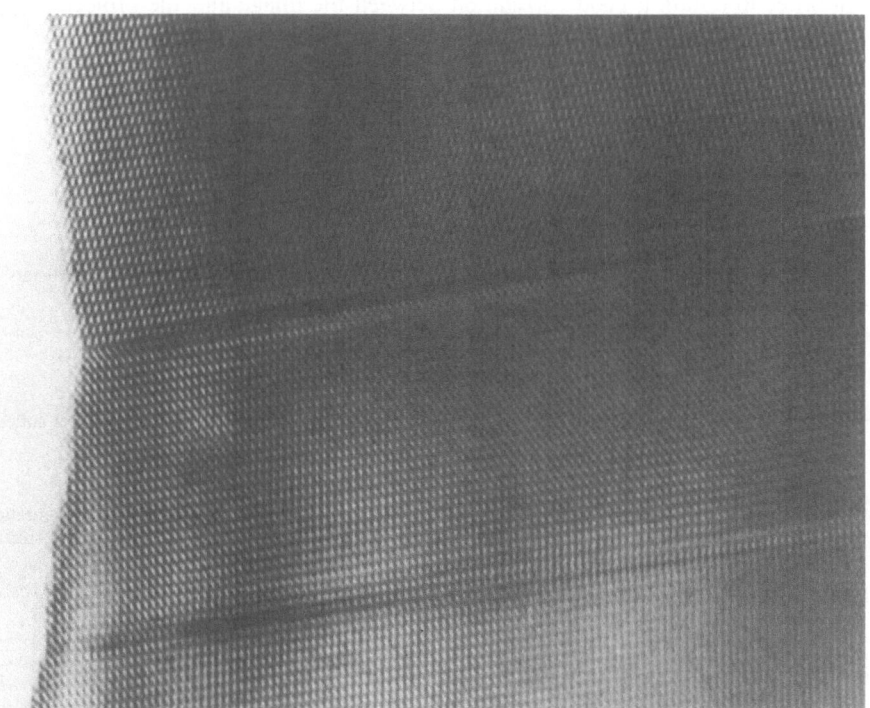

Fig. 6.   HREM image of an incoherent twin boundary in m–ZrO_2.

3.3   Complementary Diffraction Investigations

Selected area diffraction investigations provide important information on the periodicity along the common axis in the case of periodic defects (e.g., periodic twin boundaries). This is the only possible technique which permits any type of reconstruction along the atom columns with regard to the common axis. This method was recently used to complement high–resolution electron microscopy for determining the structure of grain boundaries [1,2,8].

4.   The Potential of High–Resolution Electron Microscopy in Materials Science

HREM studies can be used for identifying the structure of defects which possess exact periodicity parallel to the direction of the incoming electron beam. Such defects encompass domain boundaries and twins, tilt grain boundaries and dislocations.

Domain boundaries and various types of twins are twodimensional lattice defects which possess a perfect lattice adjacent to the boundaries. The defects are formed by simple symmetry operations (displacement, mirror image). The atomic structure of the defects and lateral displacement of the two crystals can be determined by HREM. An example of a twin boundary in a lattice is given in Fig. 6, which represents an incoherent twin boundary in monoclinic ZrO_2. The lattice constant of the monoclinic ZrO_2 is 0.54 nm. The light spots in this micrograph represent open channels and the dark spots are superpositions of various Zr atoms which have been slightly displaced relative to each other.

The atomistic structure of grain boundaries in materials was recently predicted by molecular static computer simulations [see e.g., 13-15]. Detailed investigations on grain boundaries in semiconductors, metals, and ceramics [see e.g., 14-16] resulted in grain boundary structures which could be compared to results obtained by computer simulations.

The displacement and stress fields around dislocations can be described by various approximations whereby the structure of the dislocation core is probably not properly simulated by using continuum theoretical models. The real dislocation core structures can only be analyzed by HREM if the resolution in the Scherzer focus is $d_1 < 0.13$ nm. This resolution can only be achieved by the new generation of high-resolution, high-voltage electron microscopes.

5. Case Study: The Nb/Al_2O_3 Interface

The applications of modern engineering materials such as metals and ceramics often require two different components (metals and ceramics) to be bonded. The resultant interfaces must typically sustain mechanical and/or electrical forces without failure. Consequently, interfaces exert an important, and sometimes controlling, influence on performance in such applications as composites, electronic packaging systems used in information processing, thin film technology and joining [17]. A knowledge of the atomistic structure of such interfaces is a prerequisite for an understanding of their properties.

Several methods are capable of generating well defined metal/ceramic interfaces. At the most fundamental level, ultra clean, flat surfaces readily bond at moderate temperatures and pressures [18]. Interfaces can also be produced by internal oxidation of metallic alloys [19], where small oxide particles in different metals (Nb, Pd, Ag,...) are formed by oxidation of a less noble alloying component such as Al, Cd, etc. Interfaces produced by the internal oxidation process usually show a well-defined low-energy crystallographic orientation relationship between the two components and were thus used as model systems. A third method for manufacturing metal/ceramic interfaces is by evaporation of metals onto clean ceramic surfaces. This method allows control over both substrate material/orientation and overlayer composition and by this well defined interfaces can be obtained [20].

Nb/Al_2O_3 serves as an excellent "model" system since Nb and Al_2O_3 possess nearly the same thermal expansion coefficients and most thermodynamic quantities (solubility, diffusion data, etc.) are well established for both components. Nb/Al_2O_3 composites are used in different applications such as Josephson junctions and as components for structural materials. To date, only a few detailed studies have been reported concerning the atomistic structure of Nb/Al_2O_3 interfaces formed after diffusion bonding [21-23], internal oxidation [19,24], and after thin film deposition [25-27]. The studies were all performed by high resolution electron microscopy (HREM). Orientation relationships (OR) were evaluated from diffraction studies, either by X-rays, or by selected area diffraction (SAD) patterns obtained in a transmission electron microscope (TEM). The OR between Nb and Al_2O_3 is determined by the manufacturing route [26]. While the OR is preset for interfaces prepared by diffusion-bonding, topotaxial or epitaxial OR develops during internal oxidation and epitaxial growth, respectively. During internal oxidation a <u>topotaxial</u> relationship forms bet-ween Nb and Al$_2$O$_3$ [19,24] so that close-packed planes of both systems are parallel to each other, i.e.:

$$(0001)_S \parallel (110)_{Nb} \text{ and } [01\bar{1}0]_S \parallel [001]_{Nb} \text{ (S = sapphire)}^* \tag{1}$$

Epitaxial growth of very high quality single-crystalline overlayers of Nb on sapphire has been a subject of recent experiments. There exists experimental evidence [27] that for most sapphire surfaces a unique three-dimensional <u>epitaxial</u> relationship bet-ween Nb and Al$_2$O$_3$ develops which is given by two sets of zone axes:

$$(0001)_S \parallel (111)_{Nb} \text{ and } [10\bar{1}0]_S \parallel [1\bar{2}1]_{Nb} \tag{2}$$

The aim of the present studies was to characterize by HREM the atomistic structures of Nb/Al$_2$O$_3$ interfaces formed by MBE growth of Nb layers on sapphire substrates with different orientations. The observations will be compared with results published in the literature.

5.1 Experimental Details

The Nb layers were fabricated in a MBE growth chamber equipped with electron beam sources for evaporating refractory metals. The sapphire substrates were parallel to $(0001)_S$, $(1\bar{1}00)_S$ and $(1\bar{2}10)_S$, respectively. The substrates were preheated and cleaned by annealing [28]. A special technique was used to obtain transmission electron microscopy (TEM) cross-section samples of the Nb/Al$_2$O$_3$ interface [27].

The HREM studies were performed at the Atomic Resolution Microscope (ARM) at the National Center for Electron Microscopy (NCEM), Berkeley, CA, USA. The microscope was operated at 800 kV [8]. The actual constrast transfer function (CTF) of the instrument is shown in Fig. 3b. The point-to-point resolution of the instru-ment is in the range of 0.16 nm. The CTF is mostly determined by the energy spread of the instrument ($\sim$ 15 nm). As a result of the chromatic aberration the CTF is damped as shown in Fig. 3b.

5.2 Experimental Results

The OR between Nb and Al$_2$O$_3$ was evaluated from selected area diffraction (SAD) patterns taken from different Nb films on sapphire substrates in three different orien-tations. For all films a unique OR was evaluated which is characterized by the fol-lowing coinciding planes :

$$(0001)_S \parallel (111)_{Nb} \tag{3}$$

and coinciding directions:

$$[2\bar{1}\bar{1}0]_S \parallel [1\bar{1}0]_{Nb} \quad \text{(direction A)} \tag{4}$$
$$\text{or} \quad [10\bar{1}0]_S \parallel [1\bar{2}1]_{Nb} \quad \text{(direction B)} \tag{5}$$

From crystallographic symmetry arguments the angle between the directions A and B results in 30°. Fig. 7a and 7b illustrate the relative OR of the two crystals. From eq.(3) it follows that the normals on the plane are parallel:

$$[0001]_S \parallel [111]_{Nb}.$$

Thus, the three-fold axes of the two crystals, i.e. Nb and Al$_2$O$_3$ (sapphire), are parallel.

---

*The orientation relationship (OR) between two crystals of different lattice structure is uniquely described by one coinciding plane (in both lattices) and one set of coinciding directions in that plane.

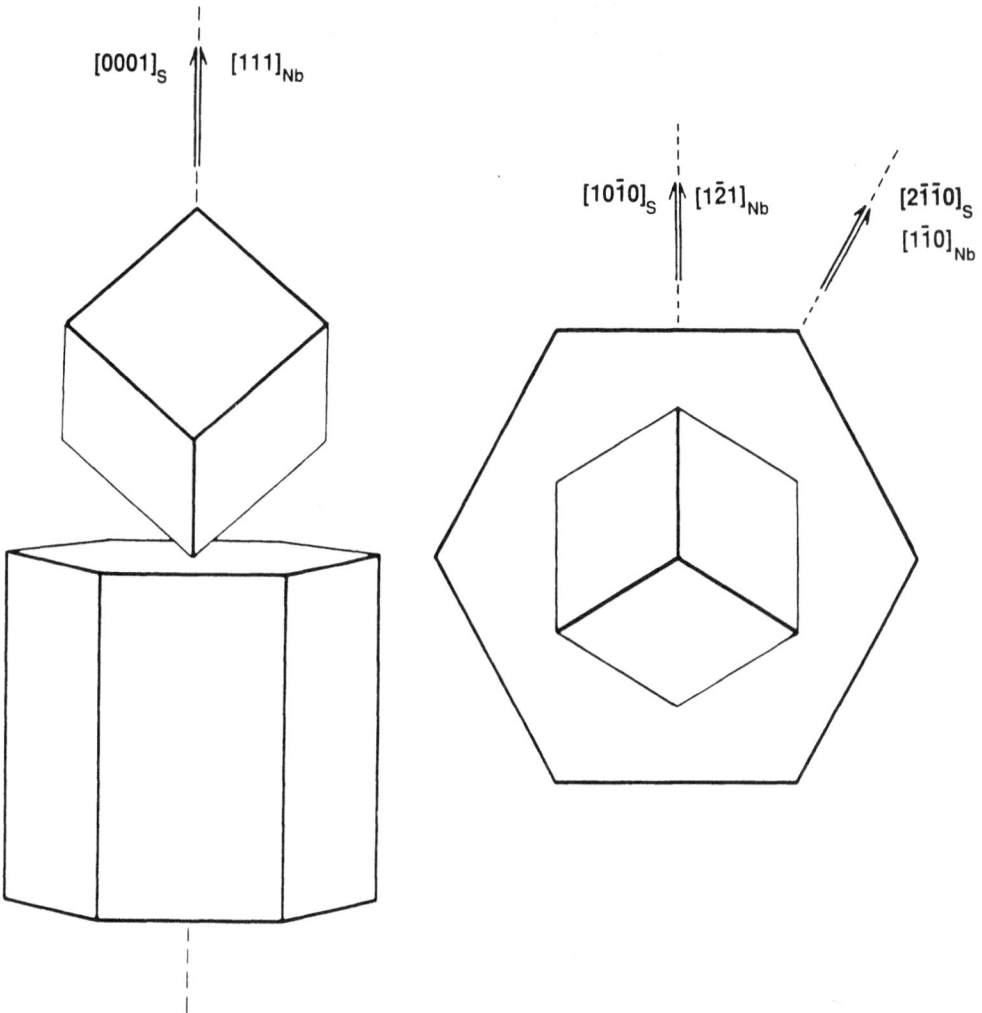

Fig. 7. Orientation relationship between sapphire substrates and niobium over-layers. a) The three-dimensional orientation relationship $[0001]_S \| [111]_{Nb}$ (S=sapphire) holds for all substrate orientations. b) Orientation relationship for different directions within the $(0001)_S \| (111)_{Nb}$ plane.

The OR described by eq.(3 – 5) is independently identified at epitaxially grown Nb/Al₂O₃ interfaces for substrate surfaces parallel to: $(0001)_S$, $(1\bar{1}00)_S$, and $(1\bar{2}10)_S$. In this paper only results for $(0001)_S$ substrates are reported. Observations for the other substrate orientations are reported elsewhere [27].

Direct lattice imaging of near interface regions allows the determination of the atomistic structure of the interface as well as the analysis of defects associated with the interface, such as misfit dislocations, etc. To obtain interpretable HREM images the electron beam should be incident along high symmetry directions in both crystals and should be parallel to the plane of the interface. A three-dimensional analysis of the structure requires HREM images taken under different directions of the incident electron beam with respect to the interface orientation. These conditions are fulfilled if the electron beam is parallel to direction A and B, respectively (eq.(4,5), see Fig. 7b). High resolution electron micrographs were taken from the same interface in both directions by simply tilting the specimen inside the ARM.

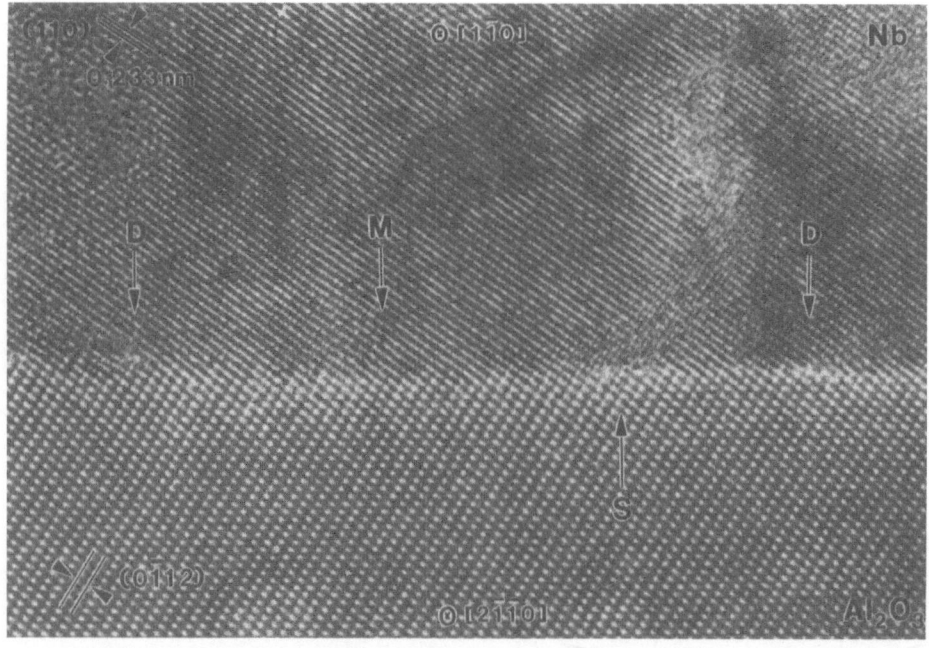

Fig. 8.   High resolution images of a Nb/Al₂O₃ interface. Direction of incoming electrons parallel to $[1\bar{1}0]_{Nb}$ and defocus value $\Delta f = -70$ nm (Scherzer value - 55 nm). Lattice planes can clearly be identified in both, sapphire and Nb. Foil thickness ~ 10 nm. At the interface regions of good matching (M) and poor matching (D) alternate. S: step in the substrate.

Fig. 8 shows an overview of a large area of a near interface region. The defocus of the objective lens is slightly more negative ($\Delta f = -70$ nm) than the Scherzer defocus ($\Delta f = -55$ nm). At this defocus the atomic distance corresponding to the (200) planes with d = 0.165 nm becomes clearly visible [17]. Lattice planes can readily be identified in Al₂O₃ and Nb. The foil thicknesses of Nb and Al₂O₃ are identical. In Nb regions of good matching (M) and poor matching (D) alternate at the interface. Steps can also be identified (S). The region of good matching (Fig. 8, M) is imaged at a higher magnification in Fig. 9. Fig. 9a shows the interface with the electron beam parallel to direction A. Nb and Al_2O_3 possess the same thickness and lattice planes transfer continuously from Nb to Al_2O_3. Fig. 9b is a micrograph of the same interface viewed along orientation B. Only $(10\bar{1})$ lattice planes with a spacing of 0.233 nm are visible in the Nb crystal in regions of good matching M. The (222) lattice planes (perpendicular to $(10\bar{1})$) possess a spacing of 0.095 nm which is beyond the resolution limit and information limit of the ARM. Therefore only $(10\bar{1})$ lattice fringes are visible. In both orientations (Figs. 9a and 9b) a perfect match of the Nb and Al_2O_3 lattice at the interface is visible.

The misfit of 1.9% between the $(11\bar{2})_{Nb}$ and $(0\bar{1}10)_S$ planes perpendicular to the interface is accomodated by localized defects (misfit dislocations) in regions of poor matching (Fig. 8). These localized misfit dislocations at the interface are spaced periodically, about one every 50 lattice planes. A higher magnification of region D of Fig. 8 is shown in Fig. 10. Fig. 10a represents the images parallel to direction A. A misfit dislocation can clearly be identified. The Burgers vector is $1/2$ $[110]_{Nb}$. No "stand-off" distance [19,24] of the misfit dislocation from the interface can be identified.

The projection of the extra plane of the misfit dislocation is parallel to the electron beam in direction A . For direction B (Fig. 10b) the extra Nb plane is inclined at an angle of 30°. There is no fit of corresponding planes in Nb and Al_2O_3 along the projection of the inclined plane of the misfit dislocation which is clearly visible in Fig. 10b. The projected length, x, can be measured:

$$x = (17 \pm 2)\ d_{110} \qquad (d_{110} = 0.233 nm)$$

and with the known tilt angle the thickness of the foil, t, can be evaluated:

$$t = (8.0 \pm 1.0)\ nm.$$

The sapphire surface will usually not be atomically flat over large distances. Steps are expected if, e.g., small deviations exist between the surface plane and the $(0001)_S$ basal plane of sapphire prior to thin film evaporation. Such surface steps can be recognized at the interface (Fig. 8 and 11). Careful inspection (especially on micrographs imaged under a grazing angle) revealed that the step height of the sapphire is always a multiple of the distance of the O^{2-} layers within the Al_2O_3 structure parallel to $(0001)_S$. This distance corresponds to $(0006)_S$ planes with d = 0.2165 nm. The corresponding lattice spacing in Nb is $d_{111} = 0.1906$ nm and is 11.9% smaller than the O^{2-} spacing. Due to this mismatch, the Nb lattice is distorted adjacent to steps. The recognizable distortion extends up to about 10 nm into the Nb film.

5.3   Discussion

The observed orientation relationship (OR) for Nb films on differently oriented sapphire substrates is equivalent to results described in the literature. The three-dimensional orientation relationship between Nb and Al_2O_3 has been confirmed for the three different orientations of sapphire substrates: $(0001)_S\|(111)_{Nb}$ and $[10\bar{1}0]_S\|[1\bar{2}1]_{Nb}$. Knowles et al. [25] demonstrated by HREM that the same OR holds for (001) niobium films deposited epitaxially on $(1\bar{1}02)R$ planes of sapphire. The characteristic misorientation from the "three-dimensional OR" is caused by small locally misorientated regions in the immediate vicinity of the Nb/Al_2O_3 interface.

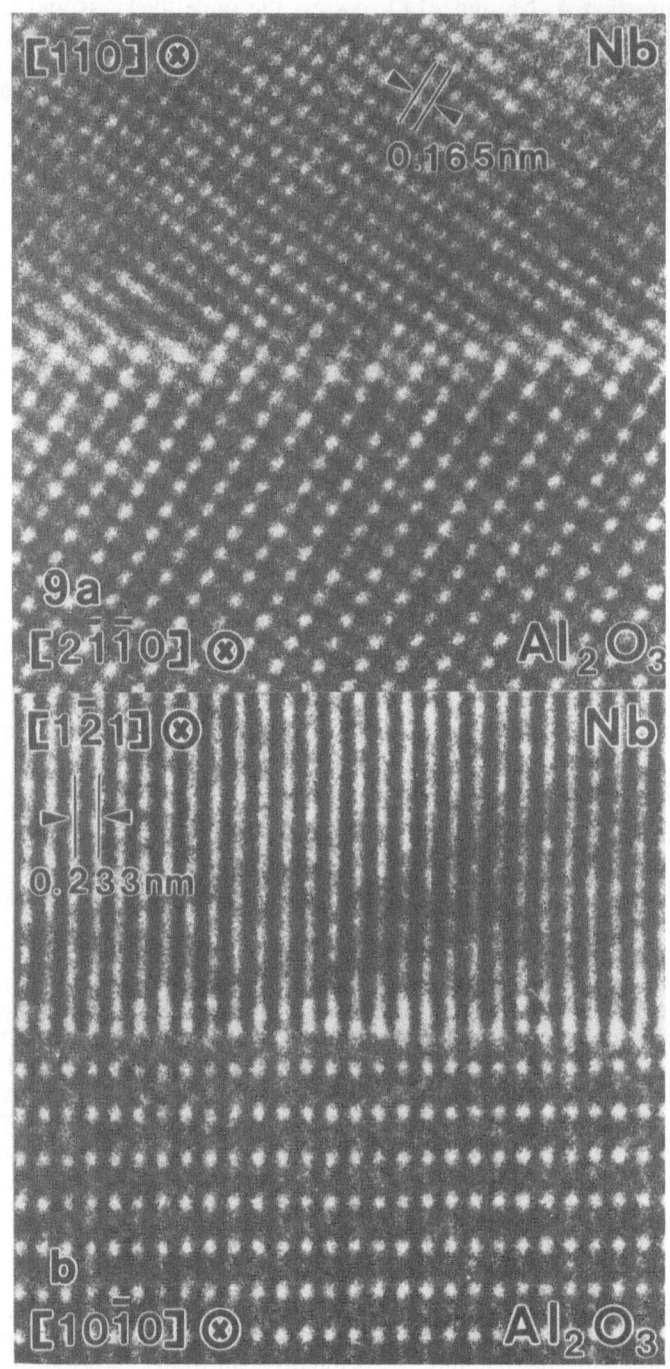

Fig. 9.   High resolution image of Nb/Al$_2$O$_3$ interface. Region of good matching (M).
a) Direction A with $[2\bar{1}\bar{1}0]_S \| [1\bar{1}0]_{Nb}$
b) Direction B with $[10\bar{1}0]_S \| [1\bar{2}1]_{Nb}$.

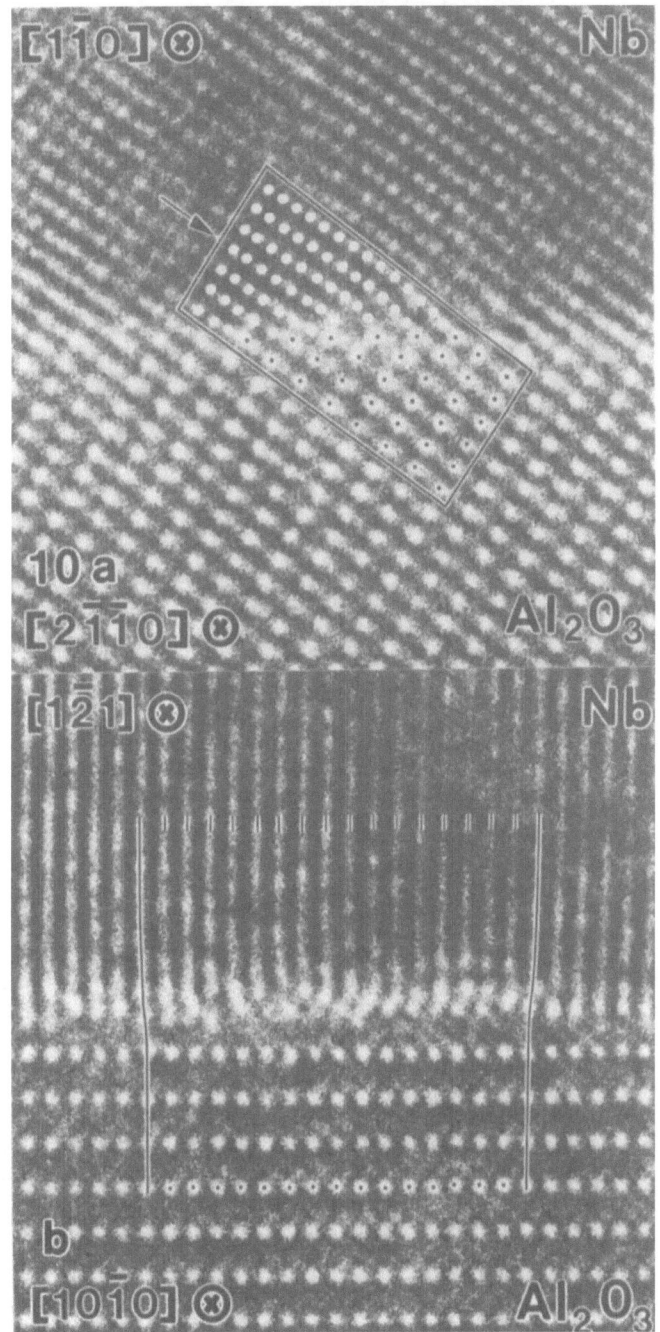

Fig. 10. High resolution image of Nb/Al$_2$O$_3$, region of poor matching. A misfit dislocation forms with no stand-off distance. a) Direction A; the core of the
misfit dislocation can readily be identified. b) Direction B; the core of the
dislocation line is inclined with respect to the electron beam. The region of
no matching of corresponding lattice planes is marked. This region corresponds to the projection width of the additional lattice plane.

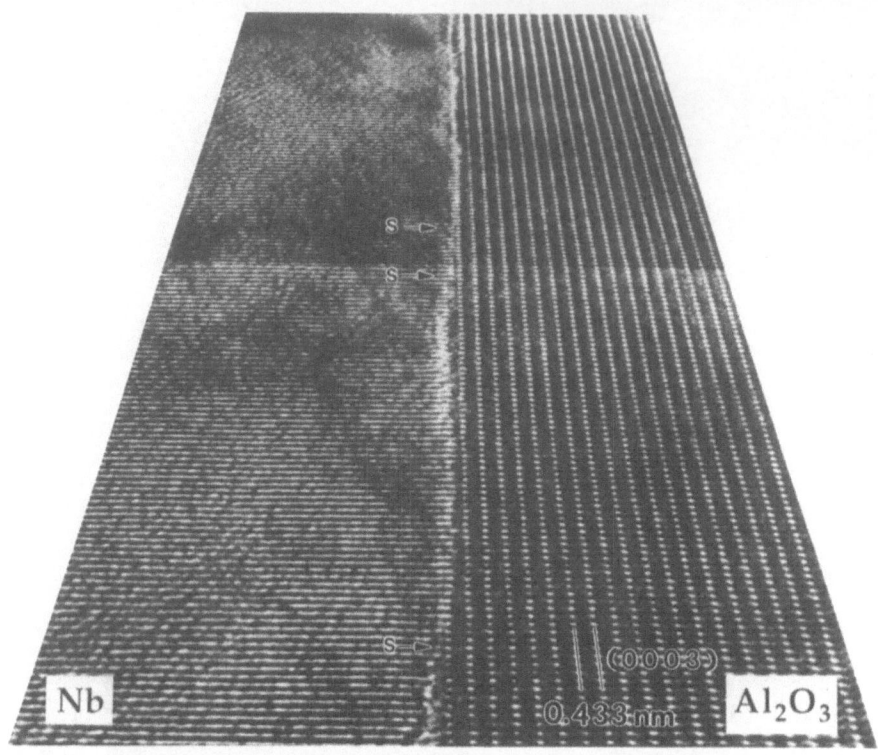

Fig. 11. "Scheimpflug image" [25] of HREM images in direction B $(0001)_S$ substrate. Steps can be observed (S). The step height corresponds to the (0006) plane distance.

Qualitatively the Nb film formation on $(0001)_S$ substrates could be described as follows. The sapphire substrate was annealed at $5 \cdot 10^{-8}$ Pa at 1500 K in a dynamic vacuum. The annealing leads to a reconstruction of the α-Al_2O_3 surface resulting in a $(\sqrt{31} \times \sqrt{31})R \pm 9°$ structure [28]. At the surface the Al:O ratio is $\approx$ 1:1, thus the sapphire is not stoichiometric at the surface. The RHEED studies performed at α-Al_2O_3 surfaces prior to deposition of the Nb film indicate that a similar reconstruction is present. In addition, atomistic steps (facets) exist at the surface of sapphire due to small deviations between the surface orientation and the $(0001)_S$ basal plane. Those steps may act as nucleation centers for the Nb film. The growth conditions are selected such that two-dimensional crystal growth occurs. The reconstructed, Al rich surface still possesses 3-fold symmetry with characteristic distances equivalent to atomistic distances of $(111)_{Nb}$ planes. For symmetry reasons it is expected that the first deposited monolayer of Nb atoms has an OR which is dictated by the symmetry of the sapphire surface leading to $(0001)_S \| (111)_{Nb}$. – No twins could be identified in the (nearly) perfect Nb film. It must be assumed that a unique atomistic relationship exists between the sapphire surface and the monoatomic Nb layer. There exists only one set of Nb atom positions on the unreconstructed sapphire surface which leads to a twin free film (Fig. 12). Nb atoms must be located on the terminating O^{2-} layer of sapphire on top of "empty sites", (Fig. 12). A quantitative evaluation of the HREM images should verify this hypothesis.

The formation of the "core" of the misfit dislocations can be explained using Frank and Van der Merwe's model [29,30]. Subsequent Nb layers bond to the first layer by metallic bonding building up the misfit dislocation nucleated at the interface. This model may explain the growth on $(0001)_S$ plane sapphire. It is, however, difficult to explain the (general) three-dimensional orientation relationship between Nb and Al_2O_3 which develops also on other sapphire surfaces.

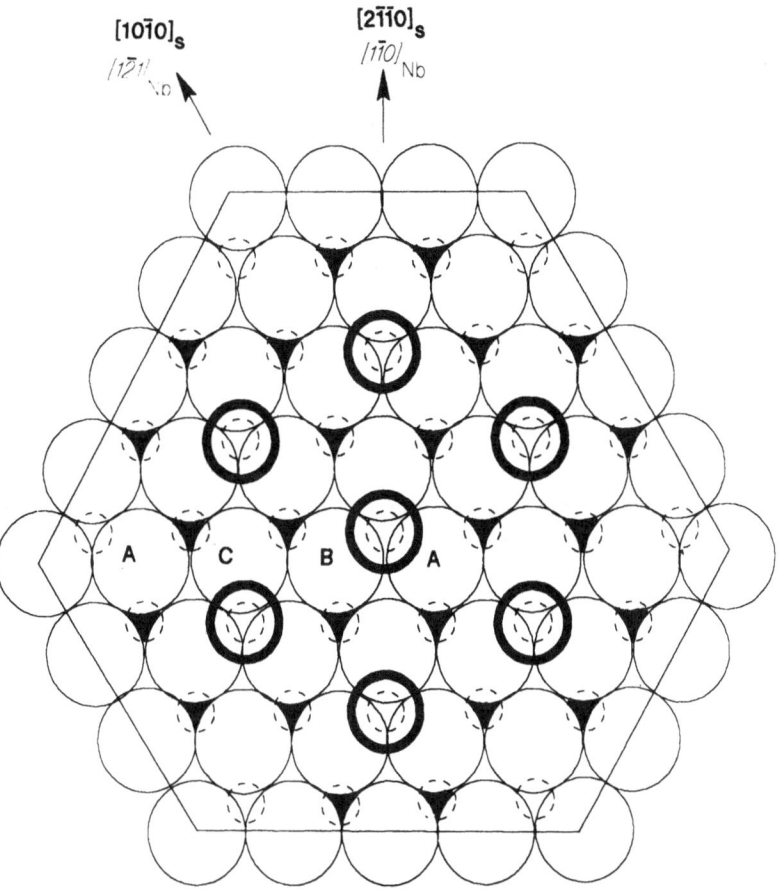

Fig. 12.   Schematic drawing of an unreconstructed $(0001)_S$ basal surface. O^{2-} ions: light large circles, Al^{3+}: dark sections of small circles, Nb atoms: bold medium sized circles. It is assumed that the O^{2-} ions form the outermost layer. The Nb atoms of the first Nb layer are positioned above the empty sites of the first Al^{3+} layer which lies immediately next to the outermost O^{2-} layer.

A careful inspection of HREM images taken under different focusing conditions of regions very close to the interface reveals that no reconstruction of sapphire close to the interface is recognizable. It must be assumed that the excess Al atoms of the reconstructed (0001) surface [28] diffuse into the Nb, which has a sufficiently high solubility for Al. Quantitative evaluation of the high resolution images will support this qualitative statement.

The misfit dislocations observed in MBE grown Nb films do not show a "stand-off" which was observed at interfaces formed by internal oxidation or diffusion bonding [19,24]. The localization of the inserted plane is well defined (Fig. 10a) and a perfect match of the Nb atoms on the sapphire lattice is obtained within 0.5 nm on either side of the dislocation. Since the deposition of the Nb film was performed at much lower temperatures than either the diffusion bonding or the internal oxidation this geometrical arrangement (Fig. 10a) of the misfit dislocations might not be thermally stable. The dislocations could climb from the interface to a stable stand-off distance if the samples are heat-treated or deposited at higher temperature. Such experiments would also show if the OR observed in our studies is thermally stable or if it is only formed at the given growth conditions due to the kinetics of the epitaxial growth process.

The Nb overlayers show an extremely high perfection which is typical for MBE grown films. However, stresses generated by unevenness of the substrate may extend up to 200 nm into the film and influence its properties. This interfacial roughnesses may also play a role in dislocation nucleation when the film is stressed.

6. Conclusion

This paper summarizes the possible applications of high-resolution electron microscopy of materials. It is important to have an instrument with the best possible resolution (small values of d_1). Reducing d_1 (below 0.2 nm) gives a better insight into the atomic structure of lattice defects. With the instruments available today, in particular the middle high-voltage instruments (400 kV), and with the use of precision pole pieces, it is expected to improve the resolution of instruments to about 0.15 nm. With this resolution and by avoiding all interference effects of the microscope, grain boundaries and phase boundaries can be obtained. Results on the atomistic structures of the Nb/Al_2O_3 interface demonstrate the power of HREM.

7. Acknowledgements

The author acknowledges helpful discussions with Dr. J. Mayer and Dr. F. Ernst. Mrs. G. Poech processed patiently and efficiently the manuscript.

8. References

[1]   Spence JCH (1988) Experimental High-Resolution Electron Microscopy (2nd Ed.), Oxford University Press, New York and Oxford
[2]   Busek P, Cowley J, Eyring L (1988) High-Resolution Transmission Electron Microscopy, Oxford University Press, New York and Oxford
[3]   O'Keefe MA (1985) Electron Image Simulation: A Complimentary Processing Technique, Electron Optical Systems, SEM, Inc., AFM O'Hare (Chicago) IL, pp. 209-220
[4]   Stadelmann PA (1987) Ultramicroscopy 21: 138-147
[5]   Hren JJ, Goldstein JI, Joy DC (eds.) (1979) Introduction to Analytical Electron Microscopy, Plenum Press, New York and London
[6]   Joy DC, Romig Jr. AD, Goldstein JI (eds.) (1979) Principles of Analytical Electron Microscopy, Plenum Press, New York and London
[7]   Egerton RF (1986) Electron Energy-Loss Spectroscopy, Plenum Press, New York and London
[8]   Hetherington CJD, Nelson EC, Westmacott KH, Gronsky R, Thomas G, (1989) Mat. Res. Soc. Symp. Proc. 139:277
[9]   Busek PR (ed.) (1985) Ultramicroscopy 18:1-4
[10] Mitchell TE, Davies PK (eds.) (1988) J. Electron. Micr. Tech. 8
[11] Barbier j, Hiraga K, Otero-Diaz LC, White TJ, Williams TB, Hyde BG (1985) Ultramicroscopy 18:211-234
[12] Merkle KL (1987) Mat. REs. Soc. Symp. Proc. 82:383-402
[13] Bourret A (1985) J. de Physique 46:C4-27 - C4-38
[14] Rühle M, Balluffi RW, Fischmeister H, Sass SL (edt.) (1985), Proc. Intern. Confernce on the Structure and Properties of Internal Interfaces, J. de Physique 46:C4

[15]   Raj R, Sass SL (eds.) (1988) Interface Science and Engineering, J de Physique 49:C5

[16]   Aucouturier M (ed.) (1990), Intergranular and Interphase Boundaries in Materials, J. de Physique 51:C1

[17]   Rühle M, Evans AG, Ashby MF, Hirth JP (eds.) (1990) Metal/Ceramic Interfaces, Pergamon Press, Oxford

[18]   Fischmeister H, Mader W, Gibbesch B, Elssner G (1988) Mat. Res. Soc. Symp. Proc. 122:529

[19]   Mader W (1989) Z. Metallkunde 80:139

[20]   Flynn CP (1990) in: Metal/Ceramic Interfaces, (eds. M. Rühle, A.G. Evans, M.F. Ashby and J.P. Hirth), Pergamon Press, Oxford, p.168

[21]   Florjancic M, Mader W, Rühle M, Turwitt M (1985) J. de Physique 46:C4-129

[22]   Mader W, Rühle M (1989) Acta metall. 37:853

[23]   Mayer J, Mader W, Knauss D, Ernst F, Rühle M (1990) Mat. Res. Soc. Symp. Proc. 183, in press

[24]   Kuwabara M, Spence JCH, Rühle M (1989) J. Mat. Res. 5:972

[25]   Knowles KM, Alexander KB, Somekh RE, Stobbs WM (1987) Inst. Phys. Conf. Ser. (EMAG 87) 90:245

[26]   Mayer J, Mader W, Phillipp FO, Flynn CP, Rühle M, Inst. Phys. Conf. Ser. (EMAG 89) 98:349

[27]   Mayer J, Flynn CP, Rühle M (1990) Ultramicroscopy 33, in press

[28]   Arbab M, Chottimer GG, Hoffmann RW (1989) Mat. Res. Soc. Symp. Proc. 153:63

[29]   Frank FC, van der Merwe JH (1949) Proc. Roy. Soc. London A198:205

[30]   Matthews JW (1975) in: Epitaxial Growth Part B (ed. J.W. Matthews), Academic Press, p. 559

ATOMISTIC SIMULATIONS OF SURFACES AND INTERFACES

Stephen M. Foiles

Theoretical Division
Sandia National Laboratories
Livermore, CA 94551

INTRODUCTION

The purpose of this article is to describe the capabilities that have been developed in recent years to calculate the structure and energetics of surfaces and interfaces on an atomic scale using approximate descriptions of the total energy. In particular, the goal is to be able to readily determine the energetics and structures of systems which may require hundreds or thousands of atoms to model and to study explicitly thermal effects by the use of molecular dynamics and Monte Carlo simulation techniques. While a great deal of progress has been made in recent years in *ab-initio* total energy calculations, it is still true that there are a wide range of problems for which one must resort to the use of approximate total energy expressions. In the last several years, a group of similar approaches have been developed for this purpose which represent a substantial improvement over previous approximations such as pair potential interaction models. In this article, these new approaches will be summarized and some illustrative examples of their use will be described.

The paper is the organized as follows. First, the various approximations for the total energy are discussed, including a description of the justification of these methods, the similarities between the methods and the physical effects that are included in these approaches which were ignored in previous treatments. This discussion is followed by some specific examples of the application of these techniques to the structure and thermodynamic properties of surfaces. The examples are by no means be exhaustive, but rather a few specific cases demonstrating both the consequences of the coordination dependence of the interactions and the consequences of the cooperative behavior of the atoms are described. These examples include the reconstruction of fcc(110) surfaces, the interaction and diffusion of metal adatoms on metal surfaces and surface phonon dispersion relations. In the last part of the paper, some applications of these approaches to grain boundaries are summarized. These include comparison of calculated grain boundary structures in Al with high resolution electron microscopy, trends in the energetics of grain boundaries, the possibility of roughening transitions in high angle grain boundaries and finally composition variations near grain boundaries in alloys.

It is important to note that the intent of this paper is not to provide a complete review of this field. There has been too much work in this area to be described here. The intent is to provide a few illustrative examples of the current capabilities of these methods. It is also important to note the goal of these calculations. The approximate nature of the techniques means that the absolute energies computed are not as reliable as could be determined using first-prin-

Equilibrium Structure and Properties of Surfaces and Interfaces
Edited by A. Gonis and G.M. Stocks, Plenum Press, New York, 1992

ciples calculations. The value of these methods is to determine trends, possible mechanisms, thermal effects and to provide a simple conceptual framework for understanding some aspects of the energetics.

MODELS OF THE TOTAL ENERGY

The basic input used in atomistic computer simulations of surfaces and interfaces is some computationally efficient method of modeling the total energy of the solid as a function of the atomic position of the various atoms. There have been significant advances in recent years in the quality of these methods with the advent of a class of models described by Carlsson[1] as pair functional approaches. In this section, a few comments will be made about pair interaction models that have been used in the past. This will be followed by a summary of the various pair functional methods that have been put forth and a discussion of the physical motivation behind the different models.

The subject of interatomic potentials has been discussed by various authors. Some recent review articles relevant to the issues discussed here have been written by Carlsson[1], Raecker and DePristo[2], Foiles[3], Nørskov[4], and Daw, Foiles and Baskes[5].

Pair Potential Models

The simplest approximation for the total energy of a solid is to assume that the total energy can be represented by a sum over pairwise interactions between atoms[6]. In particular, the total energy is then written as

$$E = \frac{1}{2}\sum_{ij} v\left(R_{ij}\right).$$ (1)

In this expression, R_{ij} is the distance between atoms i and j. (In this paper we will observe the convention that sums over atom indices will exclude terms with i=j.) The appeal of this expression is its conceptual and computational simplicity. Equation (1) does provide a reasonable description of systems composed of closed shell atoms such as the solid noble gasses. However, equation (1) does not provide a good description of metallic bonding. Some manifestations of this can be seen by considering the vacancy formation energy and elastic constants that this simple model predicts. The unrelaxed vacancy formation energy in this model must be equal to the sublimation energy for any choice of the pair interaction v. (The energy associated with allowing the atoms to relax from their ideal lattice positions is a small correction.) For the noble gas solids, this relationship is approximately true. For example, $E_v^f/E_{sub} = 0.95$ for solid Ar. (E_v^f is the vacancy formation energy and E_{sub} is the sublimation energy.) However, the values for the noble metals range from 0.23 for Au to 0.37 for Cu. Another consequence of equation (1) is the Cauchy relationship for the elastic constants, i.e. $C_{12} = C_{44}$ for any central force pair potential. Again, this is a reasonable approximation for solid Ar where $C_{12}/C_{44} = 1.1$, however it is a poor approximation for metallic systems where, for example, $C_{12}/C_{44} = 3.7$ for Au. Thus it is clear that this simple approximation is not adequate to describe metallic systems.

For the case of simple metals that can be described by weak pseudopotentials, it is possible to develop an approximation for the total energy which consists of two parts[7]. (Moriarty has also carried out similar treatments for transition metals[8]. The general results are similar but the analysis is much more involved due to the need to treat the strong potential of the d electrons.) The first contribution is a structure-independent but density-dependent energy. The second contribution is a sum over density dependent pair interactions. The basic idea behind these derivations is to start with a uniform electron gas or jellium and then add the individual ionic cores and compute the interaction of the ions and the electron gas in perturbation theory. If the perturbation expansion is truncated at second order, one obtains the description in terms of pair potentials. These arguments suggest that the above description needs to be supplemented by a density dependent energy. This leads to the approximation

$$ E = NE_0 \left(\frac{N}{V}\right) + \frac{1}{2} \sum_{ij} v(R_{ij}) . \qquad (2) $$

In this expression, N is the number of atoms, V is the total volume and E_0 is the structure-independent part of the total energy. It represents both the energy of the electron gas as well as the first-order interaction of the ions with the electron gas. This general form solves the problem with the elastic constants and the vacancy formation energy discussed above[9,10]. However, this expression is only valid for perturbations around a uniform system. Thus it is not appropriate for the treatment of free surfaces. (As a practical matter, the volume associated with atoms at a free surface is not a well defined property.) Even for internal interfaces, such as grain boundaries, the local density at the boundary is typically less than that of the bulk and so the applicability of equation (2) can be questioned.

Pair potential models have been used to study a wide variety of properties such as point defect structure, energy and migration and the structure of dislocations and grain boundaries. There are two conceptual shortcomings of these approaches for calculations with inhomogeneities, though. The first is that the structure-independent term, E_0, contains a large portion of the energy, but equation (2) provides no way to model the variation of those contributions to the energy at an inhomogeneity. The other is that the derivation of equation (2) show that the pair term depends on the environment. As usually applied, this variation in v is generally ignored.

Pair Functional Models

In the last few years there have been a variety of approaches developed to address the problems mentioned above. These methods include the Finnis-Sinclair N-body potentials[11], the embedded atom method of Daw and Baskes[12,13] and effective medium theory developed by Nørskov, Jacobsen, Puska[14] and coworkers. These different methods have different physical motivations but all fall into a class of models that Carlsson[1] has described as pair functional models. In these models the total energy is expressed in the form

$$ E = \sum_i F\left(\sum_{j \neq i} f(R_{ij})\right) + \frac{1}{2} \sum_{ij} v(R_{ij}) . \qquad (3) $$

The physical meaning ascribed to the various terms differs between the methods. However, there are certain consequences of having an energy expression of the form of equation (3) that do not depend on the physical motivation. In particular, one introduces a certain class of many-body interactions. It is the inclusion of these many body interactions that has lead to the success of these approaches in cases where the use of simple pair interactions has failed. In the following sections, the physical interpretation and derivation of the different models will be summarized. This will be followed by a discussion of the general consequences of using models based on equation (3). In particular, we will see how this leads to the observation that increased coordination implies weaker bonds and longer bond lengths.

Finnis-Sinclair N-body potentials

Finnis and Sinclair[11] introduced a simple model of the energetics of central transition metals. The motivation for this model is best understood in terms of a tight-binding description of the bonding in the central transition metals. In this picture, the emphasis is placed on the d-electrons which are localized to atomic-like orbitals. In a tight-binding picture, the total energy is assumed to consist of two parts. The first is the sum of the one-electron energies computed from the tight-binding analysis. This is supplemented by a repulsive pair potential which incorporates the core-core interactions as well as the double counting terms in the total energy. The idea behind the N-body potentials is to provide a simple model for the sum of the one electron energies.

The effect of placing the atoms into a solid is to take the atomic d electron energy and to spread it into a band. For systems with approximately half filled d bands, the bonding states are occupied and the energy of these states is lowered relative to the atomic level. The amount that the energy of a state is lowered should scale with the band width. Thus one expects that the cohesive energy associated with the d electron should scale with the d-band width.

The next step is to find a simple model of the d-band width. This follows from noting that the width is proportional to the square root of the second moment of the density of states. The second moment, μ_2, is

$$\mu_2 = \text{Tr}\langle i, \alpha| H^2 |i, \alpha\rangle = \text{Tr}\sum_{j, \beta} \langle i, \alpha| H |j, \beta\rangle\langle j, \beta| H |i, \alpha\rangle. \tag{4}$$

In this expression, the indices i and j refer to atoms and the greek indices refer to the d orbitals localized on each atom. Thus the second moments can be computed by considering all of the possible paths from a given atom to its neighbors. The important point is that the sum over all the orbital indices gives a simple radial function. Thus the second moment of the density of states can be computed from a sum over neighbors of the form

$$\mu_2^i \propto \sum_j f(R_{ij}) \tag{5}$$

where the function f(R) in equation (5) represents the result of summing the terms in equations (4) over the orbital indices. The subscript i has been added to indicate that we are considering a local band width defined on each atom.

The total energy expression that results from this reasoning is given by the form

$$E = A\sum_i \sqrt{\sum_j f(R_{ij})} + \frac{1}{2}\sum_{ij} v(R_{ij}). \tag{6}$$

This expression is interpreted as follows. The first term represents the d-electron band energy. The sum over j gives the estimate of the second moment and the total band energy is then obtained by taking the square root of the local band width on each atom and summing the result over all the atoms in the system. The factor A is an adjustable parameter. The second term in equation (6) is the sum over the repulsive pair potentials. Note that equation (6) is in fact of the pair functional form defined by equation (3). In the Finnis-Sinclair model, the function F is just the square root, and the function f(R) is interpreted as the square of the tight-binding hopping matrix elements between the atoms.

The Finnis-Sinclair has been used to study a variety of problems in the structure of bcc metals including elastic properties[11], bulk phonon dispersion relations[15], simple point defect properties[16] and thermal expansion[17]. A similar method has been proposed by Rosato, et al.[18] for the description of the fcc metals. This method has been applied to point defect properties, bulk phonons, surfaces and finite temperature behavior of the solid.

Embedding Energies

Both the embedded atom method and the effective medium theory of metals are based on the concept of an embedding energy. As will be shown, the use of this concept also leads to approximations for the total energy which are of the pair functional form. The physical motivation, though, is very different from that discussed above.

The basic idea behind the concept of an embedding energy is the following. As a first approximation, an atom placed in a metallic environment does not feel the detailed structure of the nearby metals atoms due to the effective screening in metallic systems. It may then be possible to model the interaction of the atom with the metal by the interaction of the atom with a simpler system. In defining the embedding energy, the chosen model-host is taken to be a uniform electron gas. This is thought to be a good model since the interaction of the valence electrons with a uniform electron gas should be similar to that in the real metal.

This idea has been used to study the interaction of light impurity atoms in and on metals by a variety of authors. These include the quasiatom method of Stott and Zaremba[19] and the effective medium theory due to Larsen and Nørskov[20], Nørskov and Lang[21] and Nørskov[22]. Some of the general features of these approaches will be discussed here before going to the case of metallic systems.

The uniform electron gas is completely defined by the electron density. Thus if the interaction of the impurity atom is modeled by the interaction with a uniform electron gas, the energy of the impurity in the model host just depends on the density of the electron gas. This energy is defined to be the embedding energy. It is the difference between the energy of the impurity atom in an electron gas and the sum of the energy of the atom and the energy of the electron gas. This quantity has been computed for various impurity atoms by Puska, et al.[23] within the local density approximation. The results for some of the light elements are shown in Figure 1 . The qualitative behavior of the embedding energy falls into two classes. Inert closed shell elements, such as He and Ne, have a purely repulsive interaction with the electron gas. The chemically reactive elements have a minimum at some optimal density and then the energy increases for higher densities.

Nørskov[22] has derived an expression for the energy of an impurity atom based on the embedding energy. This approach is referred to as effective medium theory. The expression that Nørskov obtains for the impurity energy is

$$\Delta E = \Delta E_{eff}^{hom} (\bar{n}) + \Delta E^{core - core} + \Delta E^{hyb}. \tag{7}$$

The first term in this expression is the effective embedding energy. It is related to the embedding energies shown in Figure 1 by a correction due to the fact that the jellium model contains a uniform compensating positive charge background and the positive charge in the real metal is localized to the nuclei. The qualitative behavior of the effective embedding energy is the same as that shown above with the minima occurring a somewhat higher densities and exhibiting stronger binding. The term n is an average of the electron density in the metal over a region near the impurity atom. The second term is a short-range core-core repulsion that is represented by a short-range repulsive pair potential. The last term, ΔE^{hyb}, is the hybridization energy. It represents the differences in the change of the band structure of the metal and the change of the band structure of the electron gas. The hybridization term plays an important role in the transition metals.

There are some interesting general conclusions that can be made about the interaction of impurities with metals just from the consideration of the embedding energy term. If one assumes that the embedding term is the dominant term, then the impurity will attempt to move to a location where the local electron density is the optimum electron density for that impurity, i.e. the electron density at which the embedding energy is a minimum. This has some interesting consequences. First, consider the impurity-metal bond length for adsorbate atoms. It is a general trend that at higher coordination, bond lengths are larger. This is explained in terms of the effective medium ideas as follows. The adsorbate wants to be in an optimum electron density. For higher coordination sites, the density at the impurity contributed by each metal atom should therefore be less than the amount contributed when the atom is in a low coordination site. This implies that the metal-adsorbate bond lengths should be longer for higher coordination since the density at the impurity from a given metal atom decreases as the bond length increases. The embedding concept also provides an explanation of a couple of properties of hydrogen in metals. The first is that hydrogen is trapped by any defect in the metal, such as a dislocation or grain boundary, that lowers the overall density. This is due to the fact that the optimum electron density for hydrogen is less than the lowest electron densities found in the ideal metal. Thus any defect that lowers the density will trap the hydrogen since the hydrogen can locate in a position with an electron density closer to its optimum density. Another property of hydrogen in metals is that hydrogen atoms in vacancies are not located at the center of the vacant site, but rather somewhat off-center. This is due to the fact that the electron density

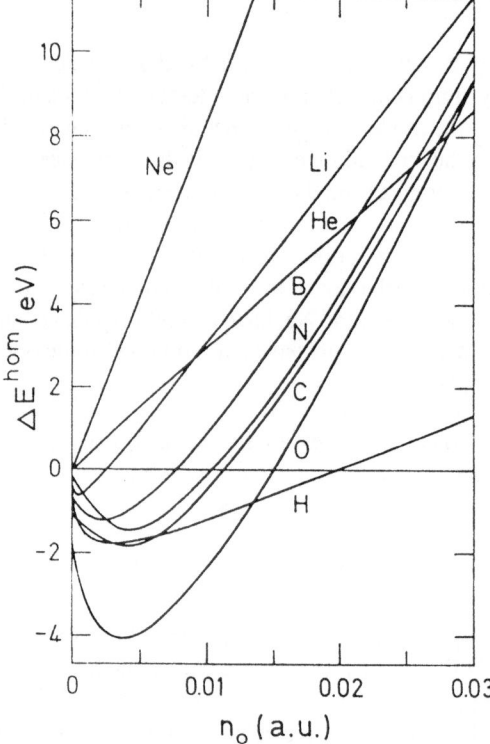

Figure 1. The embedding energy in a homogeneous electron gas for a number of elements as a function of the density of the homogeneous electron gas. The inert elements show a purely repulsive behavior while the chemically active elements exhibit a minimum in the embedding energy for some optimum density. (from Puska, *et al.*[23])

at the center of a vacancy is lower than the optimum electron density for hydrogen, so the hydrogen moves off-center to a location of higher electron density.

The effective medium theory has been used to study of the chemisorption of hydrogen[24], oxygen[25] and carbon[26] on transition metal and used to study hydrogen in metals[22,27,28,29]. A review of this work has been written by Nørskov[4].

Embedded Atom Method

Daw and Baskes[12,13] proposed a generalization of the effective medium idea which treats the metal atoms and impurity atoms on the same footing. This approach is referred to as the embedded atom method (EAM). The motivation for this work was to describe the properties of hydrogen interacting with metals. The qualitative idea is to view each atom as being embedded in a host electron gas provided by the other atoms of the system. This energy is then supplemented by a pair potential which was initially viewed as incorporating core-core repulsions. This led them to propose the following ansatz for the total energy of the system

$$E = \sum_i F(\rho_i) + \frac{1}{2} \sum_{ij} \varphi(R_{ij}) \ . \tag{8}$$

In this expression, F is the embedding energy and ρ_i is the electron density at the location of atom i due to the remaining atoms of the system. The second term is the sum over the pair interactions. In this approach the electron density of the solid is approximated by the superposition of the atomic electron densities of the constituent atoms. This leads to the simple expression for the electron density at site i,

$$\rho_i = \sum_{j \neq i} \rho_j^a(R_{ij}) \ . \tag{9}$$

Here $\rho_j^a(R)$ is the atomic electron density of atom j evaluated at the distance R from the nucleus. Note that the combinations of equation (8) and (9) leads to a energy expression which is of the pair functional form defined in equation (3). In the EAM, the various functions entering the theory are determined empirically by choosing appropriate parametrized functional forms and adjusting the parameters[30].

A more formal justification of the EAM has been developed by Daw[31]. The nature of the arguments will be summarized here. The goal is to derive the above equations from density functional theory. The derivations starts with the density functional expression for the cohesive energy of the solid[32] given by

$$E_{coh} = G[\rho] + \frac{1}{2} \iint \frac{n(r_1) n(r_2)}{|r_1 - r_2|} - E_{atom} \ . \tag{10}$$

In this expression, ρ is the total electron charge density, $G[\rho]$ is the kinetic, exchange and correlation energy functional, n(r) is the total charge density given by

$$n(r) = \sum_i Z_i \delta(r - R_i) - \rho(r) \tag{11}$$

where Z is the nuclear charge and R_i is the location of atom i. The integrals are over the positions r_1 and r_2. The last term is the energy of the isolated atoms.

95

The EAM form can be obtained if two approximations are made. First, it is assumed that the total electron density can be approximated by the superposition of atomic electron densities, namely

$$\rho(r) = \sum_i \rho_j^a (r - R_i).$$ (12)

(The effect of allowing charge redistribution is discussed by Daw[31].) There are various motivations for this approximation. From a practical standpoint it introduces significant computational simplifications. On the fundamental side, there are two arguments for this choice. First, since the total energy is a minimum at the true electron density, any errors in the electron density used in the energy functional should only appear to second order in the total energy. Second, it is known that for many of the close packed metals that the superposition of atomic electron densities is in reasonable agreement with full calculations of the electron density.

The second basic approximation is that the kinetic, exchange, and correlation functional is local in the sense that it can be written as an integral over a local function of the density and its gradients, i.e.

$$G[\rho] \approx \int g(\rho(r), \nabla\rho(r), \dots)$$ (13)

This approximation is in the spirit of the Thomas-Fermi statistical models of electronic structure but allows for the possibility of more sophisticated local energy densities. One expects this approximation to reasonably describe changes in the energetics due to changes in the overall electron density but one does not expect this model to reproduce effects that depend on the detailed band structure of the metal. The consequences of this will be discussed below.

In order to make contact with the EAM form, we define the embedding energy of atom i as the difference in kinetic, exchange and correlation energy between the atom in a uniform electron gas of density ρ_i, and the sum of energy of the isolated atom. and the electron gas. This leads to the definition

$$G_i(\rho_i) = G[\rho_i^a(r) + \rho_i] - G[\rho_i^a(r)] - G[\rho_i].$$ (14)

If one now assumes that one can determine an appropriate background electron density at each site, ρ_i, then one can write the cohesive energy as

$$E_{coh} = \sum_i G_i(\rho_i) + \frac{1}{2}\sum_{ij} U(R_{ij}) + E_{err}$$ (15)

In this expression, U(R) is the electrostatic interaction of two fixed charge density distributions. This term follows directly from the electrostatic terms of equation (10). The last term is the error made by using the embedding function. This term provides the formal means of determining the background electron densities. The optimum values of ρ_i are those that cause the error to vanish.[31] A good approximation to the solution is given by the condition

$$\int_{\Omega_i} g(\rho_i^a + \sum_{j \neq i} \rho_j^a(r - R_j)) = \int_{\Omega} g(\rho_i^a + \rho_i).$$ (16)

This condition indicates that the background density, ρ_i, should be chosen as some average of the tails of the electron densities contributed by the neighboring atoms. The choice for

ρ_i that is made in the EAM is to evaluate the electron density at the nucleus. In the limit that the background density is slowly varying, one would expect this to be a good approximation. How well the density at the nucleus represents the optimum background density as determined by the condition of vanishing E_{err} has also been investigated. The background density is computed both ways for a variety of conditions including bulk lattices in bcc and fcc crystals and fcc surfaces. Daw found that the different procedures yielded different values for the background density but that there was a linear relationship between the background densities computed in the two different manners. The existence of a unique relationship between the optimum background density and the density at the nucleus provides support for the use of the latter approximation in the EAM. The embedding functions used in the EAM are determined empirically by fitting to known properties. Any consistent difference between the density at the nucleus and the optimum background density is then just absorbed into the determination of the embedding function. However, if one is using an embedding function determined from some first principles calculation, as is done in the effective medium theory discussed below, then using the density at the nucleus is not a good choice.

The EAM has been used for a wide range of problems concerning the structure and thermodynamics of fcc metals and alloys. Some of the applications will be discussed in detail below. Some of the other applications include bulk[33] and surface[34,35] phonons, bulk thermodynamic functions and melting[36], liquid metals[37], point defect properties[13,30], grain boundary structure[38,39], alloys[30,40], segregation of binary alloys to surfaces[41] and grain boundaries[42], interdiffusion of alloys[43], fracture and mechanical properties[44], surface structure[30] and reconstruction[45,46], and ordered surface alloys[47]. In general, the agreement between the results obtained with the EAM in these various applications and experimental results, where available, has been very good. This success for a wide range of materials properties constitutes the most convincing demonstration of the utility of this approach.

Effective Medium Theory for Metals

The effective medium theory has been generalized to the case of metallic bonding by Jacobsen, et al.[14] The resulting effective medium theory for metals yields an expression for the total energy which for certain metals is approximately of the pair functional form. The basic idea is to apply the embedding energy concept to each atom of the system. This analysis has been carried out in detail by Jacobsen, et al.[14], and only the main points will be summarized here. The starting point of the derivation is that the total electron density can be approximated by a linear combination of the induced charge densities for the case of the atom in the electron gas. In particular, this gives

$$n(r) = \sum_i \Delta n_i(r) \qquad (17)$$

where $\Delta n(r)$ is the difference between the electron density for the atom in a uniform electron gas and the uniform electron density. With this ansatz, the total energy that they obtain can be written as

$$E = \sum_i E_{c,i}(\overline{n_i}) + E_{ASC} + E_{1-el}. \qquad (18)$$

The various terms in this expression will be described below. The first term is the sum of the embedding energies. The cohesive energy, E_c, is related to the embedding energy as mentioned above. The argument n_i is the background density computed by averaging the density, $n(r)$, in equation (17) over a sphere of radius s_i around each atom. Though this appears to be of the pair functional form, it is not due to certain self-consistency issues. The induced electron densities, $\Delta n_i(r)$, depends on the background density n_i for the atom i. In addition, the radius, s_i, that is used for the averaging is chosen such that the net charge inside the sphere of radius s_i for the case of the atom in the electron gas is zero. This means that the averaging radius, s_i, also depends on the background density. How this is dealt with in practice is described elsewhere[14].

The second term in equation (18) is the atomic sphere correction. In the derivation of equation (18), space is divided into spheres of radius s_i around each atom. The atomic sphere correction accounts for the errors involved with the overlap of these spheres and the regions not included in the various spheres. It is shown to be given by

$$E_{ASC} = -\frac{1}{2}\int O(r)\,n(r)\,\phi(r)\,dr \qquad (19)$$

where the overlap function $O(r)$ is equal to $(m-1)$ where m is the number of spheres that the point r is located in, $n(r)$ is the total charge electron density and $\phi(r)$ is the electrostatic potential. The overlap energy describes the electrostatic energy associated with overlapping neutral charge distributions.

The final term in equation (18) is the change in the one electron energy levels. This term represents the corrections to the energy due to the fact that the one-electron energy levels of the homogeneous electron gas do not respond to the addition of the atom in the same manner that the energy levels of real metals do. This term is argued[14] to be small for simple metals and metals with filled d bands. For applications involving these metals, this term is ignored.

In the practical application of the effective medium theory, some simplifications on the above energy expressions are made to allow for computational efficiency. First, the behavior of the averaging radius, s, and the background density are parametrized. This allows the background density to be approximated by a function of a sum over neighbors. The atomic sphere correction is also approximated by a sum of pair interactions. When these approximations are invoked and the one-electron terms are ignored, the effective medium theory is of the pair functional form.

A major difference between the embedded atom method and the effective medium theory is the manor in which the various terms entering the method are determined. In the embedded atom method, these quantities are determined empirically by fitting to known properties. In the effective medium theory, the various quantities involved are all related to quantities that can be calculated from first principles calculations.

The effective medium theory of metals has been applied to a variety of problems. These include the thermal expansion of Al[48], the reconstruction of fcc (110) surfaces[49] including adsorbate induced reconstructions[50], the surface phonons of Cu(100)[51], the premelting of Al surfaces[52], and the structure of small Cu clusters[53].

Practical Consequences of the Pair Functional Form

In this section, some of the consequences of using the pair functional type of energy expression defined in equation (3) will be described. Most of these follow from the fact that the pair functional form introduces many-body interactions. This can be seen by deriving two-, three- and higher body interactions which approximate the pair functional form. To do this, assume that the density at each site is close to some average value f_0. (I am using the terminology of the EAM here for convenience but the arguments are general.) The density at each atom, f_i, can then be written in the following form

$$f_i = f_0 + \left(\sum_{j \neq i} f(R_{ij}) - f_0 \right). \tag{20}$$

The second term will be small for an appropriate choice of f_0. Now replace the function F (see equation (3)) with its Taylor series around f_0. If the Taylor series is taken to first order, one obtains pair interactions, the second order term introduces three-body interactions, etc. The energy expression that results from terminating the series at second order[45] is

$$E \approx NE_0 + \frac{1}{2} \sum_{ij} \psi_{ij}(R_{ij}) + \frac{1}{6} \sum_{ijk} \zeta_{ijk}(R_{ij}, R_{ik}). \tag{21}$$

where the one-, two- and three-body terms are related to the pair functional expression by

$$E_0 = F(f_0) - f_0 F'(f_0) + \frac{1}{2} f_0^2 F''(f_0), \tag{22}$$

$$\psi(R) = 2f(R) F'(f_0) + f(R)(f(R) - 2f_0) F''(f_0) + v(R), \tag{23}$$

and

$$\zeta_{ijk}(R_{ij}, R_{ik}) = 3f(R_{ij}) f(R_{ik}) F''(f_0). \tag{24}$$

There are several points to be made about these expressions. First, note that the result for the total energy, equation (21), is very similar, except for the inclusion of the three body interactions, to the pair potential plus density dependent energy, equation (2), that had been used frequently before the introduction of the pair functional methods. The E_0 term here depends on f_0 which is similar to the density dependence in the pair potential formulation. Note also that the pair potential is now environment dependent, through the value of f_0, consistent with the pseudopotential perturbation theory derivations of the pair potential expression. (The nature of the dependence will be discussed below.) The two-body interaction is shown in Figure 2

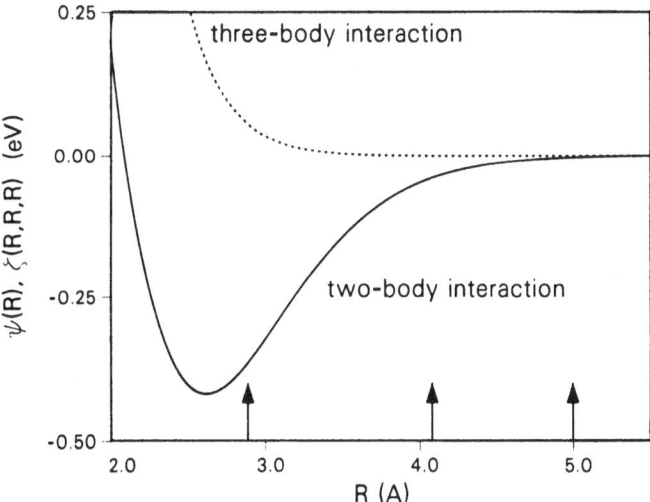

Figure 2. The effective two-body interaction, ψ, and the effective three-body interaction, ζ, derived from the EAM energy expression (see equations (23) and (24)) evaluated for Au. The three-body interaction is computed for the case of a central atom with two atoms a distance R away. The arrows indicate the separations of the first, second, and third neighbors in the bulk lattice.

Note that the qualitative shape of the two-body interactions is similar to what has been used in the pair potential treatments of metals.

In order to understand how the effective pair interaction changes with the environment, i.e. f_0, we need to know that the sign of F" is positive. This can be seen from the various motivations for the pair functional form. In the Finnis-Sinclair potential, the curvature of F follows directly from the choice of the square root behavior and the fact that the band width energy should be negative. For the embedding based arguments, the first-principles calculations of the embedding functions by Puska[23] and the model calculations by Daw[31] both yield embedding functions with positive curvature. An empirical manifestation of the curvature can been found in the elastic constants. As shown by Daw and Baskes[13], the Cauchy discrepancy, $C_{12} - C_{44}$, is determined by the curvature of the embedding function. Most of the fcc metals where one expects these methods to work well have a positive Cauchy discrepancy which implies a positive curvature of the embedding function. (For those metals, such as Ir, which have a negative Cauchy pressure, it is not correct to conclude that the embedding function has a negative curvature. The negative Cauchy discrepancy is an indication that the energetics of that metal are not well described by a pair functional method.)

The change of the effective pair interaction with environment can now be qualitatively determined. As the overall coordination of an atom decreases, the appropriate f_0 also decreases. Since the curvature of F is positive, this implies that F' will become more negative. This in turn implies that the first term of equation (23) will be more attractive. The conclusion of this reasoning is that as one lowers the coordination of an atom., the effective pair interactions will be more attractive and will have shorter bond lengths. This is consistent with the general trend that lower coordination leads to stronger bonds and shorter bond lengths. An extreme example of this is the difference between the bond energy and bond lengths of metallic dimers and bulk metals. Some of the consequences of this trend will be explored in the examples discussed below.

The qualitative behavior of the three body term is also determined by F". Since F" is positive, the three-body term is purely repulsive. Further, since it goes as the product of two density terms, the effective three-body interaction is fairly short-ranged. Figure 2 shows the three-body term for the case where the two atoms j and k are equidistant from the atom i. The repulsive three-body term will be seen below to have important consequences in the understanding of reconstructions and preferred adatom geometries.

While the pair functional form does introduce many-body interactions, it is important to note that they are of a restricted type. In particular, they do not have any explicit angular dependence. For example note that the three body interaction obtained in equation (24) only depends on bond lengths and does not depend on the bond angle of the three atoms. This lack of angular dependence is one of the main short-comings of the pair functional methods which needs to be addressed in future developments before systems where the angular terms are important can be treated. These systems include, for example, semiconductors, and central (partially filled d band) transition metals.

There is an arbitrariness in the pair functional form that should be realized whenever one is comparing different methods or different parametrization of the same basic method. Note that if the function F is linear, that the pair functional form reduces to a simple pair potential expression. A consequence of this is that the addition of a linear contribution to F can always be identically compensated for by a modification of the pair potential term. Specifically, the following transformations of the functions F and v leave the total energy, as computed from equation (3), unchanged for any value of the parameter a;

$$F(s) \rightarrow F(s) + as, \tag{25}$$
$$v(R) \rightarrow v(R) - 2af(R). \tag{26}$$

As a consequence of this, models may seem to be quite different but in fact be very similar. For example, variation of the parameter a can change a model with a purely attractive pair interaction into one where the pair interaction has an attractive minimum and not change any of the results that would be obtained.

Types of Calculations Performed

The energetic models discussed above provide a method of calculating the total energy and forces for an arbitrary arrangement of the atoms. There are three basic types of calculations that can then be performed; energy minimization (molecular statics), molecular dynamics, and Monte Carlo. These techniques will not be discussed in detail here, just the basic ideas will be presented to make it clear what can be done within each approach. A discussion of molecular dynamics and Monte Carlo simulations can also be found in the article by Prof. Car in this volume

The simplest calculations are the energy minimization calculations, sometimes also called molecular statics. In these calculations, the positions of all of the atoms are adjusted until a local energy minimum is found. These calculations are useful for determining the zero-temperature structure of a defect such as a surface and for obtaining the energies of various structures so that the lowest energy structures can be determined. One important point to note about these calculations is that the numerical methods generally search for local minima in the energy versus position surface. There is no guarantee that the global minimum has been found. Thus care is needed to differentiate metastable states from the ground state.

The next type of calculations are molecular dynamics calculations. In these calculations the classical motion of the atoms in time is computed by solving Newton's equation of motion. These calculations can be used either to determine the detailed dynamical process or to compute thermodynamic quantities by averaging over time. The time scale that can be simulated, though, is rather short; simulations can generally run only tens to hundreds of picoseconds. Thus "slow" processes, such as diffusion, can not be directly simulated in this manner. Also, these simulations treat the motion of the atoms classically. For light impurities it may be necessary to explicitly consider the quantum mechanical nature of the motion.

The last type of calculations are equilibrium Monte Carlo simulations. The goal of these simulations is to compute thermodynamic averages by generating a sampling of the appropriate statistical ensemble. An advantage of these calculations is that they can compute thermodynamic properties for systems where the physical equilibration time are longer than those that can be simulated using molecular dynamics. An example of this is the ability to study the segregation in a binary alloy to a defect. In a MC simulation, one can directly change the chemical identities of atoms to sample statistical ensemble. Of course, this is not the physical process by which atomic species interdiffuse so these simulations do not give insight into the kinetic aspects of the problem.

APPLICATIONS TO SURFACES

Three different applications of the EAM to the properties of surfaces will be presented. The first is a discussion of the reconstruction of the fcc(110) surfaces. This will demonstrate the importance of the coordination dependence to obtain the correct reconstruction and how atomistic simulations can provide insight into order-disorder transitions. Next the energetics of metal adatom clusters on metal surfaces and the diffusion of adatoms will be presented. Again this will show the importance of the many-body interactions as well as the importance of substrate relaxation. It will also demonstrate the different possible diffusion mechanisms that can occur for metal atoms diffusing on metal surfaces. Finally, the calculations of the surface phonon dispersion relations will be presented. This provides a comparison of the predictions of the EAM for the change in the force constants at a surface with experiment.

Reconstruction of fcc(110) surfaces

It is well known that the (110) surfaces of the fcc transition metals Au, Pt and Ir undergo a reconstruction to a structure with 1x2 symmetry. A variety of experiments[54] and total energy calculations[55] have shown that the structure is the "missing-row" structure in which alternate close packed rows of atoms are removed. The calculations have included work based on the EAM by Foiles[45] and using the effective medium theory by Jacobsen and Nørskov[49]. The order-disorder transition for the Au and Pt surfaces has also been investigated using the EAM by Daw and Foiles[46]. Here the various calculations performed using the EAM will be summarized. It should be noted that more reliable first-principles calculations[55] of the total energy and structure of the reconstructed surface have also been performed. The value of calculations based on the EAM is that they were useful in the early stages of the research to eliminate various structures that were proposed, it provides a conceptually simple framework in which to understand the tendency to reconstruct and it provides a means by which realistic simulations of the finite temperature behavior of the surface can be performed.

The energy of a variety of different models that were proposed for the Pt(110) 1x2 reconstruction were computed by Daw[56]. The results clearly showed that the two lowest energy structures were the unreconstructed surface and the (1x2) missing row reconstruction. The energy of the missing row reconstruction relative to the energy of the unreconstructed surfaces was computed for the six fcc metals in the Ni and Cu columns of the periodic table using the EAM by Foiles[45]. It was found that for Ni and Cu the unreconstructed surfaces is lower in energy by 1.4 and 1.2 meV/A^2, respectively, that for Pd and Ag the energies of the two surfaces are very close in energy with the missing row reconstruction lower by 0.6 and 0.5 meV/A^2, and for Pt and Au the surfaces the missing row reconstruction is lower in energy by 2.9 and 2.3 meV/A^2. (Similar results were obtained by Jacobsen and Nørskov using the effective medium theory[49].) These results agree with the experimental observations that Au and Pt reconstruct and that Ni and Cu do not reconstruct. They disagree with experiment for the case of Pd and Ag but the computed energy difference for those cases is small. There is an interesting trend in the calculated results. Namely, the tendency to reconstruct increases as one moves down the periodic column. It is important to note that the important aspect of these results are not the specific numbers obtained above for the differences in the surface energy since these diferences are smaller than the expected reliability of methods such as the EAM. The important results are that the reconstructed and unreconstructed surfaces are very close in energy and the trend of the energy difference between the different elements.

This trend and the reason why the missing row reconstruction is favored in some cases can be understood in terms of the effective two- and three- body interactions discussed above[45]. The calculations show that the difference in energy computed between the reconstructed and unreconstructed surfaces does not depend strongly on the inclusion of the surface relaxations. One can therefore hope to understand the energetics by counting the number of neighbors in the unrelaxed structure. This analysis shows that the number of nearest neighbor pairs per unit area is the same for the unreconstructed and missing row surfaces. The differences between the two surfaces are that the unreconstructed surface has more second neighbor pairs of atoms per unit area and it also has more trios of atoms where two of the atoms are nearest neighbors than the missing row reconstruction. As can be seen from Figure 2, the energy associated with the second neighbor pairs is attractive while the energy associated with the trios of atoms is repulsive. The first conclusion from these observations is that if one uses a pair interaction model where the pair interaction is attractive at second neighbors, than one will always predict that the unreconstructed surface is lower in energy. It is the inclusion of the repulsive three-body interactions that leads to the prediction of the missing row reconstruction. *Therefore, the many-body interactions contained in the pair functional form of the energy are essential to correctly predict the existence of the missing row reconstruction of fcc(110) surfaces.* The relative energetics of the unreconstructed and missing row surfaces is now seen as a competition between the strengths of the second neighbor attractive pair interactions, which favor the unreconstructed surface, and the strength of the repulsive three-body interactions. The latter is related to the curvature of the embedding function and so the Cauchy discrepancy of the elastic constants as discussed above such that a larger Cauchy discrepancy implies stronger three-body repulsions. It is interesting to note that in both of these columns of the periodic tables the Cauchy discrepancy is largest for Au and Pt which are the metals which are observed to undergo the missing row reconstruction.

Another aspect of the missing row reconstructions of the fcc(110) surfaces is the observation via LEED measurements that the surfaces undergoes an order-disorder transition at elevated temperatures. This transition occurs for Au[57] at 650 K and for Pt[58] at 960 K. This transition has been simulated using Monte Carlo techniques using the EAM by Daw and Foiles[46]. (Roelofs, *et al.*[59] have also studied the order-disorder transition using interactions developed from a modified EAM method.) In these simulations three types of atomic displacements were considered: atoms are allowed to make small displacements from there current position to account both for relaxations and thermal vibrations, atoms can jump to positions near other ideal lattice sites and finally atoms can jump to positions near ideal lattice sites on the other surface of the slab. (The calculations are performed on a slab of metal that is repeated periodically in the plane of the surface.) The last two types of steps allow the system to rearrange in accord with thermal equilibrium. In particular it is possible for the system to order into either the unreconstructed surface or the missing row reconstruction by moving half of a monolayer of atoms from one surface of the slab to the other surface. The missing row reconstruction was monitored by evaluating the structure factor at the wavevector for the 1x2 reconstruction. The simulations found the structure factor drops rapidly around a temperature of 570 K. This is in excellent agreement with the experimental critical temperature especially when one recalls that the only input into the calculations are bulk materials properties. Similar calculations for Pt predict an order-disorder critical temperature of 750 K again in reasonable agreement with experiment. In particular, the critical temperature is predicted to be higher for Pt than for Au in agreement with experiment.

In addition to correctly predicting the critical temperature, these simulations can provide insight into the nature of the order-disorder transition. Figure 3 shows a snapshot of the (110) surface at a temperature somewhat above the critical temperature. This shows the nature of the disordering that occurs. The top layer atoms still form rows along the close packed [110] directions as occurs in the ideal missing row reconstruction. The order breaks down in the placement of the rows. In the ideal missing row structure, these rows occupy alternate troughs of the (110) surface, while above the critical temperature the alternation in the occupation of the troughs breaks down.

Another insight that can be gained from the Monte Carlo simulations is that the detailed atomic motion makes a significant contribution to the critical temperature. This can be seen in two ways. First, one can perform a Monte Carlo simulation where the atoms are constrained to sit on ideal lattice sites and simply jump between lattice sites. The other approach is to treat the interaction of the surface atoms within a lattice gas model and determine the energy of the lattice gas interactions by comparing with the energy of pairs and trios of adatoms on the surface where the adatoms are located on ideal lattice sites. The critical temperatures deduced by these two methods are both about 1100 K. *Thus the inclusion of the atomic displacements changes the prediction of the critical temperature by about a factor of 2!* This difference arises from two different physical effects. The first is the effect of the local relaxations on the overall interaction energy of the surface atoms. The magnitude of this effect can be seen from computing the binding of a pair of adatoms separated by a nearest neighbor distance on a unreconstructed surface. If the atoms are constrained to ideal lattice sites, the binding energy of these two atoms is computed to be 0.45 eV while if the adatoms and substrate are allowed to relax, this binding energy is reduced to 0.28 eV. Thus the inclusion of adatom and substrate relaxation is critical to the quantitative calculation of the interaction of atoms on a surface.

The second effect which is missing when the atoms are frozen on lattice sites is the contribution to the order-disorder transition of vibrational entropy. Implicit in a lattice gas model is the assumption that the vibrational entropy is independent of the atomic arrangement. This assumption can be tested by computing the vibrational entropy in a harmonic approximation for different atomic arrangements on the surface. The vibrational part of the free energy has been computed for two surfaces: the ideal missing row surface and a surface where one atom has been moved from a row to an adjacent trough. At high temperatures the difference in the vibrational free energy is $\Delta F_{vib} = -1.4 \, k_B T$. The sign of this effect is such as to favor the breaking up of the rows and so lowers the transition temperature. The qualitative point is that when the atom is moved out of the row, its freedom to vibrate is increases which increases its entropy and so lowers the free energy.

The above results have shown how the EAM or similar techniques can be useful in studying surface reconstruction. They provide a simple method of determining the plausible structures, they can provide insight into the driving force behind the reconstruction, and they can shed light on the finite temperature behavior. It is also important to note the important role that the many-body interactions played in these calculations and the importance of allowing for the full motion of the atoms in the determination of the finite temperature behavior.

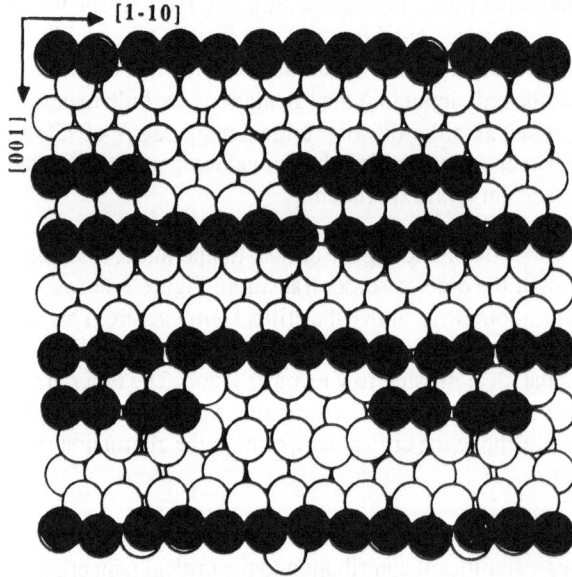

Figure 3. Snapshot of a randomly chosen configuration from the Monte Carlo simulation of the Au(110) surface showing behavior typical of a temperature slightly above the critical temperature. All atoms are Au atoms with the adatoms darkened for illustration. (from Daw and Foiles[46]

Metal Adatoms on Metals Surfaces

The interactions and diffusion of metal adatoms on metal surfaces is interesting both because it provides a measurable manifestation of the energetics of the adatoms and because the behavior of adatoms on surfaces is a crucial element in modeling surface processes such as crystal growth. Field Ion Microscopy (FIM) experiments have yielded interesting information about the structure of adatom clusters and about adatom diffusion. For example, the stable configurations of small cluster of Ni, Pd, Pt and Ir adatoms on W(110) surfaces are linear chains[60]. However, larger clusters of Ni and Pd form close-packed islands and larger clusters of Pt and Ir form intersecting linear chains. For Ir adatoms on Ir(001)[61], linear chains along the <110> directions are observed and for Pt adatoms on Pt(001)[62] the structure oscillates between linear chains and close-packed islands as the size of the cluster increases to six atoms.

The EAM has been used to determine the structure of adatom clusters of Ni, Pd, and Pt adatoms on the Pt(001) surface by Schwoebel, *et al.*[62] and by Wright, *et al.*[63]. In the paper by Schwoebel, *et al.*, it was shown that the EAM correctly predicts that the stable adatom structures for Pt on Pt(001) are that the 3 and 5 atom clusters form lines along the <110> direction and that 4 and 6 and larger adatom clusters form close packed islands. These results are in agreement with the FIM observations. The work of Wright, *et al.*[63] showed that similar results are obtained for calculations for Pd adatoms on Pt(001) except that the 5 atom cluster will also form a close packed island. FIM results for Pd on Pt(001) agree with these predictions except for the three adatom cluster where both linear and island clusters are observed, though the island geometry is preferred[64]. The generally good agreement between the EAM and observation is encouraging.

The interesting aspect of these results is the origin of the different cluster geometries. The existence of linear arrangements of the adatoms is somewhat surprising because in general fcc metals like to form close packed structures and the linear arrangement has lower coordination than the island geometries. If one views the interaction between the adatoms in a pair interaction lattice gas model, the observation that linear chains can be more stable implies that there must be an attractive interaction between adatoms on nearest neighbor sites separated by $(a/2)[110]$ in order to stabilize any type of cluster. However, the interactions between atoms located on second neighbor sites separated by $a[100]$ must be repulsive in order to make the close-packed clusters, which contain these interactions, higher in energy than the linear clusters.

The nature and origin of the effective interactions between the adatoms has been examined by Wright, *et al.*[63]. In order to separate the different contributions to the interactions, the effective pair and trio interactions between the adatoms are calculated on a frozen substrate. The results are shown in Figure 4. Note that the first neighbor interactions are attractive and the second neighbor interactions are repulsive. This change in sign of the adatom interactions between first and second neighbors can be understood qualitatively as follows. There are two contributions to the adatom interactions. The first is the direct attraction between the adatoms. This is expected to be attractive at both first and second neighbor distances but substantially smaller at second neighbors (see Figure 2). The second contribution is a substrate mediated interaction. Formally, for each substrate atom that is shared by two adatoms, a three-body interaction, ζ, is added (see equation (24)). Since the three-body interaction is repulsive, it leads to an effective repulsion between the adatoms whose strength is proportional to the number of substrate atoms that the adatoms share. Note that first neighbor adatoms share two substrate atoms and the second neighbor adatoms share one substrate atom. The alternation of the net interaction of the adatoms can now be understood. At nearest neighbors the direct attraction of the adatoms dominates and the adatom interaction is repulsive. At second neighbors, the direct interaction has decreased substantially but the substrate mediated repulsion has only dropped by a factor of 2. Thus the substrate mediated repulsion can dominate at second neighbors.

Another way of viewing the substrate mediated repulsion is through the idea of bond strength saturation. As the coordination of an atom increases, the strength of subsequent bonds that it forms decreases. When two adatoms share a substrate atom, the bonds between the adatom and the substrate atom are weaker than if the adatoms were separated. Thus the adatom-substrate bonding is weaker when adatoms share substrate atoms and this leads to an effective repulsive contribution to the interaction of the adatoms.

Another important contribution to the energetics of the adatom clusters is the relaxation of the substrate atoms. Schwoebel, et al.[62] found that the calculated cluster geometries only agreed with experiment if substrate relaxations are included in the calculations. The difference in energy between the adatoms on a frozen substrate and on a fully relaxed substrate has been calculated by Wright, et al.[63] and the results for Pt on Pt(001) are shown in Figure 5 The relaxation energy associated with the linear chains is greater than that for the close packed islands. This can be understood qualitatively as follows. The relaxations are significant near the edge of the clusters. Thus one expects the relaxation energy to scale roughly with the length of the edge of the cluster. For a linear chain, the length of the edge and so the relaxation energy goes as the number of atoms in the cluster. For close-packed structures the length of the edge and so the energy to go as the square root of the number of adatoms. Thus the relaxation energy should be larger for the linear chains as calculated.

The above discussion shows that the energetics of adatom clusters is controlled by a variety of factors. The direct interaction of the adatoms leads to attraction at short distances. In addition, there is a substrate mediated repulsion between the atoms due to bond saturation effects. Finally, the energy associated with the relaxation of the substrate plays a major role in stabilizing linear clusters relative to close-packed one.

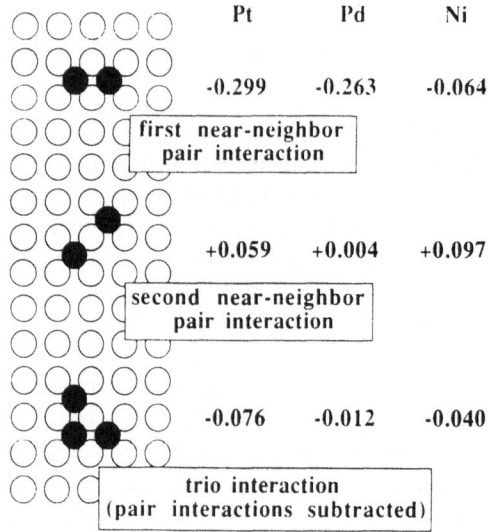

Figure 4. The important adatom interactions, in eV, on a frozen substrate. The pair interactions are simply the pair binding energies. The triplet interactions is computed by subtracting the relevant pair interactions from the triplet binding energy (from Wright et al.[63]).

The diffusion of adatoms on fcc(001) surfaces has recently been shown to have some interesting behavior. FIM experiments on the diffusion of Pt on Pt(001)[65,68] and Ir on Ir(001)[66] show that the diffusing adatom only visits half of the surface sites., i.e. those four-fold surface sites that are separated by a<100> displacements. This implies that the diffusion hop must be between second neighbor sites separated by a<100>, since if the adatoms diffused to nearest neighbor surface sites all of the four-fold surface sites would be occupied. This result is surprising if you view the diffusion of the adatom as a single atom moving on a corrugated potential surface provided by the substrate. In this view, diffusion to the nearest neighbor sites is over a two atom bridge site which should be lower in energy than diffusion over a top site as is required to diffuse directly to the second neighbor site. The resolution of this was proposed by Feibelman[65,67] based on calculations for Al using the local density functional theory based Green's function scattering method. These calculations show that the diffusion mechanism is a concerted exchange mechanism where the adatom moves into an adjacent substrate site and the substrate atom moves to the four-fold site. This path is competitive because the Al atoms are able to form strong short covalent bonds during the exchange process. Liu, et al.[69] have computed the energetics of the exchange mechanism for Pt using the EAM and found that the exchange mechanism is lower in energy than diffusion over the two-fold bridge site. This is consistent with the experiments. If they use a Lennard-Jones pair potential, they find that the exchange mechanism is much higher in energy than diffusion over the bridge site. This shows that the inclusion of the coordination dependence of the bond strengths and bond lengths, as is done in the pair functional methods, is crucial to correctly understand the diffusion process.

Surface Phonons

The pair functional methods approximately incorporate the change in the strength of the interactions as the environment is changed. The surface phonon dispersion relations are a manifestation of the surface interactions. Thus, the ability of the pair functional methods to describe surface phonon modes is a critical test of the ability of these methods to describe the change in the interactions at a surface. The force constants that result from pair functional methods differ from simple pair interaction models in two ways. First, the effective pair interactions are different at the surface than in the bulk, and second, there are explicit many-body

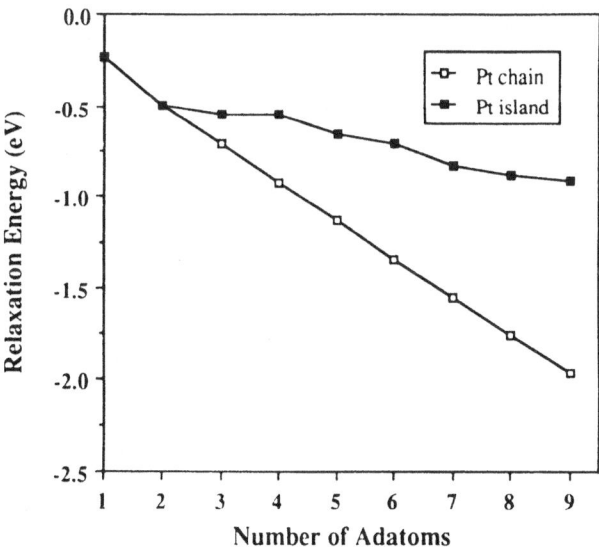

Figure 5. Relaxation energy associated with unfreezing the substrate for linear chains and close-packed island clusters of Pt adatoms on Pt(001) as a function of cluster size. The energies are in eV (from Wright, et al.[63]).

contributions to the force constants. This can be seen explicitly by examining the force constant tensor[70] between two atoms,

$$K_{ij} = U''_{ij} \hat{R}_{ij} \hat{R}_{ij} + U'_{ij} (I_{ij} - \hat{R}_{ij} \hat{R}_{ij}) + B_{ij}, \tag{27}$$

where R_{ij} is the vector from atom i to atom j, $\hat{R}_{ij}$ is the corresponding unit vector, and U(R) is the effective pair potential one deduces at first order, i.e.

$$U_{ij}(R) = F'(\sum_{j \neq i} f(R_{ij})) f(R_{ij}) + F'(\sum_{i \neq j} f(R_{ij})) f(R_{ij}) + v(R_{ij}). \tag{28}$$

The first two terms of equation (27) are formally equivalent to what one obtains for the force constant between two atoms interacting with the pair potential U(R). Note that here, though, the U(R) is environment dependent. The last term of equation (27) is the explicit many-body term which is given by

$$B_{ij} = F''_i f'(R_{ij}) g_i \hat{R}_{ij} + F''_j f'(R_{ij}) g_j \hat{R}_{ij} + \sum_{k \neq i,j} F''_k f'(R_{ik}) f'(R_{jk}) \hat{R}_{ik} \hat{R}_{jk} \tag{29}$$

with the definitions $F''_i = F''(\sum_{l \neq i} f(R_{il}))$, and $g_i = \sum_{j \neq i} f'(R_{ij}) \hat{R}_{ij}$.

The consequences of these expressions for the force constant can be appreciated if one compares the predictions of a simple central force model with the pair functional models for the specific case of two nearest neighbor atoms separated by (a/2)[011] in an unrelaxed (100) surface. Taking the surface normal to be the x direction, one obtains the following force constant matrix for the case of central force pair interactions

$$K = \begin{bmatrix} 0 & 0 & 0 \\ 0 & -a & -a \\ 0 & -a & -a \end{bmatrix}. \tag{30}$$

Here the value of the parameter a depends on the pair interaction. This form is in fact the same as that obtained in the bulk. The expression that is obtained for the pair functional methods is

$$K = \begin{bmatrix} c' & d & d \\ -d & -a' & -b' \\ -d & -b' & -a' \end{bmatrix}. \tag{31}$$

The primes denote terms that are different at the surface than in the bulk and for the bulk d=0. Note that the pair functional form yields a much richer description of the force constants at a surface and that the change in the force constants occur naturally in the theory.

The surface phonon dispersion of Cu(100) surfaces has been computed using the EAM by Nelson, et al.[70] and the results are compared to experimental dispersions obtained by Wuttig et al.[71]. The comparison of the modes obtained in a calculation of a 28 layer slab with the experimental data is shown in Figure 6. Note that the calculations correctly predict that the S_1 mode is softened more relative to the bulk edge than is the S_4 mode. The quantitative agreement is quite good. The largest disagreement between the calculations and experiment is for the S_4 mode. This is largely due to errors in the determination of the bulk phonon frequencies. This can be seen in Table 1 which presents the differences between the surface mode frequencies and the edge of the bulk modes at the high symmetry points. This shows that the shift in the surface modes from the bulk modes is calculated to within 0.05 THz. This excellent agreement provides convincing evidence that the pair functional models realistically describe the change in the force constants that occur at surfaces. The change in the force constants deduced from the EAM calculations are a 15% softening of the intralayer force constants and a 15% stiffening of the interlayer force constants.

Calculations have also been performed for the (111) surfaces of Cu and Ag[72]. Over the past several years there has been considerable interest in the force-constant changes on (111) noble metal surfaces. Measurements of the surface phonon dispersions have been performed with He scattering and EELS experiments. The main point of controversy is the amount of softening of the intralayer force constants necessary to account for the observed position of the longitudinal resonance mode. Values of the softening which range from 15% to 70% have been proposed. In their study of the Ag(111) surface, Bortolani, $et\ al.$[74] used a force-constant model with central and angular interactions including up to second neighbors. The surface force constants were determined by fitting the measured inelastic He-scattering cross sections. The procedure produced a 48% softening of the surface nearest neighbor radial intralayer force constants for Ag and Cu. On the other hand, using a much simpler nearest-neighbor central potential model, Hall $et\ al.$[75], reproduced the observed Cu(111) surface phonon spectrum

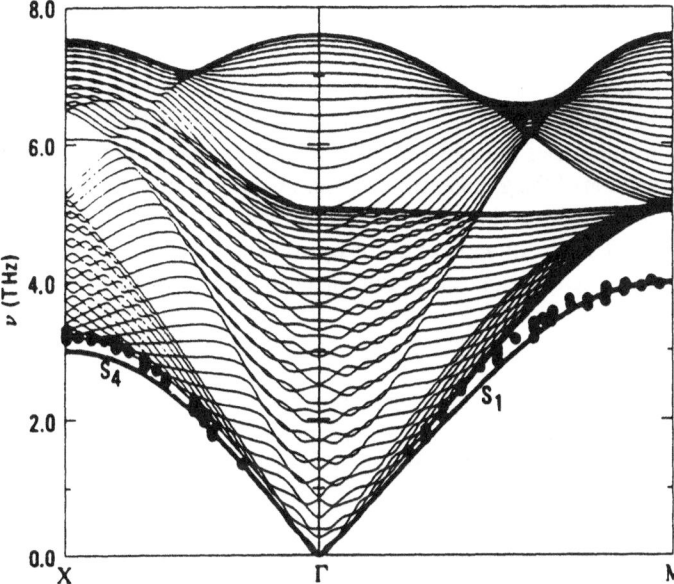

Figure 6. Comparison of the experimental[71] and theoretical[70] phonon frequencies for a Cu(100) surface. Only modes of even symmetry are shown and the experimental data is plotted as solid circles.

Table 1. Difference between the surface mode frequencies and bulk edge for Cu(100) as calculated[70] using the EAM and from experiment[71].

	calculated	experiment
S_1 mode at M		
ν(THz)	3.97	4.05
$\Delta\nu$(THz)	1.03	1.08
S_4 mode at X		
ν(THz)	2.99	3.24
$\Delta\nu$(THz)	0.18	0.13

and the inelastic-electron-scattering intensities with only a 15% reduction. (This value is similar to that found using the EAM for Cu(100).)

Detailed analysis of the EAM results for the (111) surfaces of Cu and Ag have been performed by Nelson, et al.[72]. The results are in good agreement with experiment and show that the force constants are modified at the surface by 10-15%. Also, it was shown that many features of the dispersion curves are due to avoided crossings of surface modes. The existence of these avoided crossings may have contributed to the confusion in the analysis of the experimental data. The surface phonons on the Cu(111) surface have also been investigated using effective medium theory with similar conclusions regarding the change in the force constants[73].

Similar work on surface phonons using the EAM have been carried out by Ningsheng, et al.[76]. Also, surface phonon lifetimes have been studied using molecular dynamics and the EAM[77]. The "glue model", which is another model of the pair functional type, is shown to predict anomalously high-frequency modes of the missing row reconstruction of the Au(110) surface[78].

APPLICATIONS TO GRAIN BOUNDARIES

The pair functional methods, and in particular the EAM, have been applied to the calculation of properties of internal interfaces. A few of the applications to the properties of grain boundaries will be described below. These include some detailed comparisons of the atomic structure of grain boundaries to experimental observations, calculations of the trends of the energetics as a function of the grain misorientations, the possibility of roughening transition in high angle grain boundaries and finally the segregation of the constituents of a binary alloy to the grain boundary. This list of applications is by no means complete, but is intended to give the reader a flavor of the types of calculations that can be performed.

Grain Boundary Structure

One of the most fundamental properties of a grain boundary is its structure on an atomic level. This is an interesting problem in its own right, and it provides insight into other properties such as diffusion at grain boundaries or segregation of impurities to grain boundaries. Comparison of computed grain boundary structures with experimental results is useful both because it can help in the analysis of the experimental observation and because it provides a crucial test of the model for the interatomic interactions. (If the model predicts a qualitatively incorrect structure, it is likely that results for other properties, such as mechanical strength or diffusion, will not be reliable.) Comparisons of calculations with high resolution microscopy studies have been performed for Au[79] and Al[81,82] and comparisons between calculations and x-ray diffraction have been performed for Au[80]. In general, the qualitative structures predicted in the calculations agree with the experimental observations.

Here, the work by Mills, et al.[82,83] on the structure of the Σ9 [110] symmetric tilt boundary in Al will be summarized. This example was chosen because it demonstrates some of the variety of phenomena which can occur. The experimental observations were performed using a high resolution electron microscope on a grain boundary in the symmetric Σ9 orientation. In this boundary, the boundary planes correspond to (221) lattice planes. The ideal Σ9 boundary has a 38.9^{o} misorientation between the two crystals. The measured misorientation for the experimental crystal is 39.5^{o} with an additional twist of about 0.3^{o} about the [221] direction. The observation of a tilt boundary is made to take advantage of the fact that high resolution microscopy can image the individual columns of atoms in the <110> viewing direction.

Low magnification observations of the boundary reveal a wavy boundary structure which appears to lack periodicity. Closer examination, however, shows that the boundary is actually composed of short atomically flat segments or microfacets. The microfacets are seen to have two distinct structures, which will be labeled A and B. The microfacet A has a glide-plane symmetry along the (221) plane with a relative translation of about 0.5 A in the <114> direction. By comparison, microfacet B has an apparent mirror plane symmetry across the boundary. The two microfacets are found to alternate along the boundary. The appearance of both structures supports the concept of the multiplicity of structures that has been proposed on theoretical grounds by Vitek[84].

Calculations of the structure of the boundary using the EAM have been performed using potentials for Al developed by Voter and Chen[85]. These calculations revealed four different metastable structures. These structures were obtained by considering different relative translations of the two crystals both along the tilt axis and normal to the tilt axis and by considering different terminations of the two crystals. The structures were then allowed to relax to a local minimum. In this relaxation all the atoms are free to move so relative motion of the two crystals in addition to local adjustments at the boundary are allowed. Of these four structures, two were found to have glide-plane symmetry and two have mirror plane symmetry.

The two structures with glide-plane symmetry have similar energies, 280 and 300 ergs/cm^2. They differ by a relative shift of the two crystals along the tilt axis by (a/4)[110]. The two structures appear the same when projected along the [110] tilt axis. Image simulations indicate that the two structures would also have essentially the same image. The experimental image is in good agreement with the image from either structure. Thus there is good correspondence between the calculations and the boundary images, but there still remains an uncertainty as the correct boundary structure due to the presence of the two structures with similar projected images. The energy differences between the two structures are too small for the correct structure to be conclusively determined from the calculations.

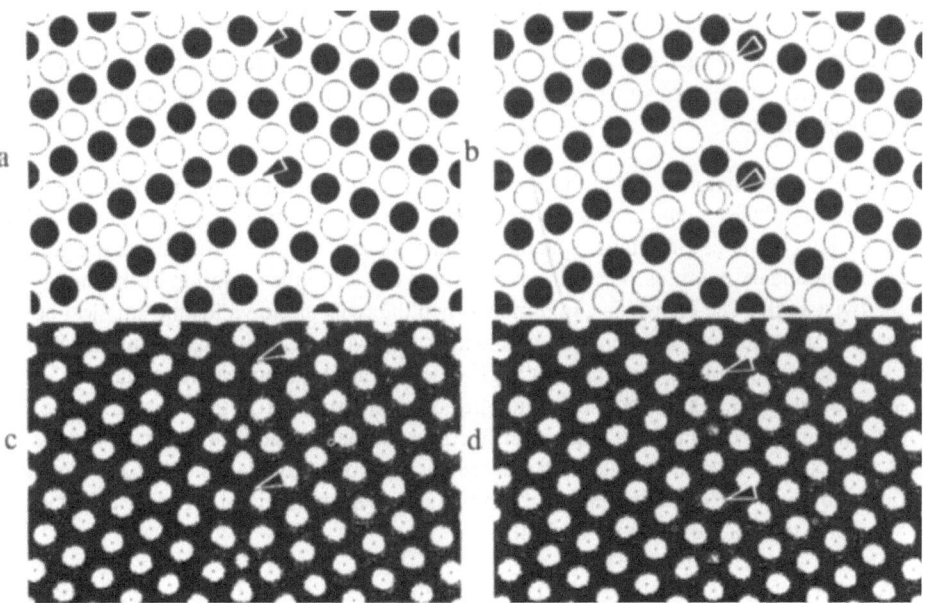

Figure 7. Atom positions of the two calculated structures with mirror plane symmetry (a and b) and the corresponding image simulations (c and d). The atom positions are superimposed as black symbols on the image simulations (from Mills, et al.[82])

The two calculated structures that contain a mirror plane are shown in Figure 7 along with the simulated images of these two structures. The energy of the structure shown in a is 435 ergs/cm^2 compared to the energy of the structure in b which is 307 ergs/cm^2. The filled and open circles indicate atoms which are in different atomic planes normal to the tilt axis. Notice that the simulated image produced by the two structures are qualitatively different. This allows the correct structure to be determined by comparison with the experimental image. This comparison shows that the simulated image in d corresponds very well with the experimental image. This shows both that the structure shown in b is the correct one and also that the positions determined in the calculations are quantitatively accurate. The fact that structure b corresponds to the observed one is in agreement with the substantially lower energy computed for this structure in the calculations. There is an interesting feature of the structure in b that should be noted. The periodicity of the boundary along the tilt axis has doubled from (a/2)[110] to a[110]. The position of the atoms in one column "zig-zag" as one moves down the tilt axis. This "zig-zag" destroys the true mirror plane symmetry across the boundary but yields a structure which appears the have a mirror plane in the image simulations.

This example has demonstrated several phenomena that can occur at grain boundaries. The existence of two different microfacets demonstrates the possible multiplicity of structures. The variety of structures found in the calculations demonstrate the necessity of exploring a wide range of relative displacements when determining grain boundary structures. Finally, the observed structure for microfacet B demonstrates the fact that the periodicity of the grain boundary can be larger than that of the corresponding coincident site lattice which gives the smallest periodicity allowed by symmetry.

Grain Boundary Energies

The systematics of the grain boundary energy have been studied extensively in calculations by Wolf[86,87] using both EAM potentials and Lennard-Jones potentials. (The qualitative trends obtained are independent of the potential model used.) In discussing the energetics, Wolf uses the following description of the five degrees of freedom that determine the macroscopic grain boundary orientation. Four of the degrees of freedom are specified by the crystallographic orientation in each crystal of the normal to the boundary plane. The final degree of freedom is a twist angle corresponding to a relative rotation of the two crystals around the boundary plane normal.

There are two major trends in the energetics that Wolf has found. First, boundaries whose boundary planes are parallel to dense packed planes have lower energies. This is equivalent to stating that large interplanar spacing for planes parallel to the boundary plane give lower grain boundary energies. The other major trend relates to the variation of the energy with the twist angle. The energy is a minimum when the twist angle yields the smallest possible unit cells. For the case of symmetric boundaries, this would correspond to pure tilt boundaries. The energy rises rapidly as the twist angle is varied until it then reaches a plateau. These trends can be seen in the grain boundary energies of symmetric <110> tilt boundaries shown in Figure 8. Note the deep trough in the energy corresponding to the (111) boundary plane. This is an example of the lower energy for dense packed boundary planes. Also, the overall variation with twist angle can be seen. Similar plots have been obtained for other orientations[86] and the qualitative behavior is similar.

Another observation made by Wolf[88] is that the dislocation model of the energetics of grain boundaries proposed by Read and Schockley[89] can be applied to high as well as low angle boundaries. For low angle boundaries, the boundary can be viewed as an array of dislocations. The energy of the boundary should then be the sum of the core energies of the dislocations plus the strain energy associated with these dislocations. This picture has been assumed to be valid only in the limit where the dislocations are well separated, i.e. for misorientations of less than around 10⁰. In the calculations by Wolf, it was shown that the calculated energy of grain boundaries fits the form suggested by the dislocation model even for high angle boundaries. This results indicates that the dislocation picture may retain validity even in the case were the dislocation cores are not well separated.

<u>Grain Boundary Roughening</u>

There is a great deal of interest in phase transitions that may occur at internal interfaces. In this section, the possibility of a roughening transition, analogous to those seen on solid surfaces, is considered for a high angle twist boundary. The possibility of roughening transitions at low angle boundaries has been discussed previously by Rottman[90] as well as experimental evidence of phase transitions at internal interfaces[91]. The present work was prompted by the x-ray studies of the structure of the Σ5 (001) twist boundary in Au[80,92]. The results here suggest that further studies of the temperature dependence of the x-ray scattering may be interesting.

In this work[39], a mechanism for interface roughening is demonstrated for the Σ5 twist boundary. The energetics of various structural defects at the boundary are computed. It is found that certain local defects, which correspond to local motion of the boundary normal to the boundary plane, have low energy. The energetics and interactions of the defects are mapped onto a lattice model equivalent to the solid-on-solid (SOS) model[93] used to describe surface roughening. Based on this model, the grain boundary should undergo a roughening transition at or near experimentally accessible temperatures. The possibility of interfacial disorder due to the presence of different structural units at the boundary has been suggested previously by Oh and Vitek[94] and by Majid and Bristowe[95]. Theoretical work by Najafabadi, et al.[96] has suggested the possibility of a different structural transition in this boundary as a function of temperature. They consider the vibrational entropy associated with different possible structures and predict a first order transition from the CSL structure to one related by a rigid shift of the two crystals.

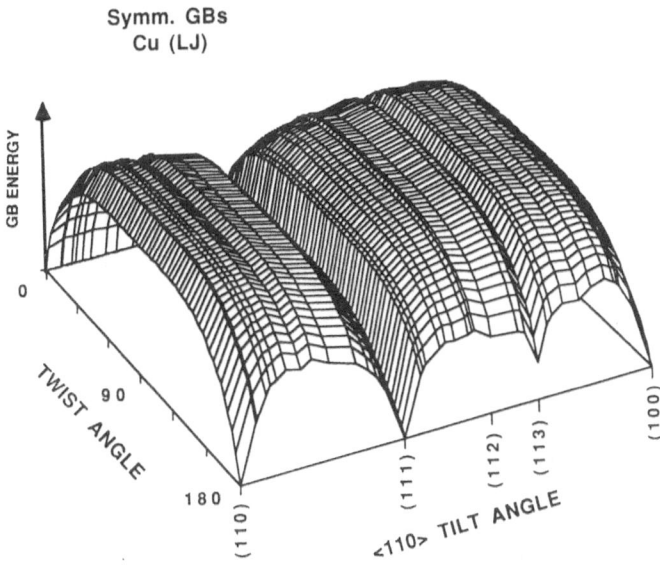

Figure 8. The energy of symmetric <110> tilt boundaries computed using the Lennard-Jones pair potentials. The energetic trends are discussed in the text. (from Wolf[86])

The lowest energy structure found for the Σ5 twist boundary using the EAM is the coincident site lattice (CSL) structure. This structure is shown in Figure 9. A metastable defect exists in this CSL structure. Consider the nearest neighbor square of atoms denoted by plusses in the center of Figure 9 which are in the plane adjacent to the boundary. These atoms are rotated counterclockwise by 36.8^0 to positions close to the ideal lattice sites in the other crystal. This has the effect of moving the position of the interface by one atomic layer in the CSL cell. The energy required to produce this defect is computed to be 0.035 to 0.114 eV depending on the exact EAM parameters used[39]. These energies are such that these defects can be produced in thermal equilibrium.

To develop a statistical model of these defects, it is useful to have a more systematic picture of the defects. Consider the coincident site lattice which is the lattice of all points common to both crystals if they are shifted such that the two crystals have a point in common. The coincident site lattice in this case is a body centered tetragonal (BCT) lattice. The fcc lattice can be constructed from the BCT lattice plus one of two five atom bases. The two possible bases, called A and B are related by a 36.8^0 rotation and yield the two crystals on either side of the boundary. The grain boundary can now be considered as the BCT lattice with A or B basis units at each site such that far from the boundary the lattice is all A units or B units. The problem is now reduced to a lattice problem of how to place A and B basis units on the lattice in the vicinity of the grain boundary.

This problem can be simplified further to a two dimensional problem by the following observation. Consider a given column of the BCT lattice along the direction normal to the boundary. For each time the type of basis unit changes, the energy required will be approximately the grain boundary energy times the area of the CSL cell. This energy is much greater than thermal energies. Thus in thermal equilibrium, the type of unit on the sites in a given column should change exactly once in that column. The interface structure can then be described by a two-dimensional array of heights denoting the position in each column where the basis unit changes. This picture is analogous to that used in modeling the surface roughening transition. There the height denotes the last occupied lattice site in each column and the requirement that there is only one point in each column where the occupation changes is called the "solid-on-solid" condition.

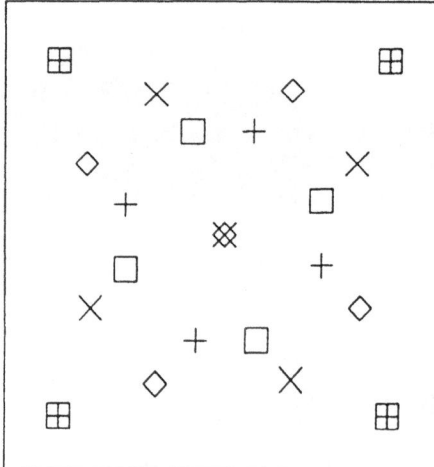

Figure 9. Relaxed position of the Σ5 (001) twist boundary. The two atomic planes on either side of the interface are shown and the symbols represent the planes in the order square, diamond, plus and x. (from Foiles[39]).

In order to study the statistical mechanics of the interface, a model of the energetics in terms of the height differences between the columns will be defined. It is convenient to define the height of each column in units of the interplanar spacings. In these units, the height of a given column will always be odd or even, Note that since each column ends on a lattice site of the BCT lattice, the height of adjacent columns must differ by one. EAM calculations indicate that structures where adjacent columns have a height difference greater than one are unstable. Thus the height difference of adjacent columns in this model will be constrained to be +/- 1. With this constraint, the height of second and third neighbor columns can differ only by 0 or +/- 2. An energy will be assigned for each height difference between second and third neighbor columns. (First neighbor columns do not have to be considered explicitly since the height difference is always 1.) This leads to the model hamiltonian

$$ H = \frac{1}{2} \sum_{i, d_2} J_2 |h(i) - h(i + d_2)| + \frac{1}{2} \sum_{i, d_3} J_3 |h(i) - h(i + d_3)| \tag{32} $$

where $h(i)$ is the height of column i, d_2 and d_3 are the displacements to the second and third neighbor columns and J_2 and J_3 model the interaction between columns of unequal height.

The parameters J_2 and J_3 have been determined by comparison of EAM results for the energetics of various different height distributions. The transition temperature was then determined by Monte Carlo simulations of equation (32). The results depended strongly on the EAM functions used with the transition temperature predicted to be either 100 K or 500 K.[39] The important point in these calculations, though, is not the detailed prediction of the critical temperature. The significant result is that an explicit atomistic mechanism has been presented which could lead to a grain boundary roughening transition and that the transition temperature may be in a region where it can be observed experimentally.

Segregation of Binary Alloys at Grain Boundaries

The local equilibrium composition of an alloy is different in the vicinity of a defect, such as a surface or a grain boundary, than in the bulk material. This segregation can have a significant affect on the properties of the defect. Foiles has applied the EAM in conjunction with Monte Carlo simulations to calculation of these effects at surfaces[3] and at grain boundaries[42]. Here, the application of this technique to the segregation of Ni-Cu alloys to (100) twist boundaries will be described. This approach has also been used by Seki, et al.[97] to study twist boundaries in Pt-Au alloys, and by Bacher, et al.[98] to study the segregation of Au to the interphase boundary between the Cu-rich and Ag-rich phases of a Cu-Ag alloy.

The thermal equilibrium of the grain boundary is computed using Monte Carlo simulations techniques applied to a slab of atoms containing a grain boundary. These simulations allow for both changes in the positions of the atoms as well as for changes in the local composition. In particular, the simulations incorporate two types of steps. First, the atoms are allowed to make displacements from their current positions. This incorporates both lattice relaxations effects (strain energy) as well as vibrational contributions to the thermodynamics. In addition, since all atoms are free to move the two crystals can displace relative to each other. The second type of step is to transmute the chemical identity of the atoms. (The simulations are performed in an ensemble where the total number of atoms and the chemical potential difference between the two elements is held fixed.) This procedure for allowing compositional rearrangement leads to rapid convergence of the simulations but means that the simulations provide no information about the kinetics of the segregation process. This simulation method has been described in more detail elsewhere[3,41].

The equilibrium segregation has been computed for the Ni-Cu alloy system at a temperature of 800 K. The relative chemical potentials of the two elements were chosen to correspond to bulk alloy compositions of 10%, 50% and 90% Cu. The boundaries studied are the $\Sigma 5$, $\Sigma 13$ and $\Sigma 61$ twist boundaries. These boundaries are created by rotating the two crystals around a [100] axis by 36.9^0, 22.6^0, and 10.4^0, respectively. The area of the grain boundary is held fixed during the simulations at a value consistent with the lattice constant determined from zero pressure simulations of the bulk at the same temperature and chemical potential difference.

The results for the overall composition of each of the three planes adjacent to the boundary are presented in Table 2. In all cases the grain boundary region is enriched in Cu and the change in concentration is confined to the region within 3 to 4 atomic planes of the boundary. In addition, Table 2 lists the net expansion normal to the boundary. This is defined here as the difference in the distance between two planes on opposite sides of the boundary for the system with the grain boundary and the distance for the same number of interlayer spacings at the bulk lattice constant. In all cases, there is an expansion of the boundary region and the amount of expansion is greater than can be accounted for by the increased concentration of Cu at the boundary. (Cu has a somewhat larger (3%) lattice constant than Ni.) This expansion at the boundary is a general feature of grain boundaries. The systematics of the expansion at grain boundaries as a function of the geometrical parameters of the boundary has been discussed by Wolf[88].

There are two trends which are apparent from the results in Table 2. First, the segregation is strongest for the higher angle boundaries. This is consistent with the results for Pt-Au boundaries[97] where this variation was studied in some detail. Second, the segregation is strongest for the Ni-rich alloy and weakest for the Cu-rich alloy. This latter observation is consistent with calculations of the dilute segregation energies performed for the $\Sigma 5$ boundary. The dilute segregation energy is computed by comparing the energy of the impurity at a position near the grain boundary versus its energy in the bulk. For the case of a Cu impurity in Ni, the Cu is bound by 0.22 eV to the coincident sites of the boundary which comprise 1/5 of the boundary sites and is bound by 0.13 eV to the 4 equivalent non-coincident sites. For the case of a Ni impurity in Cu, the Ni is repelled from the boundary plane by 0.07 eV for both types of sites. These energies indicate that the segregation will be strongest for dilute Cu in Ni as observed in the full simulations.

The segregation at the boundary in the concentrated alloys, though, cannot be determined simply by these dilute segregation energies. If one ignores interactions between the sites at the grain boundaries, the above energies predict an average concentration of 48% Cu

Table 2. Composition of the first three planes adjacent to the grain boundary computed by Monte Carlo simulations using the EAM. The net expansion of the grain boundary normal to the interface is also listed. The statistical uncertainty in the compositions is ±1%. (from Foiles[41])

	First plane	Second plane	Third plane	Net expansion
Ni-Cu (10 at. %) $\Sigma 5$	74 at. % Cu	22 at. % Cu	11 at. % Cu	0.60 Å
Ni-Cu (10 at. %) $\Sigma 13$	62 at. % Cu	24 at. % Cu	11 at. % Cu	0.35 Å
Ni-Cu (10 at. %) $\Sigma 61$	41 at. % Cu	27 at. % Cu	15 at. % Cu	0.44 Å
Ni-Cu (50 at. %) $\Sigma 5$	78 at. % Cu	48 at. % Cu	42 at. % Cu	0.42 Å
Ni-Cu (50 at. %) $\Sigma 13$	74 at. % Cu	52 at. % Cu	44 at. % Cu	0.29 Å
Ni-Cu (50 at. %) $\Sigma 61$	66 at. % Cu	57 at. % Cu	51 at. % Cu	0.30 Å
Ni-Cu (90 at. %) $\Sigma 5$	95 at. % Cu	90 at. % Cu	90 at. % Cu	0.32 Å
Ni-Cu (90 at. %) $\Sigma 13$	95 at. % Cu	90 at. % Cu	89 at. % Cu	0.27 Å
Ni-Cu (90 at. %) $\Sigma 61$	92 at. % Cu	91 at. % Cu	90 at. % Cu	0.20 Å

in the boundary plane for the case of the $\Sigma 5$ NiCu(10%) boundary using the segregation expression derived for this case by McLean[99]. This is substantially smaller than the value of 74% obtained in the simulations. The sense of this difference is consistent with the fact that the Ni-Cu alloy system is a clustering alloy. Thus the enhancement of the Cu concentration due to the presence of the boundary is complemented by the tendency of the Cu atoms to cluster.

In addition to the average composition of each plane, the simulations can determine the composition variations with each plane. In particular, it is interesting to see how the variations in the composition correlate with the positions of the dislocation cores that comprise the boundary. This has been examined for the $\Sigma 61$ boundary where the decomposition into an array of screw dislocations is clear. In Figure 10, the projected atomic positions of the two planes on either side of the boundary are shown for a randomly chosen configuration from the simulation. The boundary clearly breaks up into regions of good match separated by a square array of poor match. These areas of poor match are the screw dislocations. The average concentrations in the first three planes are also shown for the central unit cell. (The screw disloca-

Figure 10. a) Atomic positions projected on the boundary plane for the $\Sigma 61$ boundary in the NiCu(10%) alloy. The first two atomic layers on either side of the boundary are shown. The open circles are Ni atoms and the filled circles are Cu. The screw dislocation network is located in the regions of poor match between the two crystals. b-d) Contour plots of the Cu concentration as a function of position in the plane of the first (b), second (c) and third(d) atomic layers from the boundary. these plots correspond to the central unit cell of a. The numbers indicate the compositions at extremums in atomic percent Cu and the contour spacing is 4.4% (from Foiles[41])

tions are located along the diagonal lines that connect the midpoints of the sides of this cell.) For planes two and three planes away from the boundary, the Cu concentration is highest in the center and at the corners. These areas correspond to the regions of good match between the crystals. This at first unexpected because one expects the Cu to segregate to the defected area. It can be understood qualitatively as follows. There is a net expansion of the boundary as discussed above. However, in the regions of good match, one would expect that the bulk lattice spacing would be preferred. Thus, these regions are in effect under local tensile strain and Cu is enhanced in regions of tensile strain. (This is not primarily a size effect. The Cu is "softer" than the Ni and so the system can lowers its energy by moving the Cu to strained regions.) The composition variation in the plane adjacent to the boundary is different and does not have a simple explanation.

SUMMARY

In this article, new methods for the approximate calculation of the total energy of a large number of atoms in a metallic envoronment have been described. These methods are similar in that they are of the pair functional form or at least approximately so. The methods discussed include the Finnis-Sinclair N-body potentials, the embedded atom method and effective medium theory. The main advantage of these methods over simple pair potential models is that they incorporate the variation in the strength of interactions with changes in the bonding environment. This is equivalent to the assertion that these methods incorporate a certain class of many-body interactions. These many-body interactions do not include explicit angular dependence, though. Thus these methods should not be applicable to systems such as semiconductors and central transition metals, where angular interactions are expected to be important. It is also important to note that the motivations behind the various methods suggest that while they will describe gross coordination dependence, they do not include effects due to details of the band structure or to the formation of covalent bonds.

Various applications of these methods to the properties of surfaces and interfaces have been presented. These applications do not by any means represent a comprehensive survey of the applications of these methods. These applications demonstrated the manifestations of the coordination dependence of the interactions on various surface properties including reconstruction, adatoms and surface phonons. In addition, some applications of these methods to the properties of grain boundaries are presented which give a flavor of the types of calculations that can be performed. These applications also demonstrate the utility of these simplified techniques. These calculations frequently either involved large unit cells which can not be handled by *ab initio* techniques. In addition, some of the examples show that the explicit inclusion of the atomic scale relaxation and motion of all of the atoms is important to an accurate description of many of these phenomena.

While a great deal of progress has been made in recent years with the advent of the various pair functional methods described here, there is a need for further advances in the area of model interatomic potentials. The current methods are reliable for a very limited range of metals, specifically simple metals and filled d band transition metals. There is a need for potentials with similar reliability for other materials of scientific and technological importance. These include semiconductors, central transition metals, and oxides as well as potentials for systems which combine these materials classes to study such properties as metal-ceramic interfaces. These problems are the subject of current research by several research groups.

ACKNOWLEDGEMENTS

This work was supported by the U. S. Department of Energy, Office of Basic Energy Sciences, Division of Materials Sciences.

REFERENCES

1   A. E. Carlsson, in *Solid State Physics*, edited by H. Ehrenreich, H. Seitz and D. Turnbull, vol. 43, (Academic Press, New York, 1990), p. 1.

2   T. J. Raeker and A. E. DePristo, International Reviews in Physical Chemistry 10, 1 (1991).

3   S. M. Foiles, in *Surface Segregation and Related Phenomena*, edited by P. A. Dowben and A. Miller, (CRC Press, Boca Raton, 1990).

4   J. K. Nørskov, Reports on Progress in Physics 53, 1253 (1990).

5   M. S. Daw, S. M. Foilesand M. I. Baskes, in preparation.

6   R. A. Johnson, J. Phys. F3, 295 (1973).

7   W. A. Harrison, *Pseudopotentials in the Theory of Metals*, (Benjamin, New York, 1966).

8   J. A. Moriarty, Phys. Rev. B38, 3199 (1988); J. A. Moriarty, Phys. Rev. B42, 1609 (1990).

9   R. A. Johnson, Phys. Rev. B6, 2094 (1972).

10   M. I. Baskes and C. F. Melius, Phys. Rev. B20, 3197 (1979).

11   M. W. Finnis and J. E. Sinclair, Philos. Mag. A50, 45 (1984).

12   M. S. Daw and M. I. Baskes, Phys. Rev. Lett. 50, 1285 (1983).

13   M. S. Daw and M. I. Baskes, Phys. Rev. B29, 6443 (1984).

14   K. W. Jacobsen, J. K. Nørskov and M. J. Puska, Phys. Rev. B35, 7423 (1987).

15   R. Rebonato and J. Q. Broughton, Philos. Mag. Lett. 55, 225 (1987).

16   C. C. Matthai and D. J. Bacon, Philos. Mag. A52, 1 (1985); J. M. Harder and D. J. Bacon, Philos. Mag. A54, 651 (1987); R. Rebonato, D. O. Welch, R. D. Hatcher and J. C. Billelo, Philos. Mag. A55, 655 (1987).

17   M. Marchese, G. Jacucci and C. P. Flynn, Philos. Mag. Lett. 57, 25 (1988).

18   V. Rosato, M. Guillope and B. Legrand, Phil. Mag. A59, 321 (1989).

19   M. J. Stott and E. Zaremba, Solid State Comm. 32, 1297 (1979); Phys. Rev. B22, 1564 (1980); Can. J. Phys. 60, 1145 (1982).

20   D. S. Larsen and J. K. Nørskov, J. Phys. F 9, 1975 (1979).

21   J. K. Nørskov and N. D. Lang, Phys. Rev. B21, 2131 (1980).

22   J. K. Nørskov, Phys. Rev. B26, 2875 (1982).

23   M. J. Puska, R. M. Nieminen, and M. Manninen, Phys. Rev. B24, 3037 (1981).

24   P. Nordlander, S. Holloway, and J. K. Nørskov, Surf. Sci. 136, 59 (1984).

25   B. Chakraborty, S. Holloway, and J. K. Nørskov, Surf. Sci. 152/153, 660 (1985).

26   K. W. Jacobsen, J. K. Nørskov, Surf. Sci. 166, 539 (1986).

27   J. K. Nørskov, F. Besenbacher, J. Bottinger, B. B. Nielsen, A. A. Pisarev, Phys. Rev. Lett. 49, 1420.

28   P. Nordlander, J. K. Nørskov, and F. Besenbacher, J. Phys. F 16, 1161 (1986).

29   J. K. Nørskov and F. Besenbacher, J. Less Common Metals 130, 475 (1987).

30   S. M. Foiles, M. I. Baskes, and M. S. Daw, Phys. Rev. B33, 7983 (1986); Phys. Rev. B37, 10387 (1988).

31   M. S. Daw, Phys. Rev. B39, 7441 (1989).

32   P. Hohenberg and W. Kohn, Phys. Rev. 136, 864 (1964).

33   M. S. Daw and R. D. Hatcher, Solid State Comm. 56, 697 (1985).

34   J. S. Nelson, M. S. Daw and E. C. Sowa, Phys Rev B40, 1465 (1989).

35   J. S. Nelson, E. C. Sowa and M. S. Daw, Phys Rev. Lett. 61, 1977 (1988).

36   S. M. Foiles and J. B. Adams, Phys. Rev. B40, 5909 (1989).

37   S. M. Foiles, Phys. Rev. B32, 3409 (1985).

38   S. M. Foiles, Acta Metall 37, 2815 (1989); S. M. Foiles, M. I. Baskes and M. S. Daw, in *Interfacial Structure, Properties and Design*, edited by M. H. Yoo, W. A. T. Clark, and C. L. Briant (Materials Research Society, Pittsburgh, 1988).

39 S. M. Foiles, in *Atomic Scale Calculations of Structure in Materials*, edited by M. S. Daw and M. A. Schlüter, (Materials Research Society, Pittsburgh, 1990).

40 S. M. Foiles and M. S. Daw, J. Mater. Research 2, 5 (1987); S. M. Foiles, in *High-Temperature Ordered Intermetallic Alloys II*, edited by N. S. Stoloff, C. C. Koch, C. T. Liu and O. Izumi (Materials Research Society, Pittsburgh, 1987).

41 S. M. Foiles, in *Surface Segregation and Related Phenomena*, edited by P. A. Dowben and A. Miller, (CRC Press, Boca Raton, 1990) and references therein; S. M. Foiles, Phys. Rev. B32, 7685 (1985).

42 S. M. Foiles, Phys. Rev. B40, 10639 (1989); A. Seki, D. N. Seidman, Y. Oh, and S. M. Foiles, Acta Metall., in press; M. J. Mills, S. H. Goods, S. M. Foiles and J. R. Whetstone, in *High Temperature Ordered Intermetallic Alloys*, edited by L. Johnson, D. P. Pope and J. O. Stiegler, (Materials Research Society, Pittsburgh, 1991); S. M. Foiles and D. N. Seidman, in *Atomic-Level Properties of Interface Materials,* edited by D. Wolf and S. Yip, (Chapman and Hall, London, 1991).

43 J. B. Adams, S. M. Foiles and W. G. Wolfer, J. Mater. Research 4, 102 (1989).

44 M. S. Daw, M. I. Baskes, C. L. Bisson and W. G. Wolfer, in proceedings of *Modelling Environmental Effects on Crack Growth Processes,* edited by R. H. Jones and W. W. Gerberich, (The Metallurgical Society, 1985); M. S. Daw and M. I. Baskes, in Proceedings of the NATO Advanced Study Workshop on *Physics and Chemistry of Fracture*, edited by R. M. Latanision and R. H. Jones (Martinus Nijhoff, Dordrecht, 1986); M. I. Baskes, S. M. Foiles and M. S. Daw, J. de Physique C5, 483 (1988); R. G. Hoagland, M. S. Daw, S. M. Foiles and M. I. Baskes, J. Mater. Research 5, 313 (1990).

45 S. M. Foiles, Surf. Sci. 191, L779 (1987).

46 M. S. Daw and S. M. Foiles, Phys. Rev. Lett. 59, 2756 (1987).

47 S. M. Foiles, Surf. Sci. 191, 329 (1987).

48 P. Stoltze, K. W. Jacobsen and J. K. Nørskov, Phys. Rev. B36, 5035 (1987).

49 K. W. Jacobsen and J. K. Nørskov, in *The Structure of Surfaces II*, edited by J. F. van der Veen and M. A. Van Hove, (Springer-Verlag, Berlin, 1988), p. 118.

50 K. W. Jacobsen and J. K. Nørskov, Phys. Rev. Lett. 60, 2496 (1988).

51 P. D. Ditlevsen and J. K. Nørskov, Journal of Electron Spectroscopy and Related Phenomena 54, 237 (1990).

52 P. Stoltze, J. of Chemical Physics 92, 6306 (1990).

53 O. B. Christensen, K. W. Jacobsen, J. K. Nørskov and M. Manninen, Phys. Rev. Lett. 66, 2219 (1991).

54 W. Moritz and D. Wolf, Surf. Sci. 163, L655 (1985); G. L. Kellogg, Phys. Rev. Lett. 55, 2168 (1985); I. K. Robinson, Phys. Rev. Lett. 50, 1145 (1983); L. D. Marks, Phys. Rev. Lett. 51, 1000 (1983); M. Copel and T. Gustafsson, Phys. Rev. Lett. 57, 723 (1986).

55 D. Tomanek, H.-J. Brocksch and K. H. Bennemann, Surf. Sci. 138, L 129 (1983); H.-J. Brocksch and K. H. Bennemann, Surf. Sci. 161, 321 (1985); J. W. Davenport and M. Weinert, Phys. Rev. Lett. 58, 1382 (1987); K.-M. Ho and K. P. Bohnen, ???

56 M. S. Daw, Surf. Sci 166, L161 (1986).

57 J. C. Campuzano, A. M. Lahee and G. Jennings, Surf. Sci. 152/153, 68 (1985).

58 J.-K. Zuo, Y.-L. He, G. C. Wang, and T. E. Felter, J. Vac. Sci. Technol. ???.

59 L. D. Roelofs, S. M. Foiles, M. S. Daw and M. I. Baskes, Surf. Sci. 234, 63 (1990).

60 D. R. Tice and D. W. Basset, Thin Solid Films 20, S37 (1974); D. W. Basset, Thin Solid Films 48, 237 (1978); H. W. Fink and G. Ehrlich, Surf. Sci. 110, L611 (1981).

61 P. R. Schwoebel and G. L. Kellogg, Phys. Rev. Lett. 61, 578 (1988).

62 P. R. Schwoebel, S. M. Foiles, C. M. Bisson and G. L. Kellogg, Phys. Rev. B40, 10639 (1989).

63 A. F. Wright, M. S. Daw and C. Y. Fong, Phys. Rev. B42, 9409 (1990).

64 G. L Kellogg, private communication (?).

65 G. L. Kellogg and P. J. Feibelman, Phys. Rev. Lett. 64, 3143 (1990).

66 C. Chen and T. T. Tsong, Phys. Rev. Lett. 64, 3147 (1990).

67 P. J. Feibelman, Phys. Rev. Lett. 65, 729 (1990).

68 G. L. Kellogg, Surf. Sci. 246, 31 (1991).

69 C. L. Liu, J. M. Cohen, J. B. Adams and A. F. Voter, Surf. Sci. (in press).

70 J. S. Nelson, E. C. Sowa and M. S. Daw, Phys. Rev. Lett. 61, 197 (1988).

71 M. Wuttig, R. Franchy and H. Ibach, Z. Phys. B 65, 71 (1986); Solid State Comm. 57, 445 (1986).

72 J. S. Nelson, M. S. Daw and E. C. Sowa, Phys. Rev. B40, 1465 (1989).

73 P. D. Ditlevsen and J. K. Nørskov, J. of Electron Spect. and Related Phenomena 54-55, 237 (1990).

74 V. Bortolani, A. Franchini, F. Nizzoli and G. Santoro, Phys. Rev. Lett 52, 429 (1984); V. Bortolani, G. Santoro, U. Harten and J. P. Toennies, Surf. Sci. 148, 82 (1984); V. Bortolani, A. Franchini, F. Nizzoli and G. Santoro, Surf. Sci. 152, 811 (1985); G. Santoro, A. Franchini, V. Bortolani, U. Harten, J. P. Toennies and C. Woll, Surf. Sci. 183, 180 (1987).

75 B. M. Hall, D. L. Mills, M. H. Mohamed and L. L. Kesmodel, Phys. Rev. B38, 5856 (1988); M. H. Mohamed, L. L. Kesmodel, B. M. Hall and D. L. MIlls, Phys. Rev. B37, 2763 (1988).

76 L. Ningsheng, X. Wenlan and S. C. Shen, Solid State Comm. 67, 837 (1988); Phys. Stat. Sol. (b) 147, 511 (1988); Solid State Comm.69, 155 (1989).

77 A. D. Maradudin, R. F. Wallis, A. R. McGurn, M. S. Daw and A. J. C. Ladd, in *Lattice Dynamics and Semiconductor Physics*, edited by ??? (World Scientific Press, Singapore, 1990).

78 X. Q. Wang, G. L. Chiarotti, F. Ercolessi and E. Tosatti, Phys. Rev. B38, 8131 (1988).

79 K. L. Merkle, Colloque de Phys. 51 C1, 251 (1990).

80 I. Majid, P. D. Bristowe, and R. W. Balluffi, Phys. Rev. B40, 2779 (1989).

81 U. Dahmen, C. J. D. Hetherington, M. A. O'Keefe, K. H. Westmacott, M. J. Mills, M. S. Daw and V. Vitek, Philos. Mag. Lett. 62, 327 (1990).

82 M. J. Mills, G. J. Thomas, M. S. Daw and F. Cosandey, in *Atomic Scale Structure of Interfaces*, edited by R. D. Bringans, R. M. Feenstra and J. M. Gibson (Materials Research Society, Pittsburgh, 1990).

83 M. J. Mills and M. S. Daw, in *High Resolution Electron Microscopy of Defects in Materials,* edited by R. Sinclair, D. J. Smith, and U. Dahmen (Materials Research Society, Pittsburgh, 1990).

84 V. Vitek, A. P. Sutton, G. J. Wang and D. Schwartz, Scripta Metall. 17, 183 (1983).

85 A. F. Voter and S. P. Chen, Materials Research Society Proceeding 82, 175 (1987).

86 D. Wolf, J. Mater. Research 5, 1708 (1990), and references therein.

87 K. L. Merkle and D. Wolf, MRS Bulletin 14 #9, 42 (1990); D. Wolf and K. L. Merkle, in *Atomic-Level Properties of Interface Materials,* edited by D. Wolf and S. Yip, (Chapman and Hall, London, 1991).

88 D. Wolf, Scripta Metall. 23, 1713 (1989); D. Wolf, Scripta Metall. 23, 1913 (1989).

89 W. T. Read and W. Schockley, Phys. Rev. 78, 275 (1950).

90 C. Rottman, Phys. Rev. Lett. 57, 725 (1986).

91 C. Rottman, J. de Physique 49, C5 (1988).

92 M. R. Fitzsimmons and S. L. Sass, Acta Metall. 36, 3103 (1988).

93 J. D. Weeks, in *Ordering in Strongly Fluctuating Condensed Matter Systems*, edited by T. Riste (Plenum Press, New York, 1980).

94 Y. Oh and V. Vitek, Acta Metall. 34, 1941 (1986).

95 I. Majid and P. D. Bristowe, Scripta Metall. 21, 1153 (1987).

96 R. Najafabadi, D. J. Srolovitz and R. LeSar, J Mater. Research 5, 2663 (1990).

97 A. Seki, D. N. Seidman, Y. Oh, and S. M. Foiles, Acta Metall., in press.

98 P. Bacher, P. Wynblatt and S. M. Foiles, Acta Metall., in press.

99 D. McLean, *Grain Boundaries in Metals*, (Clarendon, Oxford, 1957).

ANALYTIC BOND ORDER POTENTIALS WITHIN TIGHT BINDING HÜCKEL THEORY

D. G. Pettifor and M. Aoki*

Department of Mathematics
Imperial College of Science, Technology and Medicine
London SW7 2BZ, UK

*Department of Physics
Gifu University
Gifu 501-11, Japan

1. INTRODUCTION

The need for simple, yet reliable interatomic potentials which can model the behaviour of thousands of atoms at surfaces and interfaces is widespread (see, for example, Foiles 1992). As fig. 1 illustrates schematically the accurate first-principles Local Density Functional (LDF) theory can be solved *routinely* for around fifty non-equivalent atoms (see, for example, Car 1992), whereas the less accurate, more intuitive Tight Binding (TB) approximation can handle around five-hundred non-equivalent atoms at surfaces and interfaces (see, for example, Wilson et al 1990 and references therein). The simulation of thousands of atoms requires the use of semi-empirical many-body potentials such as those developed by Daw and Baskes (1984), Finnis and Sinclair (1984), Stillinger and Weber (1985), and Tersoff (1988). However, such semi-empirical potentials have their limitations. For example, Embedded Atom or Finnis-Sinclair potentials cannot predict the c(2x2) reconstruction of the W(100) surface, the Stillinger-Weber three-body potential leaves the Si(111) surface unreconstructed, and the Tersoff bond order potential cannot model the buckling of the dimers on the Si (100) surface.

In this paper we show that it is possible to *derive* analytic expressions for the bond order potential by making a set of well-defined approximations that take us down from Local Density Functional theory to Tight Binding Hückel theory to the many-body bond order potentials as illustrated schematically by the arrows in fig. 1. Since this material has been covered in detail elsewhere (Pettifor 1990, Pettifor and Aoki 1992, Pettifor 1992), only a brief outline will be presented in this paper.

2. FROM LDF TO TB HÜCKEL

The semi-empirical Tight Binding Hückel model is the simplest   scheme

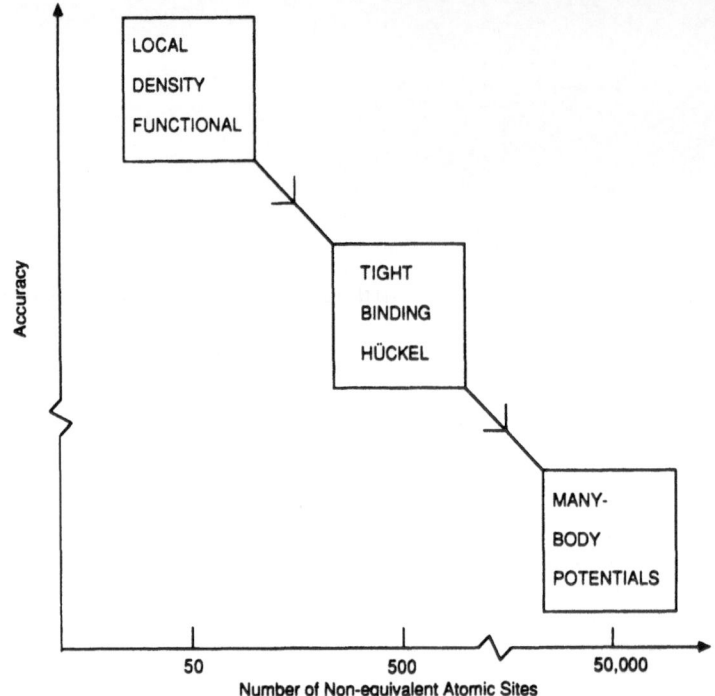

Fig. 1.   Schematic representation of the decrease in predictive accuracy
with increasing number of non-equivalent atomic sites which can be
treated computationally by different levels of approximation.   The
downward arrows linking the boxes indicate the progression from
the first principles Local Density Functional theory to Tight
Binding Hückel theory to many-body interatomic potentials through
the application of well-understood approximations.

for describing the energetics of semi-conductors and transition metals
within a quantum mechanical framework.   The total binding energy is written
in the form

$$U = U_{rep} + U_{bond} + U_{prom} \tag{1}$$

where U_{rep} is a semi-empirical pairwise repulsive contribution, namely

$$U_{rep} = \frac{1}{2} \sum_{i,j}' \phi(R_{ij}) \tag{2}$$

and U_{bond} is the covalent bond energy which results from evaluating the
local density of states $n_{i\alpha}(E)$ associated with orbital α on site i within
the two-centre, orthogonal TB approximation.   That is,

$$U_{bond} = \sum_{i\alpha} \int^{E_F} (E - E_{i\alpha}) n_{i\alpha}(E) \, dE \tag{3}$$

where $E_{i\alpha}$ is the effective atomic energy level of orbital α at site i and E_F is the Fermi energy. The third contribution in eq.(1) is the promotion energy which for the case of sp orbitals takes the form

$$U_{prom} = (E_p - E_s) \sum_i \Delta N_p^i = E_{sp} \sum_i \Delta N_p^i \qquad (4)$$

where ΔN_p^i gives the change in p occupancy on going from the free atom state to atom i in a given bonding situation. In practice, eq.(1) gives the binding energy with respect to some *reference* free atom state which usually differs from the *true* atomic ground state due to, for example, the neglect of spin-polarization or the shift in atomic energy levels arising from the renormalization of the wave functions in the bonding situation (see, for example, Sankey and Niklewski 1989).

The form of eq.(1) may be derived from first principles (see Sutton et al. 1988 and references therein). The pairwise nature of the repulsive term follows directly from the Harris-Foulkes approximation to density functional theory (Harris 1985, Foulkes and Haydock 1989), whereas the Hückel-type two-centre orthogonal form of the matrix elements may be justified *in principle* within Anderson's (1968) chemical pseudopotential theory.

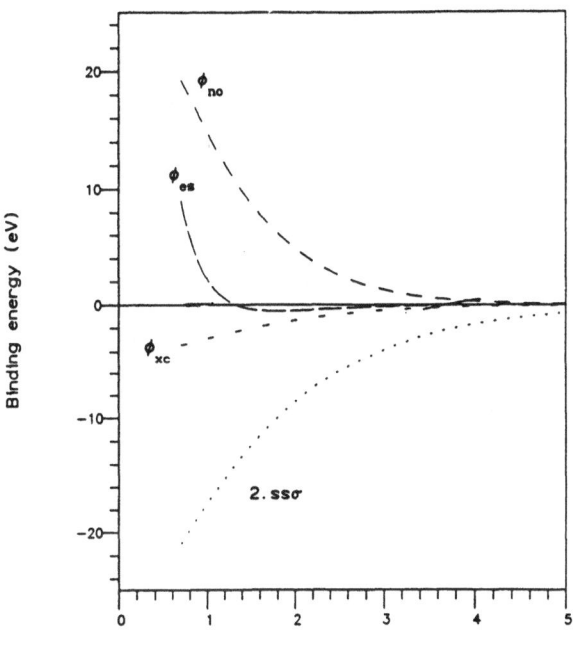

Interatomic separation (a.u.)

Fig. 2. The non-orthogonality (no), electrostatic (es), and exchange-correlation (xc) contribution to the pair potential $\phi(R)$ for hydrogen. Also shown is twice the bond integral ssσ. (Skinner and Pettifor 1991).

Skinner and Pettifor (1991) have carried this derivation through explicitly for the simplest case of s valent hydrogen by assuming a single variational orbital on each site. Fig. 2 shows that the pairwise potential $\phi(R)$ in eq.(2) comprises three terms, namely the repulsive non-orthogonality contribution ϕ_{no}, the electrostatic interaction between overlapping free-atom charge densities ϕ_{es}, and the attractive exchange-correlation contribution ϕ_{xc}. We see that at the equilibrium separation of 1·4 a.u. for the hydrogen molecule the pairwise potential $\phi(R)$ will be dominated by the non-orthogonality contribution, as expected.

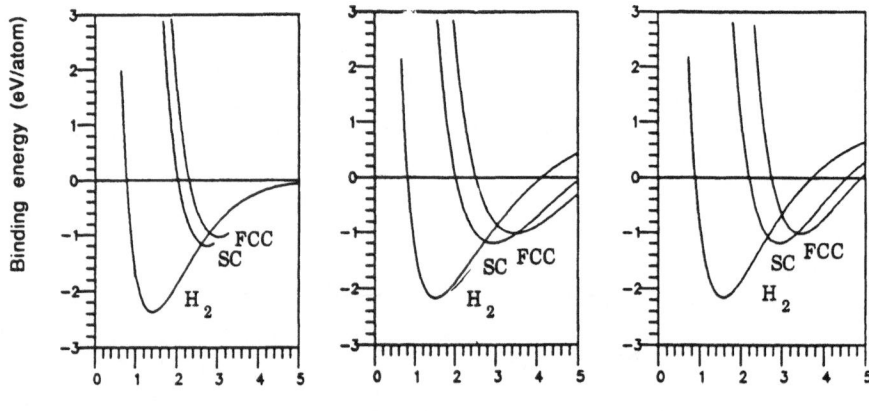

Fig. 3.   Binding energy curves for diatomic and bulk metallic simple cubic (sc) and face centred cubic (fcc) hydrogen. The left-hand panel gives accurate benchmark calculations, whereas the central and right-hand panels give the results obtained within a first nearest-neighbour TB model with environment-dependent and environment-independent orbitals respectively (Skinner and Pettifor 1991).

Fig. 3 compares the resultant TB binding energy curves for diatomic, simple cubic and face centred cubic hydrogen (right hand panel) with the accurate configuration interaction or LDF curves of Kolos and Wolniewicz (1965) and Min et al (1984) (left hand panel). The central panel allows for the orbital exponent of the atomic s orbitals to be environment dependent. We see that the use of environment independent orbitals, corresponding to the *transferable* pair potential $\phi(R)$ and hopping integral $ss\sigma(R)$ drawn in fig. 2, leads to a good description of the relative-binding energies and equilibrium nearest neighbour separations between the three different structure-types.

In practice, for other atoms besides hydrogen, the repulsive pair potential $\phi(R)$ and the two-centre hopping integrals are obtained by *fitting* to known bandstructure and Local Density Functional (LDF) binding energy curves for different structure types. For sp-bonded systems the bond integrals are assumed for simplicity to display the same functional dependence on interatomic distance $h(R)$, namely

$$
\left.\begin{array}{c}
ss\sigma(R) \\
pp\sigma(R) \\
pp\pi(R) \\
sp\sigma(R)
\end{array}\right\} = \left.\begin{array}{c}
-1 \\
p_\sigma \\
-p_\pi \\
\sqrt{p_\sigma}
\end{array}\right\} h(R) \qquad (5)
$$

where p_σ and p_π are positive constants and $sp\sigma$ is taken as the geometric mean of $|ss\sigma|$ and $pp\sigma$. This constraint of identical distance dependences $h(R)$ could be relaxed if necessary to take explicit account of the differences in behaviour of the $ss\sigma$, $pp\sigma$, and $pp\pi$ bonds (see fig. 4 of Allen et al 1986). This distance dependence is important for obtaining reasonable *bandstructure* fits over the large range of co-ordination from diamond through to fcc. For example, Papaconstantopoulos (1986) found $p_\sigma \sim 1.3$ for sp bonded diamond lattices, whereas $p_\sigma \sim 3$ for fcc lattices. The *binding energy* curves will be fitted neglecting the distance dependence of p_σ and p_π as this will lead to simpler functional forms for the many-body potentials presented in §3.

Figure 4 shows that a reasonable fit to the LDF binding energy curves of silicon (Yin and Cohen 1982) may be obtained within TB Hückel theory by using a logical rescaling method (Goodwin et al 1989). They assumed *short-range* transferable TB parameters of the form

$$
\phi(R) = A (R_o/R)^n f^n(R) \qquad (6)
$$

and

$$
h(R) = B (R_o/R)^m f^m(R) \qquad (7)
$$

where $A = 3.46eV$, $B = 1.82eV$, $R_o = 2.35\text{Å}$ (the equilibrium bond length in silicon with the diamond structure), and $f(R)$ is a smooth cut-off function, namely

$$
f(R) = \exp[-(R/R_c)^p]/\exp[-(R_o/R_c)^p] \qquad (8)
$$

where $R_c = 3.67\text{Å}$ and $p = 6.48$. The power law exponents n and m were taken as $n = 4.54$ and $m = 2$, the latter being suggested by Harrison's (1980) canonical form. We note that for simplicity $\phi(R)$ has been chosen proportional to $[h(r)]^{n/m}$. Harrison's 1983 values of p_σ and p_π were chosen, namely $p_\sigma = 1.68$ and $p_\pi = 0.48$. (Note that Goodwin et al (1989) took $sp\sigma = 1.08\ h(R)$ rather than $1.30h(R)$ given by the geometric mean in eq.(5).

The latter constraint is retained here because it simplifies the σ bond order potential dramatically; see eq. (11) of §3.) Finally, the sp atomic energy level separation was taken as 8.295eV, which is 18% larger than the free atom value of 7.03eV, in order to move the close packed binding energy curve closer to tetrahedrally co-ordinated diamond (see fig. 6.3 of Skinner 1989). Goodwin et al (1989) found that this TB parametrisation gave the geometries of small silicon clusters in good agreement with the LDF predictions of Raghavachari and Logovinsky (1985). For example, Si_3 is predicted by TB to have a bond length of 2.25 Å and a bond angle of $77.9°$ compared to the LDF values of 2.17Å and $77.8°$ respectively. In addition Wang et al (1991) have shown that the above TB parameterization yields an excellent radial distribution function for liquid silicon and a good description of point defects. However, McInnes (1992) has recently found that the 18% larger value of the sp splitting used by Goodwin et al (1989) leads to the Haneman buckling model being as stable as the Pandey π-bonded chain model for the Si(111) 2 x 1 reconstruction in disagreement with experiment.

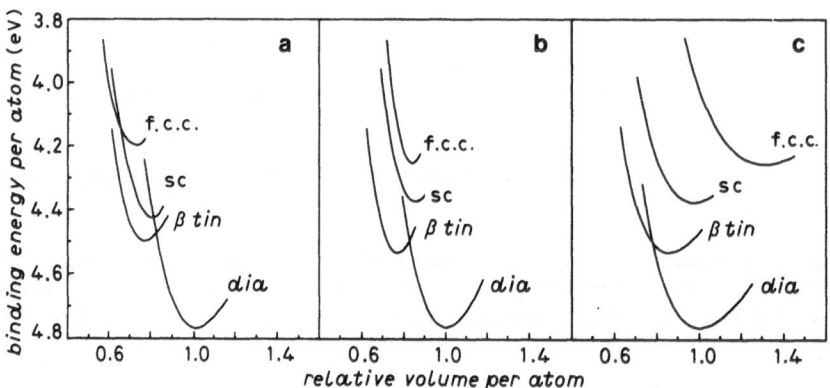

Fig. 4.   A comparison of the LDF binding energy curves of silicon (a) with the rescaled TB curves (b) and the unrescaled TB curves (c) (Goodwin et al 1989).

        This discrepancy illustrates an important point: the simple TB parametrisation of eq. (5) (in which p_σ and p_π are assumed distant independent) should be used to fit the bandstructure and binding energy curves of the ground state and *nearby* metastable phases only, rather than trying to find a *global* fit from the open diamond structure all the way through to the close-packed fcc structure (which requires distant dependent p_σ and p_π; see fig.4 of Allen et al 1986). As pointed out by Kohyama (1991) this 18% increase in E_{sp} in order to fit the twelve-fold co-ordinated fcc binding energy curve is sufficient to close the gap in the bandstructure of four-fold co-ordinated silicon. If the correct value of the sp splitting is used then the Pandey π-bonded chain is more stable than the Haneman buckling model in agreement with experiment (McInnes 1992).

3.   FROM TB HÜCKEL TO BOND ORDER POTENTIALS

The TB calculations of surface reconstructions, which have been pioneered so successfully by Chadi (1978, 1986, 1987), can in principle be simplified still further by *deriving* explicit angularly-dependent many-atom potentials. Moriarty (1990), in particlar, has shown that it is possible to obtain analytic many-atom potentials for transition metals by doing perturbation theory about a given *atomic* site. This type of potential has been used to study the surface reconstructions on bcc transition-metal (100) surfaces (Moriarty and Phillips 1991, Carlsson 1991). In this section we will derive angularly-dependent many-atom potentials for the bond order by embedding the *bond* rather than the atom within its local atomic environment (Pettifor 1989, 1990; Pettifor and Aoki 1991).

The bond energy in eq.(3) is given directly in terms of the local density of states associated with the individual *atoms*. It may be rewritten in terms of the contributions from the individual *bonds* (see, for example, Sutton et al 1988 and references therein) as

$$U_{bond} = \frac{1}{2} \sum_{i,j}' U_{bond}^{ij} \tag{9}$$

where

$$U_{bond}^{ij} = 2 \sum_{\alpha,\beta} H_{i\alpha,j\beta} \Theta_{j\beta,i\alpha} \tag{10}$$

where the pre-factor 2 accounts for spin-degeneracy. This has a particularly transparent form. H is the Slater-Koster (1954) *bond integral* matrix linking the orbitals on sites i and j together. Θ is the corresponding *bond order* matrix whose elements give the difference between the number of electrons in the bonding $\frac{1}{\sqrt{2}}|i\alpha + j\beta\rangle$ and anti-bonding $\frac{1}{\sqrt{2}}|i\alpha - j\beta\rangle$ states. For the particular choice $sp\sigma = (|ss\sigma| \, pp\sigma)^{\frac{1}{2}}$ the sp-bond energy reduces to the form (c.f. eqs.(65)-(67) of Pettifor 1990)

$$U_{bond}^{ij} = -2(1 + p_\sigma) \, h \, (R_{ij}) \, \Theta_{j\sigma,i\sigma} - 4p_\pi \, h \, (R_{ij}) \, \Theta_{j\pi,i\pi} \tag{11}$$

where the hybrid σ orbitals $|i\sigma\rangle$ and $|j\sigma\rangle$ are defined by

$$|i\sigma\rangle = (|is\rangle + \sqrt{p_\sigma} \, |iz\rangle)/\sqrt{1+p_\sigma} \tag{12}$$

and

$$|j\sigma\rangle = (|js\rangle - \sqrt{p_\sigma} \, |jz\rangle)/\sqrt{1+p_\sigma} \tag{13}$$

choosing the z-axis along $\underset{\sim ij}{R}$ and the π bond order by

$$\Theta_{j\pi,i\pi} = \frac{1}{2}(\Theta_{jx,ix} + \Theta_{jy,iy}). \tag{14}$$

Although eq.(11) gives the bond energy between a given pair of atoms i and j it is *not* pairwise because the bond order itself depends on the local atomic environment. We display this dependence explicitly by using the recursion method of Haydock et al (1972) to write the bond order as an integral over the difference of two continued fractions:

$$\Theta_{i\alpha,j\beta} = -\frac{1}{\pi} \Im m \int^{E_F} [G_{oo}^+(E) - G_{oo}^-(E)] \, dE \tag{15}$$

where $\Im m$ is the imaginary part of the bonding and anti-bonding Green's functions which are given by

$$G_{oo}^{\pm}(E) = \langle u_o^{\pm}| (E-H)^{-1} |u_o^{\pm}\rangle \tag{16}$$

$$= \cfrac{1}{(E-a_o^{\pm}) - \cfrac{(b_1^{\pm})^2}{(E-a_1^{\pm}) - \ldots}}$$

where $|u_o^{\pm}\rangle = \frac{1}{\sqrt{2}} |i\alpha \pm j\beta\rangle$. The coefficients are determined by the Lanczos recursion algorithm, namely

$$b_{n+1}^{\pm}|u_{n+1}^{\pm}\rangle = H|u_n^{\pm}\rangle - a_n^{\pm}|u_n^{\pm}\rangle - b_n^{\pm}|u_{n-1}^{\pm}\rangle \tag{17}$$

with the boundary condition that $|u_{-1}^{\pm}\rangle$ vanishes. The Hamiltonian H is, therefore, tridiagonal with respect to the recursion basis $|u_n^{\pm}\rangle$, having non-zero elements

$$\langle u_n^{\pm}|H|u_n^{\pm}\rangle = a_n^{\pm} \tag{18}$$

and

$$\langle u_{n+1}^{\pm}|H|u_n^{\pm}\rangle = b_{n+1}^{\pm}. \tag{19}$$

The dependence of the recursion coefficients on the local atomic environment about the bond ij may be obtained by using the well-known relationship between the recursion coefficients $a_n^{\pm}$, $b_n^{\pm}$ and the moments $\mu_n^{\pm} = \langle u_o^{\pm}|H^n|u_o^{\pm}\rangle$, namely

$$\mu_0^{\pm} = 1, \tag{20}$$

$$\mu_1^{\pm} = a_0^{\pm}, \tag{21}$$

$$\mu_2^{\pm} = (a_0^{\pm})^2 + (b_1^{\pm})^2, \tag{22}$$

$$\mu_3^\pm = (a_0^\pm)^3 + 2a_0^\pm(b_1^\pm)^2 + a_1^\pm(b_1^\pm)^2, \tag{23}$$

and

$$\mu_4^\pm = (a_0^\pm)^4 + 3(a_0^\pm)^2(b_1^\pm)^2 + 2a_0^\pm a_1^\pm(b_1^\pm)^2$$
$$+ (a_1^\pm)^2(b_1^\pm)^2 + (b_1^\pm)^2(b_2^\pm)^2 + (b_1^\pm)^4. \tag{24}$$

The difference between the bonding and antibonding moments may be displayed explicitly by writing

$$\mu_n^\pm = \mu_n \pm \zeta_{n+1} \tag{25}$$

where μ_n is the *average* nth moment with respect to the appropriate orbitals on site i and j, namely

$$\mu_n = \frac{1}{2}[<i|H^n|i> + <j|H^n|j>] = \frac{1}{2}(\mu_n^i + \mu_n^j) \tag{26}$$

and ζ_{n+1} is the *interference* term, namely

$$\zeta_{n+1} = <i|H^n|j>. \tag{27}$$

The ζ_n can be represented diagrammatically as in fig.5. The first diagram represents the ring term ζ_n^{ring} in which all n sites are distinct and there are no self-retracing paths, whereas the latter diagrams are dressed by self-retracing paths which must be summed over all nearest neighbour sites. We have neglected in fig.5 double-counting terms which involve hopping backwards and forwards between atoms i and j.

The σ, π or δ bond order may be written as a *linear* expansion over the interference terms (Pettifor 1989, Pettifor and Aoki 1991), namely

$$\Theta = -2\sum_{n=0}^{\infty} \chi_{on,no}(E_F) \, \delta a_n - 4\sum_{n=1}^{\infty} \chi_{0(n-1),no}(E_F) \, \delta b_n. \tag{28}$$

The response functions $\chi_{om,no}(E_F)$ are determined solely by the *average* moments μ_n, whereas the δa_n, δb_n are obtained from $\frac{1}{2}(a_n^+ - a_n^-)$, $\frac{1}{2}(b_n^+ - b_n^-)$ by retaining contributions linear in the *interference* forms ζ_n, namely

$$\delta a_0 = \zeta_2, \tag{29}$$

$$\delta b_1 = \zeta_3/(2\mu_2^{\frac{1}{2}}), \tag{30}$$

$$\delta a_1 = \zeta_4/\mu_2 - (\mu_3/\mu_2^2) \, \zeta_3 - 2\zeta_2, \tag{31}$$

etc. (c.f. eqs.(2.26)-(2.30) of Pettifor and Aoki 1991). This expression, eq.(28), as an *exact* many-atom expansion for the bond order which converges fairly rapidly (Aoki and Pettifor 1992).

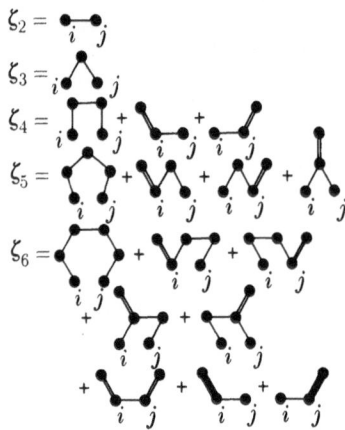

Fig. 5.   Diagrammatic representation of the interference terms ζ_n between atoms i and j (Pettifor and Aoki 1991).

The many-atom expansion for the bond order takes a particularly transparent, though approximate form for the case of *unsaturated* bonds in metals with roughly constant densities of states throughout the band. Assuming $b_n = b_1$ for all n and neglecting the normalized odd moment contributions in expansions such as eqs.(29)-(31), we find that

$$\delta a_1 = (\zeta_4 - 2\mu_2\zeta_2)/\mu_2 \qquad (32)$$

$$\delta b_2 = (\zeta_5 - 3\mu_2\zeta_3)/\mu_2^{3/2}, \qquad (33)$$

and

$$\delta a_2 = [\zeta_6 - 4\mu_2(\zeta_4 - 2\mu_2\zeta_2) - \mu_2^2\zeta_2 - 2\mu_4\zeta_2]/\mu_2^2. \qquad (34)$$

We see from fig.5, therefore, that if the double-counting terms are neglected, then the dressed diagrams cancel from the δa_n, δb_n leaving only the ring terms ζ_n^{ring}.

The bond order may, thus, be written within the Ring Approximation as

$$\Theta = 2\sum_{n=2}^{\infty} \hat{\chi}_n(N) \, \zeta_n^{ring} / b^{n-1} \qquad (35)$$

where $b = b_1$ and the reduced susceptibilities $\hat{\chi}(N)$ are given by

$$\hat{\chi}_n(N) = \frac{1}{\pi}\left[\frac{\sin(n-1)\phi_F}{n-1} - \frac{\sin(n+1)\phi_F}{n+1}\right] \tag{36}$$

with $\phi_F = \arccos(E_F/2b)$. ϕ_F is fixed by the number of valence electrons per spin per bond N through

$$N = (2\phi_F/\pi)[1 - (\sin 2\phi_F)/2\phi_F]. \tag{37}$$

Fig. 6 shows the behaviour of the first five reduced response functions $\hat{\chi}_n$ as a function of the number of valence electrons per spin per bond. We see that the number of nodes (excluding the end points) equals (n-2).

b in eq.(35) plays the role of an *embedding function*. It enters the bond order within the Ring Approximation by setting the energy scale through the writing of the response functions $\chi_{om,no} = \chi_{m+n+2}/b$ for m = n-1 or n. For the case of sp valent σ bonds, which are formed from the hybrid orbitals eqs.(12) and (13), we find (Pettifor 1990, Pettifor and Aoki 1992)

$$[b^\sigma/\langle i\sigma|H|j\sigma\rangle]^2 = 1 + p_\sigma E_{sp}^2/[(1+p_\sigma)^2 h(R_{ij})]^2$$

$$+ \sum_{k\neq ij} \frac{1}{2}[G_{ij}^\sigma(R_{ik},\theta_{jik}) + G_{jk}^\sigma(R_{jk},\theta_{ijk})] \tag{38}$$

where θ_{jik} and θ_{ijk} are the appropriate bond angles and

$$G_{ij}^\sigma(R_{ik},\theta) = [h(R_{ik})/h(R_{ij})]^2 g_\sigma(\theta) \tag{39}$$

with

$$g_\sigma(\theta) = c + d\cos\theta + e\cos 2\theta, \tag{40}$$

$$c = 1 - d - e, \tag{41}$$

$$d = 2 p_\sigma/(1 + p_\sigma)^2, \tag{42}$$

and

$$e = p_\sigma^2/2(1 + p_\sigma)^2 - p_\sigma p_\pi^2/2(1 + p_\sigma)^3. \tag{43}$$

The sp valent π bond, on the other hand, has an associated embedding function whose square is given by

$$[b^\pi/\langle i\pi|H|j\pi\rangle]^2 = 1 + \sum_{k\neq i,j} \frac{1}{2}[G_{ij}^\pi(R_{ik},\theta_{jik}) + G_{ji}^\pi(R_{jk},\theta_{ijk})] \tag{44}$$

where

$$G_{ij}^{\pi}(R_{ik}, \theta) = [h(R_{ik})/h(R_{ij})]^2 g_{\pi}(\theta) \qquad (45)$$

with

$$g_{\pi}(\theta) = C + E \cos 2\theta, \qquad (46)$$

$$C = 1 - E, \qquad (47)$$

and

$$E = \frac{1}{4} - p_{\sigma}(1 + p_{\sigma})/4 \ p_{\pi}^2. \qquad (48)$$

Fig. 7 shows the angular dependence of the embedding functions $g_{\sigma}(\theta)$ and $g_{\pi}(\theta)$ using the Goodwin et al (1989) values of p_{σ} and p_{π}. The pure s σ bond (corresponding to $p_{\sigma} = 0$) displays no angular dependence as expected, the influence of a neighbouring atom k on the ij bond strength being independent of the bond angle. On the other hand, the pure p σ bond (corresponding to vanishing ssσ and ppπ) has $g_{\sigma}(\theta)$ falling to zero for a bond angle of 90°

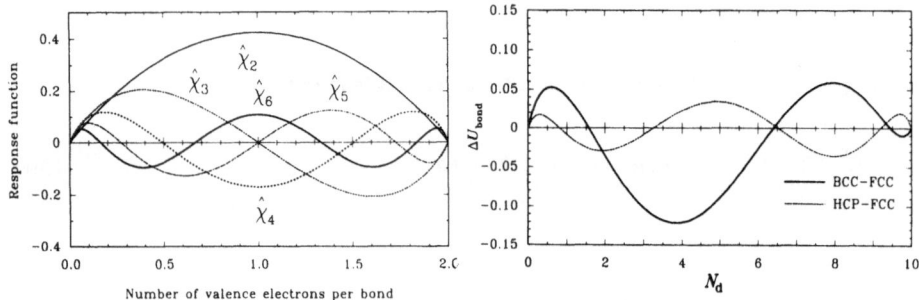

Fig. 6.   The left-hand panel shows the reduced response functions χ_n as a function of the number of valence electrons per spin per bond N. The right-hand panel shows the total bond energy difference (in units of band width) between bcc, fcc and hcp transition metals as a function of the number of valence d electrons per atom N_d for the bcc lattice (Pettifor and Aoki 1991).

since there will be no coupling to this neighbouring atom if p_{π} is zero. The sp hybrid has a minimum in $g_{\sigma}(\theta)$ around 130° and takes a value less than 0.1 for $\theta \geq 100^{\circ}$ as the hybrid orbitals, eqs. (12) and (13), have very little weight in these directions, so that neighbours may be added in this range without affecting the strength of the original σ bond. Thus, graphite and diamond with bond angles of 120° and 109° respectively will have nearly *saturated* σ bonds (for $E_{sp} = 0$). In contrast the angular dependence of the π bond with its lobes extending in planes perpendicular to the bond axis leads to *unsaturated* behaviour as any neighbour will drastically reduce the strength of the original dimeric bond, as can be seen in fig.7 where $g_\pi(\theta)$ rises to a value an order of magnitude larger than that for angularly independent s orbitals. We see in fig.7 that the angular dependence of the semi-empirical Tersoff potential mirrors that predicted by the bond order potential for the σ bond, which explains the inability of the Tersoff potential to model surface reconstructions that are driven by the π bonding.

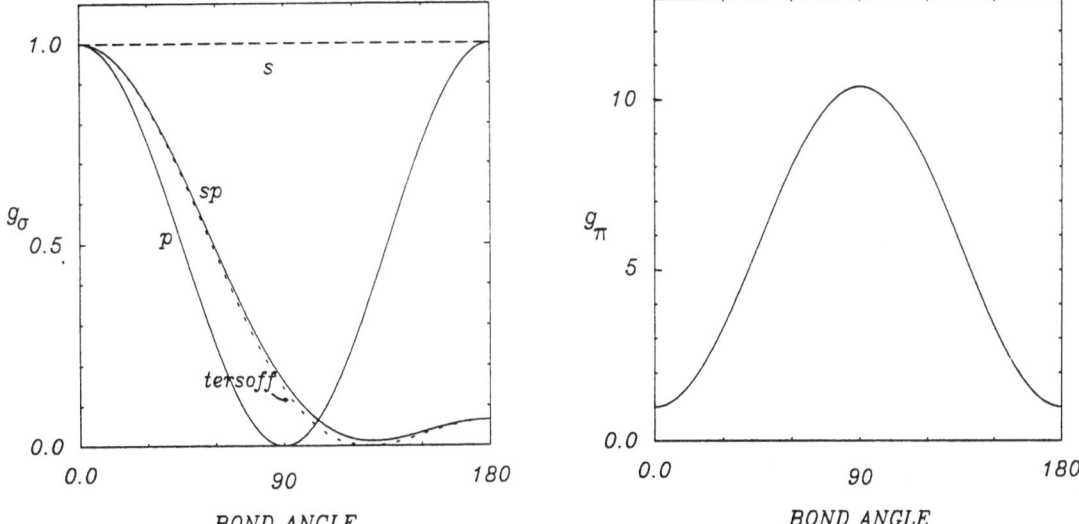

Fig. 7. The angular dependence of the predicted embedding functions $g_\sigma(\theta)$ and $g_\pi(\theta)$ with the semi-empirical Tersoff (1988) curve for comparison.

Analytic expressions for the three-atom and four-atom ring terms in eq.(35) have been given for the sp and sd valent cases (Pettifor and Aoki 1992, Aoki and Pettifor 1992). The four-membered ring term is associated with the two-node response function χ_4 (c.f. the left hand panel of fig.6). It is, thus, primarily responsible for the stability in the middle of the transition metal series of the bcc lattice compared to fcc or hcp (c.f. the right hand panel of fig.6). These bond order potentials are currently being developed to simulate elemental transition metals and intermetallics where the many-atom angular character of the bonding is known to be important. For example, the angular character of the embedding function $g_\sigma(\theta)$ in fig.7 accounts naturally for the occurence of the Zintl structure in LiAl, which cannot be accounted for by pair potentials or angularly independent embedded atom potentials (see, for example, Hafner 1989).

4. CONCLUSIONS

Classical interatomic potentials fail to describe correctly the nature of the quantum mechanical bond in metals and semi-conductors. They are often unable to model the known structures of surfaces and interfaces. We have shown that within the framework of the Tight Binding Hückel approximation it is possible to derive an *exact* many-atom expansion for the bond order. This is currently acting as the basis for further *approximations* and simplifications in order to provide user-friendly, analytic angularly-dependent potentials for atomistic simulations. The original TB description is able to account for most crystal structure trends within the periodic table or across binary compound structure maps (see, for example, Pettifor 1992). Provided the parameters are sensibly constrained the TB approximation can predict subtle surface reconstructions (see, for example, Chadi 1978, 1986, 1987). It remains to be seen how many terms must be retained in the bond order expansion so that the potentials have the same predictive capability at surfaces and interfaces.

REFERENCES

Allen, P.B., Broughton, J.Q. and McMahan A.K., 1986, *Phys. Rev.* B34: 859.
Anderson, P.W., 1968, *Phys. Rev. Lett.* 21: 13.
Aoki, M., and Pettifor, D.G., 1992, (in preparation).
Car, R., 1992, (these proceedings).
Carlsson, A.E., 1991, *Phys. Rev.* B44: 6590.
Chadi, D.J., 1978, *Phys. Rev. Lett.* 41: 1062.
Chadi, D.J., 1986, *Phys. Rev. Lett.* 57: 102.
Chadi, D.J., 1987, *Phys. Rev. Lett.* 59: 1691.
Cressoni, J.C. and Pettifor, D.G. 1991, *J.Phys.: Condens. Matter*, 3: 495.
Daw, M.S. and Baskes, M.I., 1984, *Phys. Rev.* B29: 6443.
Finnis, M.W. and Sinclair, J.E., 1984, *Phil. Mag.* A50: 45.
Foiles, S., 1992, (these proceedings).
Foulkes, W.M.C. and Haydock, R., 1989, *Phys. Rev.* B39: 12520.
Goodwin, L., Skinner, A.J. and Pettifor, D.G., 1989, *Europhys. Lett.* 9: 701.
Harris, J., 1985, *Phys. Rev.* B31: 1770.
Harrison, W.A., 1980, *Electronic Structure and the Properties of Solids.*
        (Freeman: San Francisco).
Harrison, W.A., 1983, *Phys. Rev.* B27: 3592.
Haydock, R., Heine, V. and Kelly, M.J., 1972, *J. Phys. C: Solid State Phys.*
        5: 2845.
Kohyama, M., 1991, *J. Phys.: Condens. Matter* 3: 2193.
Kolos, W., and Wolniewicz, L., 1965, *J. Chem. Phys.* 43: 2429.
McInnes, D.A., 1992, *D.Phil. thesis* (University of Oxford).
Min, B.I., Jansen, H.J.F. and Freeman, A.J., 1984, *Phys. Rev.* B30: 5076.
Moriarty, J.A., 1990, *Phys. Rev.* B42: 1609.
Moriarty, J.A. and Phillips, R., *Phys. Rev. Lett.* 66: 3036.
Papaconstantopoulos, D.A., 1986, *Handbook of the Bandstructure of Elemental
        Solids* (Plenum: New York).
Pettifor, D.G., 1989, *Phys. Rev. Lett.* 63: 2480.
Pettifor, D.G., 1990, *Springer Proc. Phys.* 48: 64.
Pettifor, D.G., in *Electron Theory in Alloy Design*, D.G. Pettifor and
        A. Cottrell, eds., (Institute of Materials: London), ch.4.
Pettifor, D.G. and Aoki, M., 1991, *Phil. Trans. R. Soc. Lond.* A334: 439.
Pettifor, D.G. and Aoki, M., 1992, in *Structure and Phase Stability of
        Alloys*, J.L. Moran-Lopez and J.M. Sanchez, eds., (Plenum: New York).
Raghavachari, K. and Logovinsky, V., 1985, *Phys. Rev. Lett.* 55: 2853.
Sankey, O.F. and Niklewski, D.J., 1989, *Phys. Rev.* B40: 3979.

Skinner, A.J., 1989, *PhD Thesis* (University of London).

Skinner, A.J. and Pettifor, D.G., 1991, *J. Phys.: Condens. Matter* 3: 2029.

Slater, J.C. and Koster, G.F., 1954, *Phys. Rev.* 94: 1498.

Stillinger, F. and Weber, T., 1985, *Phys. Rev.* B31: 5262.

Sutton, A.P., Finnis, M.W., Pettifor, D.G. and Ohta, Y. 1988, *J. Phys. C: Solid State Phys.* 21: 35.

Tersoff, J. 1988, *Phys. Rev.* B38: 9902.

Wang, C.Z., Chan, C.T. and Ho, K.M., 1991, *Phys. Rev. Lett.* 66: 189.

Wilson, J.H., Todd, J.D. and Sutton, A.P., 1990, *J. Phys.: Condens. Matter*, 2: 10259.

Yin, M.T. and Cohen, M.L., 1982, *Phys. Rev.* B26: 5668.

TOPOLOGICAL THEORY OF LINE-DEFECTS ON CRYSTAL SURFACES, AND THEIR INTERACTIONS WITH BULK AND INTERFACIAL DEFECTS

R.C. Pond and S.M. Casey

Department of Materials Science and Engineering
University of Liverpool
Liverpool, L69 3BX
United Kingdom

INTRODUCTION

Important surface processes are known to be mediated by surface defects; for example, crystal growth proceeds by the motion of steps (see Zangwill[1] for a review). Recent advances in surface science, particularly through new experimental techniques, are revealing the existence of a diversity of line defects in addition to steps, and it is anticipated that these may also significantly influence surface processes. In view of these developments, it is therefore appropriate to formulate a theory which can be used to enumerate the distinct types of defects that can arise on any given crystal surface in its unrelaxed and reconstructed conditions. This can be done on the basis of symmetry theory for unrelaxed surfaces, and also for reconstructed cases if the nature of the relaxation is known. Thus, the primary objective of this work is to present a theory which can predict the nature of all admissible defects on surfaces. In addition, we discuss the interaction of such surface defects with bulk and interfacial defects.

Defects in crystalline materials are constrained to adopt particular forms. They can be treated using topological methods, and are thereby found to be characterised by symmetry operations of the underlying crystal. A well known example of such topological characterisation is the assignment of a Burgers vector to a dislocation in the bulk of a crystal. This parameter expresses an intrinsic

Equilibrium Structure and Properties of Surfaces and Interfaces
Edited by A. Gonis and G.M. Stocks, Plenum Press, New York, 1992

geometrical property of the discontinuity, and is completely independent of the defect's core structure and line direction. In the following section we review this characterisation process, and subsequently generalise the essential idea involved to the case of surface defects. The distinctive types of defects that can arise in the bulk of a single crystal are then reviewed briefly in these terms.

In order to identify all of the admissible defects that can arise at surfaces and interfaces, it is necessary to summarise the space groups which describe the permissible symmetries of crystal surfaces and interfaces. This is done in the fourth and fifth sections of the paper, and in the sixth section we discuss defects on unrelaxed and relaxed crystal surfaces, using the (001) surface of Si as an illustration. Interfaces resemble surfaces in some respects, but, on the other hand, exist inside crystals. Therefore, in the discussion of interfacial defects in the following section, it is convenient to consider these in two categories, namely surface-like and bulk-like defects. The final section of the paper is a summary of the crystallographic origin and topological properties of all the various line defects which can arise at surfaces, in the bulk, and at interfaces, and includes an account of the interactions between defects of different types.

In the present work, we use the notation adopted in the International Tables for Crystallography[2], for the mathematical representation of symmetry operations. Using matrix form, a symmetry operator, $\mathbb{W}$, is written $(\mathbf{W},\mathbf{w})$, where $\mathbf{W}$ is the orthogonal part (e. g. rotation, or mirror reflection) and $\mathbf{w}$ represents any associated displacement (due to nonsymmorphic elements, or the location of an element, or both). Pure translation operations have the form $(\mathbf{I},\mathbf{t})$, where $\mathbf{I}$ represents the identity operation and $\mathbf{t}$ the translation vector. When referring to the ith operator in a space group, we write $\mathbb{W}_i = (\mathbf{W}_i,\mathbf{w}_i)$, and in the general case this will comprise both a translation and rotation or reflection.

CHARACTERISATION OF DISCONTINUITIES

The topological properties of a line of discontinuity in a material are intinsic geometrical aspects which are completely independent of its line direction. In crystalline materials, these properties are constrained to have particular forms because of the medium's symmetry. The object of the present section is to outline this topological characterisation using a framework which encompasses

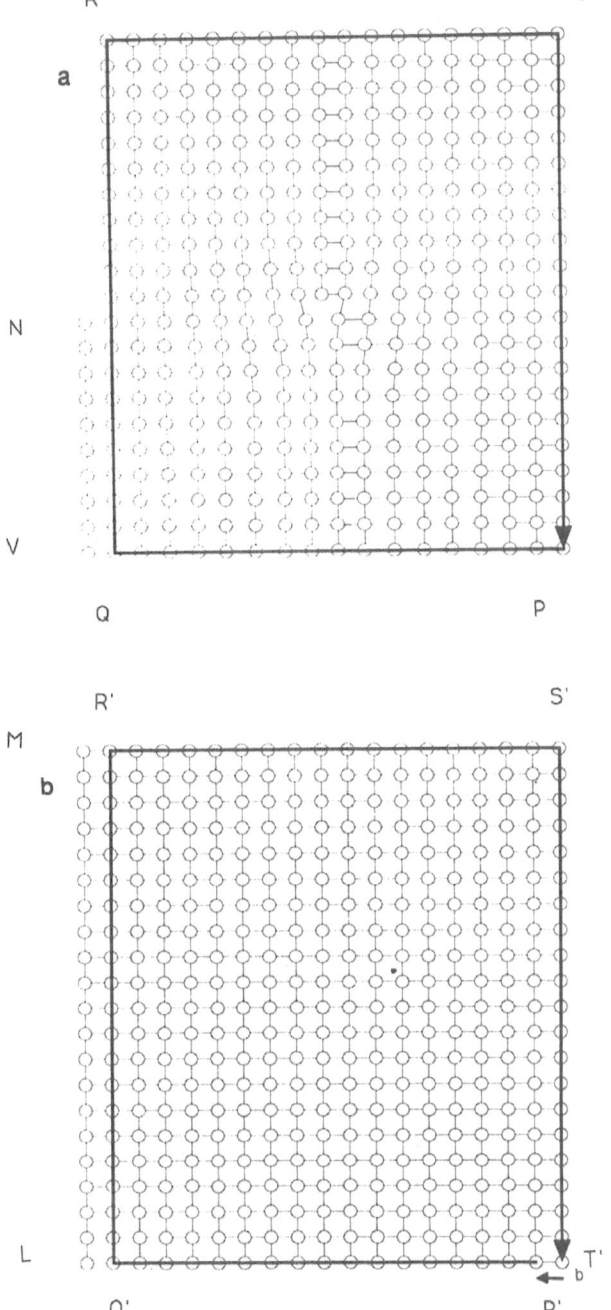

Fig. 1. A Burgers circuit, PQRSP, constructed around a
dislocation, (a), and (b), the mapping of this circuit
onto the reference lattice.

surface, bulk, and interfacial discontinuities. We begin by considering the well known procedure for characterising a dislocation in a single crystal, as depicted schematically in fig. 1. A graphical method was suggested by Frank (see[3,4,5] for reviews), whereby a circuit PQRSP is constructed around the defect, and is subsequently mapped onto the perfect crystal, P'Q'R'S'T'. The closure failure of the latter circuit, T'P', is the Burgers vector, $\mathbf{b}$, as indicated in the figure, and, in the case of perfect dislocations, this vector is a translation vector, $\mathbf{t}$, of the crystal's lattice (see[4] for the conventions used regarding line senses). An alternative procedure, leading to the identical result, is to follow Volterra[3,4,5], who envisaged the introduction of a dislocation into a crystal by making a cut, then relatively displacing the two faces of this cut, and finally rebonding them, having removed excess or introduced additional material as necessary. This characterisation procedure emphasises that $\mathbf{b}$ is the operation required to introduce the defect into the medium, i.e. the relative displacement is the operation $(\mathbf{I}, \mathbf{t})$ in this case.

Line defects in the bulk of single crystals are characterised by proper symmetry operators (i.e. no inversion operation is included), and hence the admissible types of defect are limited to dislocations, disclinations and dispirations[3,4,5]. The introduction of dislocations and disclinations (these latter are characterised by rotation operations, $(\mathbf{W}, \mathbf{0})$) into a continuum are illustrated schematically in fig. 2. In crystalline materials, disclinations can only arise in crystals exhibiting rotation symmetry, and dispirations in crystals exhibiting screw-rotation axes. However, the elastic energy per unit length is substantial in the case of disclinations and dispirations, and hence these are not normally observed except in small crystals and materials with low elastic moduli.

When a dislocation is introduced into a finite crystal, i.e. one exhibiting external surfaces like that in fig. 1 (a), a step is created, as at N. In the spirit of Volterra, we characterise this latter discontinuity by the operation required to introduce it into the surface VR, i.e. by the translation operation $(\mathbf{I}, \mathbf{t})$. Thus, although the dislocation in the bulk and the surface step are quite different configurations, the former exhibiting a distinctive strain field for example, these two discontinuities are characterised topologically by the same operation, $\mathbb{W} = (\mathbf{I}, \mathbf{t})$. Of course, the step and the dislocation are physically connected where the dislocation emerges onto the surface from the bulk, and their common characterisation expresses mathematically the conservation of topological properties in crystalline materials.

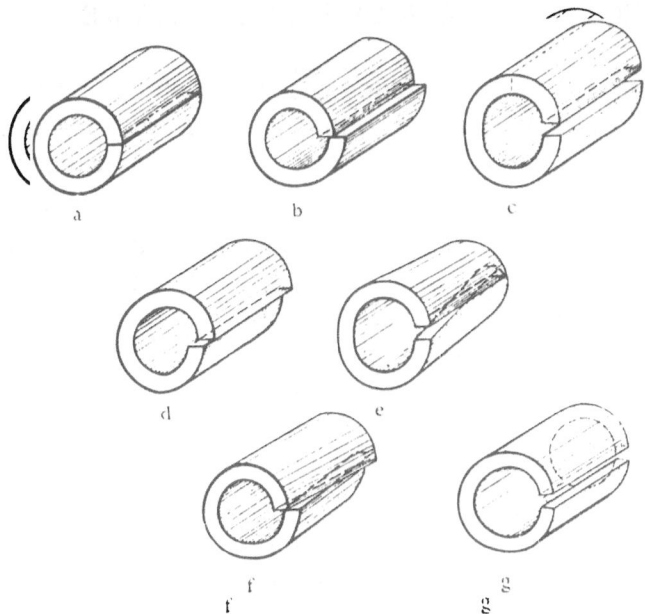

Fig. 2. The six orders of dislocations in a hollow rubber
tube (after Nabarro[5]). (a) Initial cut, (b) and (c)
edge dislocations, (d) screw dislocation, (e)
and (f) twist disclinations, and (g) wedge
disclination. (courtesy O.U.P.)

In the case of surface discontinuities like the step in fig. 1(a), it
is helpful to recognise that the operation (I,t) can also be regarded as
inter-relating two crystallographically equivalent variants in the
following sense. Consider first that the crystal depicted in fig.1(b) be
infinitely extended, and then create a surface parallel to Q'R'. This
semi-infinite object exhibits lower symmetry than the infinite crystal,
and hence, according to the Principle of Symmetry Compensation (see
Shubnikov and Koptsik[6] for example), equivalent variants can exist, and
these are inter-related by symmetry operations broken as a result of
creating the surface. Thus, in the present case, the initial semi-
infinite crystal is related to an infinity of equivalent (energetically
degenerate) variants. One member of this set, i.e. the surface parallel
to LM, is related to the initial one by the broken operation (I,t); it is
identical to the initial configuration except for its location in space.
Now imagine that the two variants coexist physically; they can be
joined contiguously everywhere except along the surface discontinuity
separating the variants. Whatever the line direction and position of this
step, the operation (I,t) inter-relates the degenerate surfaces on either
side of it, and hence characterises it topologically.

There are 230 possible space groups which exhibit three-dimensional translation symmetry, and these are therefore appropriate for classifying the symmetry of crystals. We designate such space groups Φ, and the complete list is tabulated in The International Tables For Crystallography[2]. In the case of a holosymmetric crystal, i.e. where the space group of the crystal is isomorphic to that of the lattice on which it is based, only one variant of the crystal structure exists. For example, consider the case of diamond-structure crystals, which exhibit the nonsymmorphic space group Fd3m and which are based on the fcc lattice (Fm3m); the principal symmetry operations are illustrated in fig. 3 and tabulated in Table 1 assuming an atomic site is taken as origin. The character of admissible line-defects that can arise in such crystals correspond to the proper operations in Table 1 and the translation operations defining the fcc lattice.

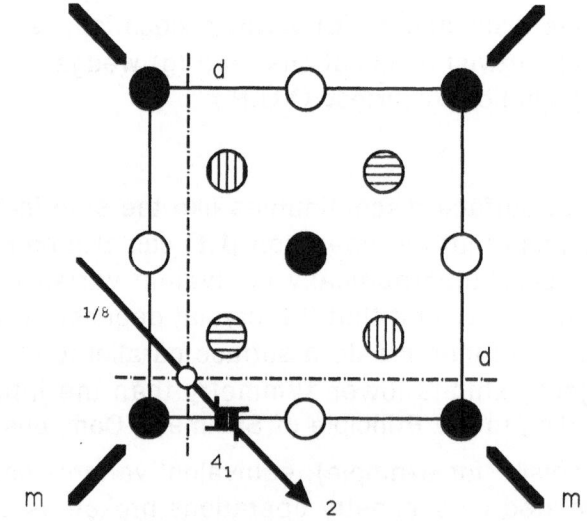

Fig. 3. Schematic illustration of the structure of Si
viewed along [001], indicating some of the
principal symmetry operations present. The
shading of the symbols represents the
heights of the atoms.

Consider next the sphalerite stucture, depicted in fig. 4, which exhibits the space group $\Phi = F\overline{4}3m$. This is an example of a nonholosymmetric crystal; although its lattice is the same as for diamond, its space group is a subgroup of that for diamond, as is indicated in Table 1.   This can be seen by considering the atomic bases

Table 1. Symmetry Operations in The Space Groups
Fd$\bar{3}$m and F$\bar{4}$3m

(Chosen origin at $\bar{4}$3m, **w** = $^1/_4$ [111]).

Symmery Operation	No. of Operations	Orientation	Matrix	
1	1	-	**(1,0)**	
$\bar{4}^+$	3	<100>	**($\bar{4}^+$,0)**	
$\bar{4}^-$	3	<100>	**($\bar{4}^-$,0)**	
2	3	<100>	**(2,0)**	F$\bar{4}$3m
m	6	{110}	**(m,0)**	
3^+	4	<111>	**(3^+,0)**	
3^-	4	<111>	**(3^-,0)**	
$\bar{1}$	1	-	**($\bar{1}$,w)**	
4_1^+	3	<100>	**(4^+,w)**	
4_1^-	3	<100>	**(4^-,w)**	
2	6	<110>	**(2,w)**	
d	3	{100}	**(m,w)**	
$\bar{3}^+$	4	<111>	**($\bar{3}^+$,w)**	
3^-	4	<111>	**($\bar{3}^-$,w)**	

in the two crystal structures. The two identical atoms in the basis of the diamond structure, with internal coordinates 0,0,0 and 1/4,1/4,1/4 and referred to here as α and β type positions, have surroundings which are related by symmetry operations such as the inversion located at 1/8,1/8,1/8. Half of the symmetry operations in Table 1, i.e. those with **w** = 1/4 [111], inter-relate α type atoms to β type, and the other half, i.e. those with **w** = 0, inter-relate α to α and, simultaneously, β to β. On the other hand, in the case of sphalerite, the two atoms in the basis are of different species and hence the operations inter-relating α type atoms to β type are suppressed. This has important consequences as far as defects are concerned. First, the set of admissible line-defect types is reduced, although the important class of dislocations is the same. Secondly, domain boundaries can arise; these are extended discontinuities which separate variant forms of the crystal based on the same lattice. The number of domains that can arise, and their inter-relation, are concisely expressed in the decomposition of the parent (diamond) structure's spacegroup with respect to that of the product (sphalerite), i.e., using right cosets

$$Fd\bar{3}m = (F\bar{4}3m) 1 \text{ u } (F\bar{4}3m) \bar{1} \qquad (1)$$

where the symbol u signifies union, and the symmetry operations are as defined in Table 1. Thus, two domains can arise inter-related by the

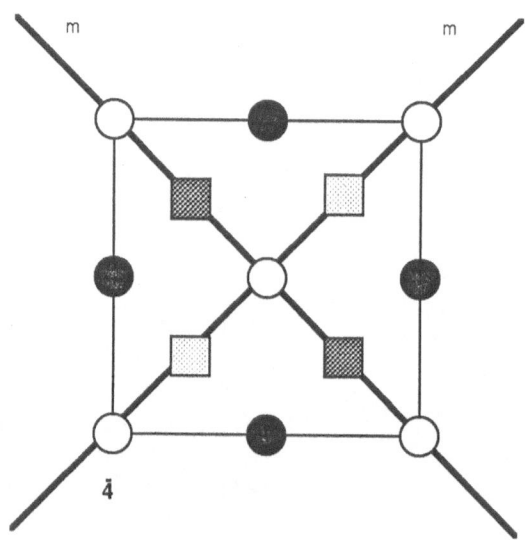

Fig. 4. Schematic illustration of the structure of
sphalerite viewed along [001], indicating
some of the principal symmetry operations
present. The circular and square symbols
represent atoms of different species, and
the shading corresponds to their various heights.

coset representative, $\bar{1}$, and hence we refer to these as inversion domains. In fact, the two domains are inter-related by any of the operations in the coset $(F\bar{4}3m)\bar{1}$, and we refer to this as the set of exchange operations, designated $\mathbb{U}^e$. An exchange operation carried out on the initial domain transforms it into the other domain, and hence can be used to characterise the domain boundary separating the coexisting variants. In the present case, we note that the set of exchange operations includes some proper operations, and hence domain boundaries in sphalerite can be terminated by line-defects, although these are not dislocations.

The discussion above shows that the range of admissible discontinuities in a single crystal is determined by the nature of the operations in its space group, and can be summarised as shown below.
(i) Dislocations; these are characterised by translation operations, (I,t), and can therefore arise in any crystal, although only those with small magnitude b are likely to be important physically.
(ii) Disclinations; these are characterised by rotation operations, $(W,0)$ and can only arise in crystals exhibiting such operations, and are only probable in very small crystals or materials with low elastic moduli.
(iii) Dispirations; these are characterised by screw-rotation operations, (W,w) and are probable in the same circumstances as disclinations.
(iv) Domain Boundaries; these can only arise in nonholosymmetric crystals, and are characterised by exchange operations (W^e, w^e); domain boundaries can be terminated by a line-defect inside a crystal only when proper operations are in this set, and must form closed volumes otherwise.

SURFACE SPACE GROUPS

A semi-infinite crystal can be imagined to be created from an infinite one by choosing the orientation and location of a surface plane, and subsequently discarding all the atoms beyond the chosen plane. In this way an unrelaxed crystal surface is created, designated in this work by its outward pointing unit normal, n. In general, this process involves dissymmetrisation, and hence variant surfaces exist. The set of discontinuities that can arise on unrelaxed surfaces can therefore be established by determining the set of variants and investigating their possible coexistence. This aspect is treated in a later section, and here we review the possible space groups of semi-infinite crystals in preparation for the discussion in the next section.

The initial infinite crystal exhibits the space group Φ, but when a surface is introduced (passing through the origin with respect to which the crystal symmetry operations are defined), only those operations which leave n invariant survive. If the ith operation in the group Φ is designated $\mathbb{W}_i$, the operations in the surface group, $\Phi(s)$, are those for which $\mathbb{W}_i n = n$, or written in full,

$$W_i n = n \qquad\qquad (2a)$$
$$n \cdot (w_i + t_j) = 0 \qquad\qquad (2b)$$

Expression 2a shows that rotation axes and mirror planes must be parallel to n, and that roto-inversion axes are not admissible. Expression 2b shows that pure translation operations must be perpendicular to n, screw-rotation axes are not admissible, and the total translation associated with a mirror glide plane must be perpendicular to n. These conclusions are summarised schematically in fig. 5.

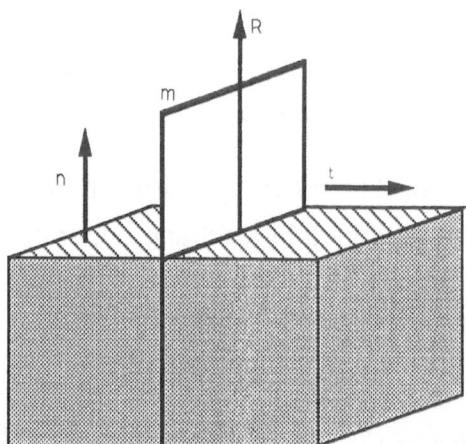

Fig. 5. Schematic illustration of the orientations of permissible symmetry elements in surface space groups; R represents possible rotation axes of order 1, 2, 3, 4 or 6.

The permissible combinations of the above operations form the space groups $\Phi(s)$, and have been tabulated by, for example, Shubnikov and Koptsik[6]. For convenience, these are reproduced here; following the suggestion of Shubnikov and Koptsik[6], the space groups are collected according to the dimensionality of the translation symmetry exhibited. Surfaces exhibiting zero, one and two dimensional translation are referred to as rosettes, bands, and layers respectively. In addition, these are designated "one-sided" to emphasise that the objects in question are different when viewed from the two sides of the surface. There are 10 groups in the set of rosettes, 7 groups in the set of bands, and 17 groups in the set of layers, and these are listed in the columns marked "ordinary" in Tables 2, 3 and 4 respectively.

Table 2. Permissible Rosette Groups for Surfaces and Bicrystals

No.	Ordinary	Antisymmetric
1	1	-
2	-	$\bar{1}'$
3	-	2'11
4	112	-
5	-	11m'
6	m11	-
7	-	2'/m11
8	-	112/m'
9	-	2'2'2
10	mm2	-
11	-	m2'm'
12	-	mmm'
13	4	-
14	-	$\bar{4}'$
15	-	4/m'
16	-	42'2'
17	4mm	-
18	-	$\bar{4}'2'm$
19	-	4/m'mm
20	3	-
21	-	$\bar{3}'$
22	-	32'
23	3m	-
24	-	$\bar{3}'m$
25	6	-
26	-	$\bar{6}'$
27	-	6/m'
28	-	62'2'
29	6mm	-
30	-	$\bar{6}'m2'$
31	-	6/m'mm

As an illustration of the assignment of space groups to unrelaxed crystal surfaces, we consider two examples. The first example is the (001) surface of Si, depicted schematically in fig. 6(a). The two shortest independent translation operations satisfying the conditions required by expression 2b are $t_1 = 1/2[110]$, and $t_2 = 1/2[1\bar{1}0]$; of the 48 operations in Table 1, only the operations 1, $2_{[001]}$, $m_{(110)}$, and $m_{(1\bar{1}0)}$ satisfy expression 2a. These operations constitute the layer group p2mm, and we note that all the operations in this group inter-relate α to α and β to β sites. The second example is the (110) surface of Si, depicted in fig. 6(b). The operations leaving this surface invariant are $t_1 = [001]$, $t_2 = 1/2[1\bar{1}0]$, $2_{[1\bar{1}0]}$, $m_{(1\bar{1}0)}$, and the (001) diamond mirror-

Table 3. Permissible Band Groups for Surfaces and Bicrystals

No.	Ordinary	Antisymmetric
1	p1	-
2	-	p1'
3	p112	-
4	-	p2'11
5	-	p2'$_1$11
6	-	p12'1
7	p1m1	-
8	pm11	-
9	-	p11m'
10	p1a1	-
11	-	p11a'
12	-	p112/m'
13	-	p2'/m11
14	-	p112/a'
15	-	p2'$_1$/m11
16	-	p12'/ml
17	-	p12'/al
18	-	p2'2'2
19	-	p2'$_1$2'2
20	pmm2	-
21	-	pm2'm'
22	-	p2'mm'
23	pma2	-
24	-	pm2'a'
25	-	p2'$_1$ma'
26	-	p2'aa'
27	-	p2'$_1$am'
28	-	pmmm'
29	-	pmam'
30	-	pmma'
31	-	pmaa'

glide plane, $d_{(001)}$. This last operation is normally expressed as $(\mathbf{m}_{(001)}, 1/4[111])$, but , by re-expressing this as $(\mathbf{m}_{(001)}, 1/4[111] + 1/2[0\bar{1}1])$, i.e. $(\mathbf{m}_{(001)}, 1/4[1\bar{1}3])$, it can be seen that expression 2b is satisfied. The operations identified above are depicted schematically in fig. 6(b), and constitute the nonsymmorphic layer group pbm2. Note that the operations 2 and $d_{(001)}$ inter-relate α and β sites on this surface.

BICRYSTAL SPACE GROUPS

In preparation for the later section where interfacial discontinuities are outlined, it is necessary to review the permissible space groups for bicrystals. As for surface, groups we use the general

Table 4. Permissible Layer Groups for Surfaces and Bicrystals

No.	Ordinary	Antisymmetric
1	p1	-
2	-	p$\bar{1}$'
3	p112	-
4	-	p12'1
5	-	p12'$_1$1
6	-	c12'1
7	-	p11m'
8	p1m1	-
9	p1a1	-
10	-	p11b'
11	c1m1	-
12	-	p12'/m1
13	-	p112/m'
14	-	p12'$_1$/m1
15	-	c12'/m1
16	-	p112/b'
17	-	p12'/a1
18	-	p12'$_1$/a1
19	-	p2'2'2
20	-	p2'$_1$2'$_1$2
21	-	p2'$_1$2'2
22	-	c2'2'2
23	pmm2	-
24	-	p2'mm'
25	-	p2'$_1$ma'
26	-	p2'aa'
27	-	p2'mb'
28	pbm2	-
29	-	p2'$_1$am'
30	-	p2'$_1$ab'
31	-	pb2'n'
32	-	p2'$_1$mn'
33	pba2	-
34	cmm2	-
35	-	c2'mm'
36	-	c2'mb'
37	-	pmmm'
38	-	pmaa'
39	-	pban'
40	-	pbmm'
41	-	pmma'
42	-	pbmn'
43	-	pbaa'
44	-	pbam'
45	-	pmab'
46	-	pmmn'
47	-	cmmm'
48	-	cmma'
49	p4	-
50	-	p$\bar{4}$'
51	-	p4/m'
52	-	p4/n'
53	-	p42'2'
54	-	p42'$_1$2'

Table 4 (cont.)

No.	Ordinary	Antisymmetric
55	p4mm	-
56	p4bm	-
57	-	p$\bar{4}$'2'm
58	-	p$\bar{4}$'2'$_1$m
59	-	p$\bar{4}$'m2'
60	-	p$\bar{4}$'b2'
61	-	p4/m'mm
62	-	p4/n'bm
63	-	p4/m'bm
64	-	p4/n'mm
65	p3	-
66	-	p$\bar{3}$'
67	-	p312'
68	-	p32'1
69	p3m1	-
70	p31m	-
71	-	p$\bar{3}$'1m
72	-	p$\bar{3}$'m1
73	p6	-
74	-	p$\bar{6}$'
75	-	p6/m'
76	-	p62'2'
77	p6mm	-
78	-	p$\bar{6}$'m2'
79	-	p$\bar{6}$'2'm
80	-	p6/m'mm

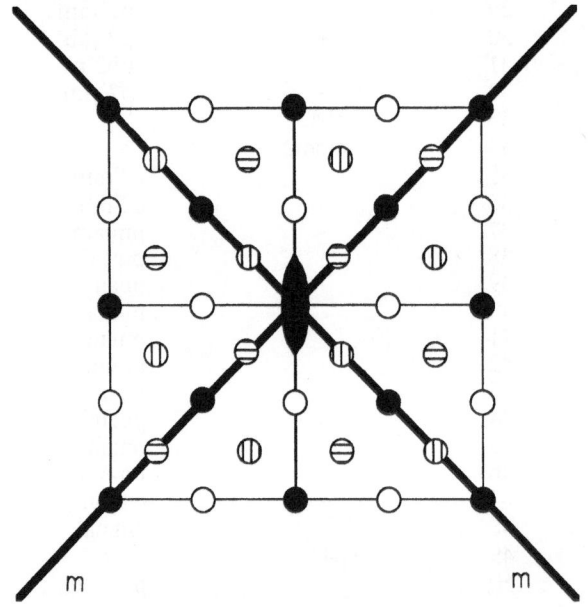

Fig. 6(a). Schematic plan view of the unrelaxed (001) surface of Si, showing the symmetry operations in the layer space group p2mm.

approach of Shubnikov and Koptsik[6], and employ the nomenclature of Pond and coworkers[7,8]. A bicrystal is taken to be a composite object and we distinguish the two crystals by designating them black (μ) and white (λ). Let the ith and j^{th} symmetry operations of these crystals be represented by $\mathbf{W}(\mu)_i$ and $\mathbf{W}(\lambda)_j$ respectively. We take the

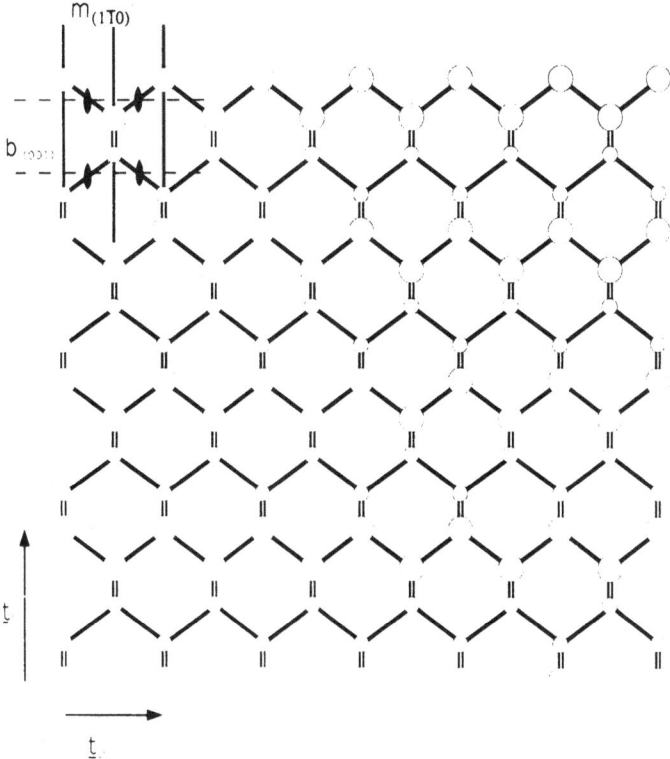

Fig. 6(b). Schematic plan view of the unrelaxed (110) surface
of Si, showing the operations present in the layer
space group pbm2.

correspondence between the coordinate frames of the two crystals to be $\mathbf{P} = (\mathbf{P}, \mathbf{p})$, where $\mathbf{P}$ is a transformation from black to white, and $\mathbf{p}$ is the relative shift of the black crystal with respect to the white. (In the specification of the crystal symmetry operators, an origin in each of the crystals is chosen; $\mathbf{p}$ is the relative position of these two points.) The unit normal to the interface, expressed in the white frame, and taken to be pointing into the white crystal, is now designated **n**. Two distinct types of symmetry operations, called coincident and antisymmetry operations, can be present in bicrystal space groups. The former can be present in the cases of both interphase interfaces and grain boundaries, whereas the latter can only occur in grain boundaries.

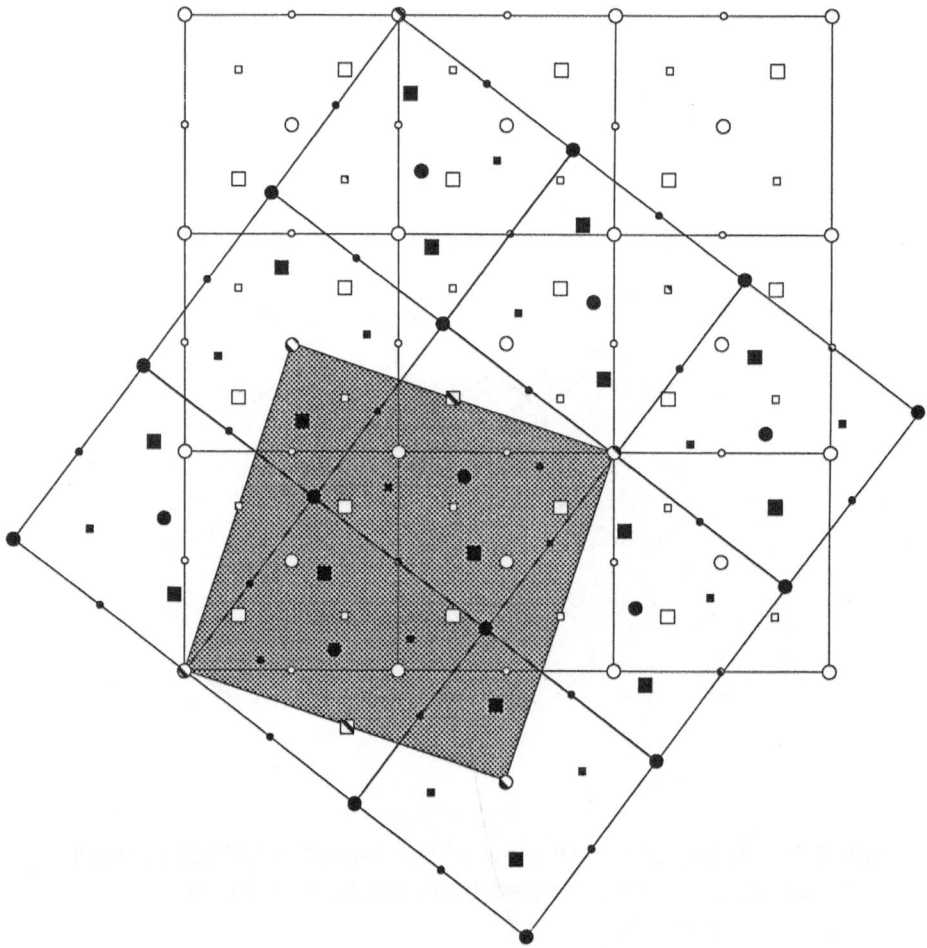

Fig. 7. Dichromatic complex for two crystals of Si rotated by 53.1^0 about their common [001]; the symmetry of the complex is $I4_1/a$ m'd'.

To introduce coincident and antisymmetry operations, it is helpful in the first instance to imagine the black and white crystals to be infinite and to interpenetrate, constituting a dichromatic complex with space group designated $\Phi(\lambda\mu)$. An example is shown in fig. 7 which shows black and white diamond structure crystals misoriented by a rotation of 53.1° about their coincident [001] direction. Coincident black and white operations, designated $\mathbb{W}(c)_k$, can be identified by re-expressing black operations in the white frame and comparing the resultants to the set of white operations, i.e. by finding solutions, if any, to the equation

$$\mathbb{W}(c)_k = \mathbb{W}(\lambda)_j = \mathbb{P}\, \mathbb{W}(\mu)_i\, \mathbb{P}^{-1} \qquad (3)$$

In fig. 7 both coincident translation and point symmetry operations are present. The former define a coincident site lattice, which is body-centred tetragonal in this case. Coincident symmetry operations simultaneously relate white sites to white ones, and black ones to black. On the other hand, antisymmetry operations inter-relate sites of opposite colour. These operations therefore correspond to those alternative descriptions of $\mathbb{P}$ which have the form of symmetry operations, and are designated $\mathbb{W}'$. Thus, to obtain this set it is necessary to inspect the alternative descriptions, which are given by $\mathbb{P}\mathbb{W}(\lambda)_i$; in the present case this leads to eight operations. The total set of coincident and antisymmetry operations exhibited by fig. 7 constitute the space group $\Phi(\lambda\mu) = I4_1/a\ m'd'$, where the anti-operations are indicated by primed symbols.

Now, imagine that a bicrystal is created from the dichromatic complex by choosing the interface plane with normal $\mathbf{n}$, and then discarding all the white atoms on one side and all the black ones on the other. We can now identify the symmetry operations which leave the bicrystal invariant. Those coincident operations, $\mathbb{W}(c)_k$, which also leave the bicrystal invariant must satisfy the conditions set out in expressions 2a and 2b. In other words, when no antisymmetry operations are present, the permissible symmetries of bicrystals are identical to those for surfaces, as listed under the "ordinary" columns in Tables 2, 3 and 4. Antisymmetry operations present in a bicrystal space group are those for which $\mathbb{W}'_i\mathbf{n} = \mathbf{n}$, or written in full,

$$\mathbf{W}'_i\mathbf{n} = \mathbf{n} \qquad (4a)$$
$$\mathbf{n}.\mathbf{w}_i' = 0 \qquad (4b)$$

Expressions 4a and b look, at first sight, very similar to 2a and b. However, it must be kept in mind that these operators involve colour-reversal in addition to a conventional symmetry operation. Putting this mathematically, we say that an operation W'_i is equivalent to operating the conventional part, W_i, followed by the inversion operation, i.e. W'_i is equivalent to TW_i. The inversion operation can be regarded as the mathematical means by which colour is reversed, since **n**, which by definition points into the white crystal, is inverted by this operation. Therefore, rather than there being a parallel between expressions 2 and 4, there is a complementarity. For example, according to expression 2 coincident mirror planes can only arise perpendicular to the interface plane, whereas expression 4 shows that antimirrors can only exist paralell to the interface. Similarly, coincident rotation axes can only be perpendicular, and antirotations can only be parallel to the interface, as summarised in fig.8.

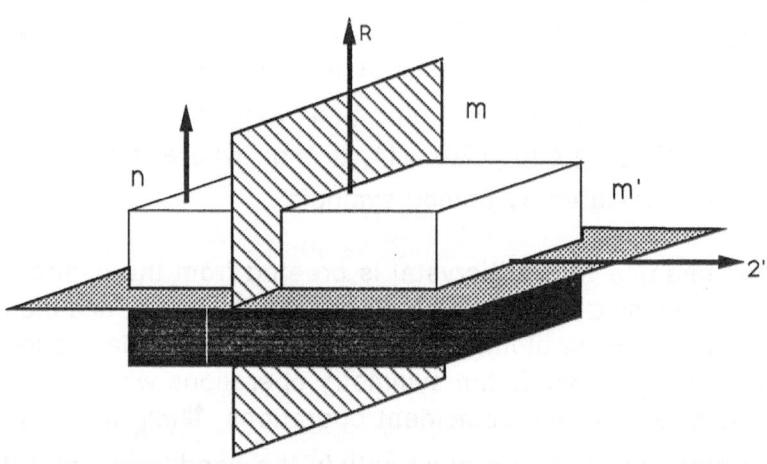

Fig. 8. Schematic illustration of the orientations of permissible symmetry elements and anti-elements in bicrystal space groups; the symbol R is the same as for fig. 5.

The possible existence of antisymmetry operations increases the number of permissible bicrystal space groups significantly. In the terminology of Shubnikov and Koptsik[6], two-sided, rather than one-sided, groups can arise, meaning that the bicrystal may not appear different when viewed from opposite sides of the interface. Thus, the "ordinary" groups listed in Tables 2, 3 and 4 are augmented by "antisymmetry" groups, leading to a total of 31 rosette, 31 band, and 80 layer groups, as listed in the Tables. An example of the assignment of a two-sided layer group to a bicrystal is depicted in fig. 9. The bicrystal is the unrelaxed configuration obtained from the dichromatic complex shown in fig. 7 with $\mathbf{n}$ parallel to [310]; the space group is $p2_1$'am'.

SURFACE DISCONTINUITIES

The characterisation of discontinuities on unrelaxed and reconstructed surfaces is described in the first two sections below. Illustrations of the range of defect types on the (001) surface of Si are treated in the third section.

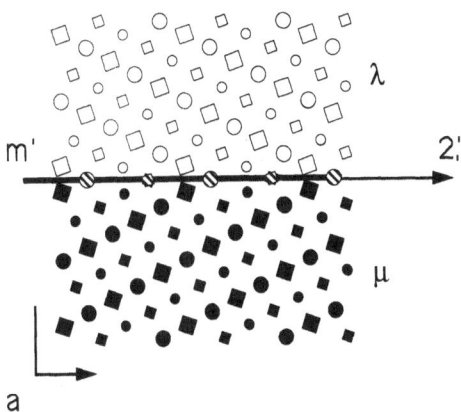

Fig. 9. Schematic cross-sectional view of a bicrystal created from the dichromatic complex shown in fig. 7. The interfacial orientation is [310], and the layer space group of the bicrystal is $p2_1$'am'.

Unrelaxed surfaces

The procedure for establishing the character of all the admissible discontinuities on a given surface, $\mathbf{n}$, is as follows. First, the space group of the infinite crystal, Φ, which comprises the operations $\mathbb{W}_i$, is decomposed with respect to that of the surface group, $\Phi(s)$, comprising the operations $\mathbb{W}_j(s)$, i.e.

$$\Phi = (\Phi(s)) \mathbb{W}_1{}^{se} \ u \ (\Phi(s)) \mathbb{W}_2{}^{se} \ u \ ....(\Phi(s)) \mathbb{W}_q{}^{se} \qquad (5)$$

This decomposition determines the number of variants, q, and the set of (coset-representative) surface exchange operations, $\mathbb{W}_k{}^{se}$, where $\mathbb{W}_1{}^{se}$ is the identity. Each operation in the set $\mathbb{W}_k{}^{se}$ characterises a surface discontinuity, and these can be classified into three principal types as described below. (i) Facet Junctions; these are characterised

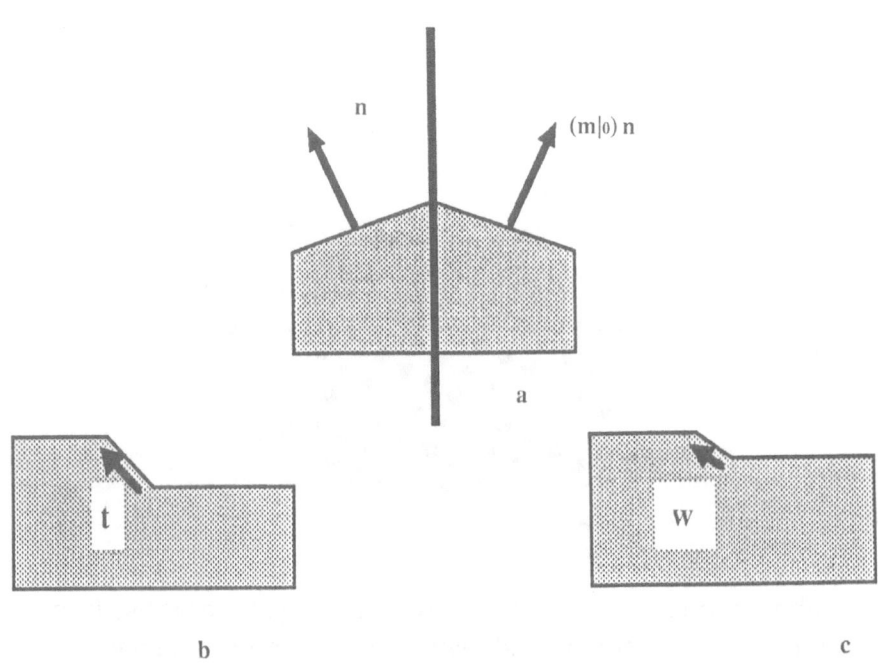

Fig. 10. Schematic illustration of surface discontinuities; (a) facet junction, (b) step, and (c) demistep.

by surface exchange operations $\mathbb{W}_k{}^{se} = (W_k{}^{se}, w_k{}^{se})$ where $W_k{}^{se}$ represents a rotation, roto-inversion, or a mirror operation. Such features separate equivalent surfaces with normals $\mathbf{n}$ and $W_k{}^{se}\mathbf{n}$, as depicted schematically in fig. 10(a).

(ii) Steps; these are characterised by translation operations, $\mathbb{W}_k{}^{se} =$

(I, t_k^{se}). They can arise on surfaces with any orientation, n, and the step height, h, is equal to $n.t_k^{se}$, as depicted in fig. 10(b).

(iii) Demisteps; these can arise only on certain surfaces of nonsymmorphic crystals (i.e. where the space group includes screw-rotation and/or mirror-glide planes). They are characterised by operations $\mathbf{W}_k^{se} = (W_k^{se}, w_k^{se})$ such that $W_k^{se}n = n$, i.e. the orientation of surface in question is left invariant by the point operation, and the step height, h, is equal to $n.w_k^{se}$, as shown in fig. 10(c). An important distinction between steps and demisteps is that the equivalent surfaces separated by the latter are inter-related by a point symmetry operation, W_k^{se}, rather than the identity as in the former case.

In the case of nonholosymmetric crystals, a domain boundary in the bulk of the crystal, characterised by $\mathbf{W}_l^{e}$, may emerge onto the surface. This operation belongs to the space group of the crystal's parent structure, but does not belong to the crystal group, Φ, or the surface group, $\Phi(s)$, and the discontinuity delineating the line of emergence is characterised by $\mathbf{W}_l^{e}$. The surface feature separates equivalent surfaces, and is either a facet junction, step, or demistep depending on the form of $\mathbf{W}^{e}$.

Reconstructed surfaces

Reconstruction at surfaces reduces the symmetry of the surface group, $\Phi(s)$. Therefore, the number of variant surface structures increases, and hence additional discontinuities can arise. Let the space group of the reconstructed surface be $\Phi(rs)$; the additional variants are given by the decomposition

$$\Phi(s) = (\Phi(rs))\mathbf{W}_1^{re} \ u \ (\Phi(rs))\mathbf{W}_2^{re} \ u.........(\Phi(rs))\mathbf{W}_v^{re} \qquad (6)$$

where $\mathbf{W}_i^{re}$ is the ith exchange operation due to surface reconstruction. These exchange operations characterise surface-domain lines, and are referred to as translation-domain lines when surface translation symmetry, is broken, $\mathbf{W}_i^{re} = (I, t_j^{re})$, and rotation- or mirror- domain-lines when point symmetry, $\mathbf{W}_j^{re} = (W_j^{re}, w_j^{re})$ is broken. Facet junctions, steps or demisteps are not intrinsically associated with these features since they correspond to operations in the unrelaxed surface group, Φ(s).

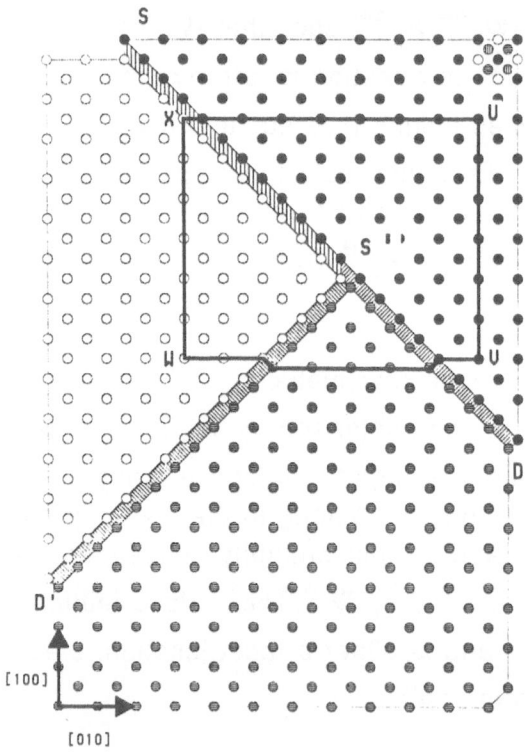

Fig. 11. Schematic plan view of a step, SS', on a (001) surface of Si decomposing into two demisteps, D and D'.

Discontinuities which can potentially arise on the (001) surface of Si are now considered in order to illustrate the treatment developed above. The unrelaxed structure of this surface is shown in fig. 6(a), and its surface space group, Φ(s), is the one-sided layer group p2mm. Decomposing the crystal space group, Fd3̄m, with respect to the surface group, we find that there are twelve variant surfaces related to the initial one by point operations, $\mathbb{w}_q^{se}$, in addition to an infinity of translation operations characterising admissible steps. The twelve variants correspond to a pair of surfaces separated by demisteps on each of the six {001} faces of a cube. Fig.11 is a schematic illustration of a step on the (001) surface characterised by $\mathbb{w}_q^{se} = (I, t_q^{se}) = (I, 1/2[01\bar{1}])$, reacting with two demisteps, characterised by $\mathbb{w}_r^{se} =$

160

$(W_r{}^{se}, w_r{}^{se}) = (4^+, 1/4[1\bar{1}1])$ and $(W_s{}^{se}, w_s{}^{se}) = (4^-, 1/4[1\bar{1}1])$. These latter four-fold screw operations are oriented along [001]; similar operations along [010] and [100] characterise 90^0 facet junctions.

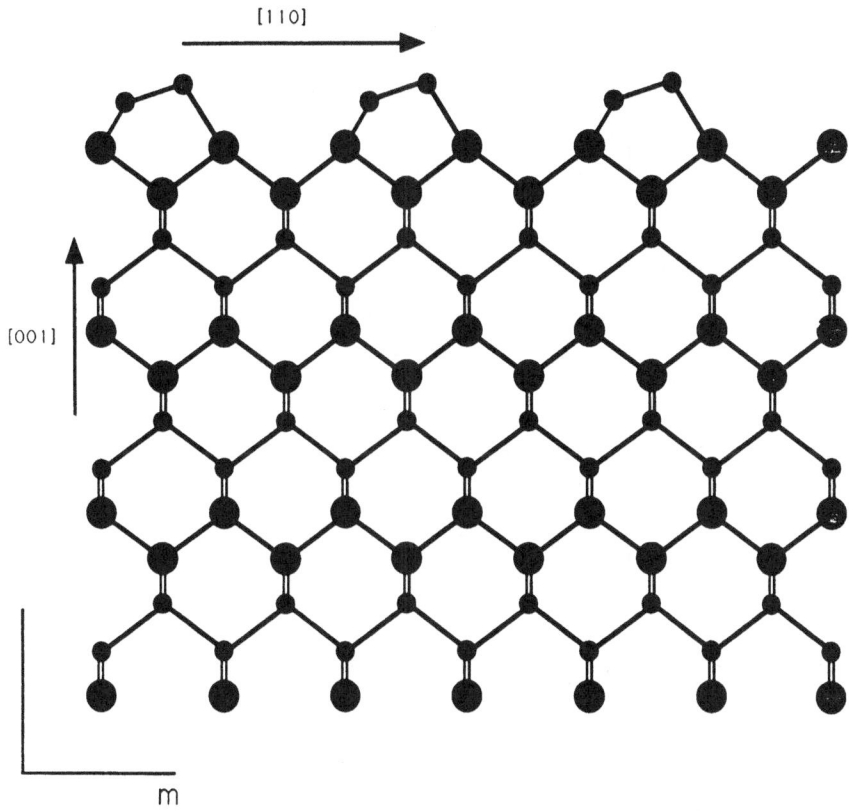

[110]

[001]

m

Fig. 12. Schematic cross-sectional view of the (2x1) reconstructed (001) Si surface, showing the residual symmetry elements.

The (001) Si surface is known to reconstruct, and the modified (2x1) structure is shown schematically in fig. 12, and the new space group is $\Phi(rs) = p1m1$. Both translation and point symmetry operations are broken by this relaxation, and hence become exchange operations, $\mathbb{w}_j{}^{re}$, characterising surface domain lines. Fig.13 is a schematic illustration showing translation-domain lines, characterised by $\mathbb{w}_j{}^{re} = (I, t_j{}^{re}) = (I, 1/2[110])$, and mirror-domain lines, characterised by $\mathbb{w}_k{}^{re} = (W_k{}^{re}, w_k{}^{re}) = (m_{(110)}, 0)$. Steps and demisteps are also shown in the figure, and it is interesting to note how the reconstruction rotates by

161

90^0 on crossing a demistep, but not across a step. Experimental observations of steps, demisteps, and translation-domain lines have been reported in the literature[9-13] , and observations of mirror-domain lines, using scanning-tunneling microscopy at low temperatures, have been reported by private communication to the authors.

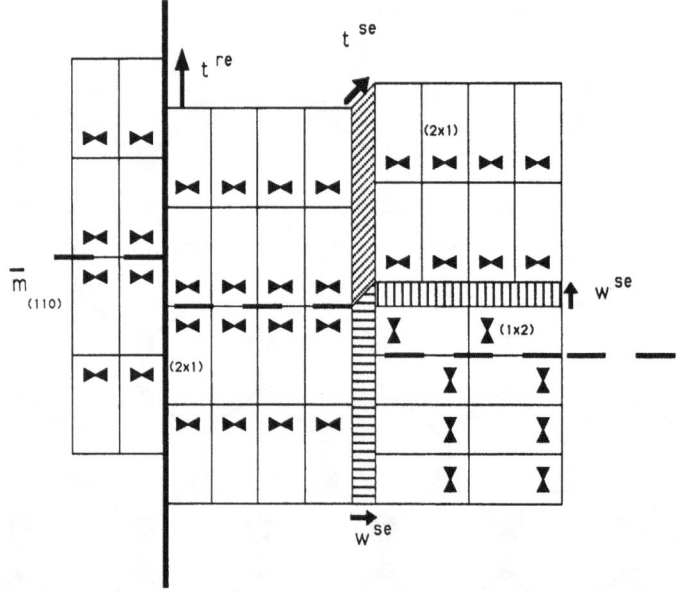

Fig. 13. Schematic plan view illustration of a reconstructed (001) Si surface showing the arrangement of (2x1) cells near a translation-domain line, a mirror-domain line, a step, and demisteps. The "bow-tie" symbols represent the presence of mirror symmetry parallel to the long axis of the (2x1) cell, and its absence parallel to the short axis.

INTERFACIAL DISCONTINUITIES

As might be expected, interfaces can exhibit features similar to both surface and bulk discontinuities, and these are considered separately below. Since detailed accounts of such features have been presented elsewhere[14], we only summarise the admissible possibilities here.

Interfacial Features Resembling Surface Discontinuities

The theoretical method for identifying surface-like discontinuities at interfaces is analogous to that described by expressions (5) and (6) for unreconstructed and reconstructed surfaces respectively. In the former case, it is necessary to decompose the space-group of the dichromatic complex, $\Phi(\lambda\mu)$, with respect to that of the bicrystal in question, $\Phi(b)$, as follows

$$\Phi(\lambda\mu) = (\Phi(b))\mathbb{W}_1{}^{be} \ u \ (\Phi(s))\mathbb{W}_2{}^{be} \ u \ ....(\Phi(b))\mathbb{W}_q{}^{be} \qquad (7)$$

The bicrystal exchange operations, $\mathbb{W}_q{}^{be}$, correspond to coincident or antisymmetry operations, $\mathbb{W}(c)_k$ or $\mathbb{W}'$ respectively, which do not leave the particular bicrystal invariant. These operations therefore characterise interfacial facet junctions, steps, or demisteps as summarised below.

(i) Interfacial facet junctions; these are characterised by rotation, roto-inversion, or mirror operations. The discontinuity separates the interface with normal $\mathbf{n}$ from the crystallographically equivalent one with orientation $\mathbb{W}_p{}^{be}\mathbf{n} = W(c)_k\mathbf{n}$, or $W'_l\mathbf{n}$. An example, derived from the dichromatic pattern illustrated in fig.7, is shown in fig.14.

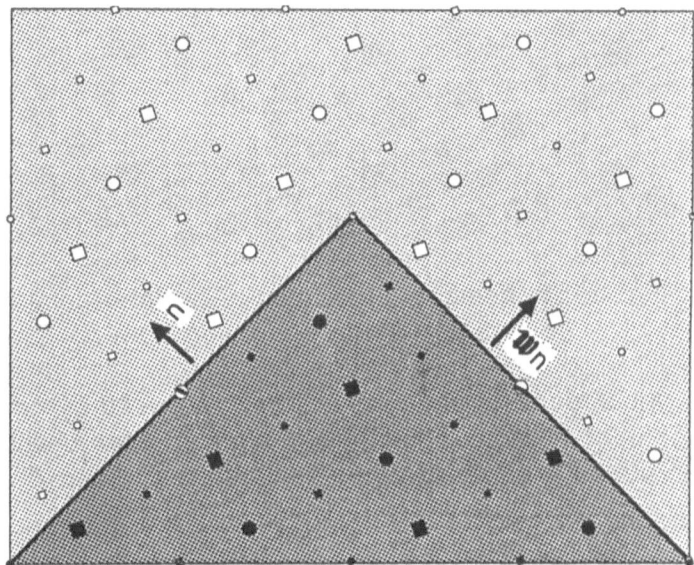

Fig. 14. Schematic illustration of an interfacial facet junction based on the dichromatic complex in fig. 7. The interfaces have {210} orientations and are related by the coincident four-fold screw axis parallel to [001].

(ii) Interfacial steps; these are characterised by coincident translation operations, $\mathbb{w}_n{}^{be} = (\mathbf{I}, \mathbf{t}(c)_k)$, and exhibit heights equal to $\mathbf{n}.\mathbf{t}(c)_k$, and an example is shown in fig.15 (antitranslations cannot arise if the adjacent crystals have different orientations).

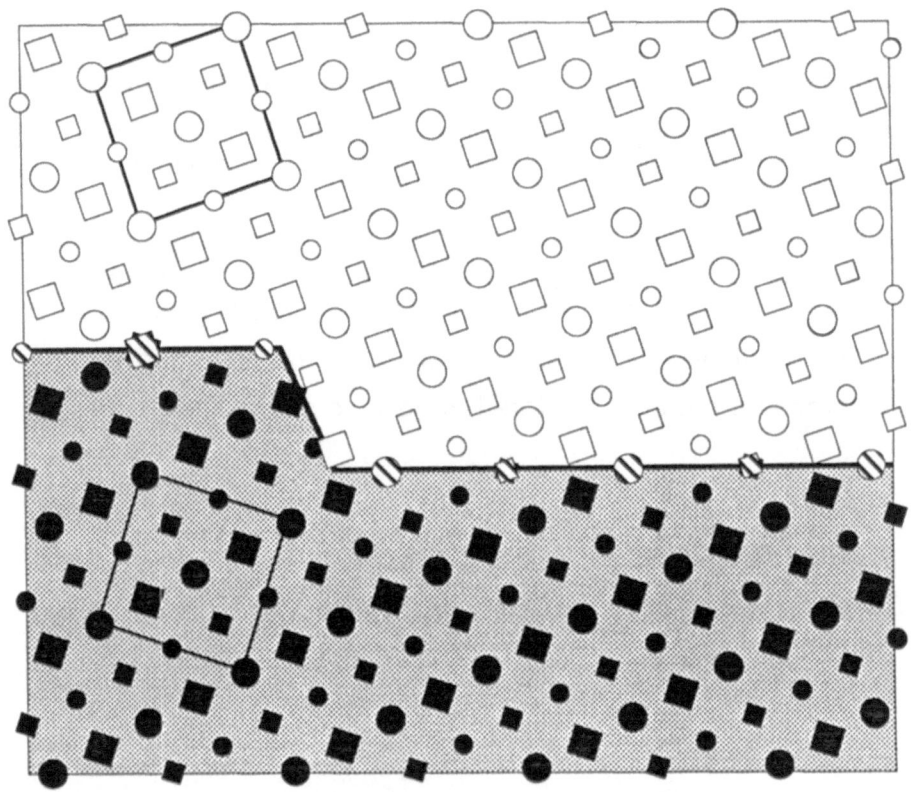

Fig. 15. Schematic illustration of an interfacial step in a (310) interface, derived from fig. 7.

(iii) Interfacial demisteps; these are characterised by nonsymmorphic exchange operations. In general they arise at facet junctions, but can occur on planar interfaces in the special cases where $\mathbf{W}(c)_k\mathbf{n} = \mathbf{n}$, or $\mathbf{W'}_I\mathbf{n} = \mathbf{n}$, and the heights are then equal to $\mathbf{n}.\mathbf{w}(c)_k$ and $\mathbf{w'}_I\mathbf{n}$ respectively. A schematic illustration is shown in fig. 16.

Reconstruction, leading to the breaking of coincident and antisymmetry operations in a bicrystal group, are known to occur at interfaces (we do not include rigid-body displacements, $\mathbf{p}$, in the present discussion since these lead to dislocations as considered later). For example, (211) twin boundaries in Ge, for which the space group of the unrelaxed structure is p2'mm', reconstruct to a centred

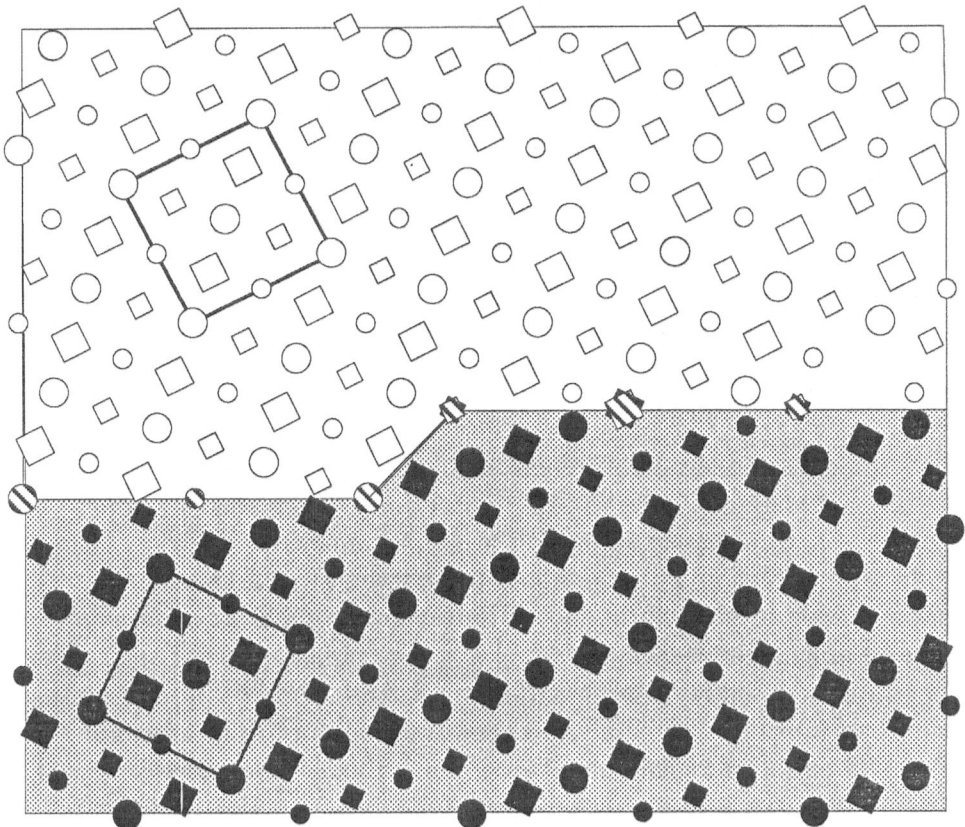

Fig. 16. Schematic illustration of an interfacial demistep in a (210)
interface derived from fig. 7, characterised by the coincident
mirror-glide plane parallel to (001).

2x1 configuration[15,16] , and dimerisation has been detected in $CoSi_2$:Si

interfaces[17] . To find the possible discontinuities resulting from
reconstruction, it is necessary to decompose $\Phi(b)$ with respect to the
group of the reconstructed bicrystal, $\Phi(rb)$, i.e.

$$\Phi(b) = (\Phi(rb))\mathbb{W}_1{}^{re} \ u \ (\Phi(rb))\mathbb{W}_2{}^{re} \ u...........(\Phi(rb))\mathbb{W}_v{}^{re} \qquad (8)$$

The exchange operations are coincident or antisymmetry operations
belonging to $\Phi(b)$ which are broken by the interfacial reconfiguration,
and characterise interfacial domain lines as summarised below.
(i) Translation-domain lines can arise when a coincident translation
operation exhibited by the unrelaxed interface is broken by

reconstruction, and are characterised by an exchange operation, $\mathbb{W}_k{}^{rb} =$

$(\mathbf{I}, \mathbf{t}_k{}^{rb})$. Although such domain lines are characterised by translation operations, it is important to recognise that these discontinuities are not dislocations; they are not introduced into the material by a rigid-body displacement, but by local relaxations. Displacement fields may arise, but will not have the special form characteristic of dislocations. (ii) Rotation-, or mirror-domain-lines can arise if the appropriate coincident or antioperation is broken by the reconstruction (as before, we exclude rigid-body displacements, $\mathbf{p}$, from this consideration). As far as the authors are aware, such symmetry breaking relaxations have not been reported in the literature.

Interfacial defects resembling bulk discontinuities

Since interfaces exist inside bulk materials, it is to be expected that discontinuities analogous to crystal defects, i.e. dislocations, disclinations, and dispirations, can arise in them. Our task now is to obtain an expression which enables the character of admissible discontinuities to be determined. A derivation of such an expression has been presented elsewhere[14]), and is only outlined here. The procedure is to follow a Volterra like approach, finding the operations which can introduce interfacial discontinuities separating variant interfacial structures. This can be conceived in an alternative manner in terms of surface discontinuities as follows. Consider first the bicrystal of interest; this can be regarded as having been created by bringing together the black and white crystal in the appropriate relative orientation and position, represented by $\mathbb{P}$, with the required surfaces having been prepared, and then bonding these together. Now, since variant forms of these surfaces can exist, we imagine a similar process in which the crystal surfaces both exhibit a discontinuity before bonding. Now, we bond together the initial black and white surfaces to regain the starting bicrystal, and find the rigid-body operation necessary to bond together the variant surfaces in such a way, if possible, that a variant interfacial structure arises. It has been shown elsewhere that these operations are given simply by the combination of the symmetry operations characterising the two surface features, $\mathbb{W}(\lambda)_j$ and $\mathbb{W}(\mu)_i$. Let the operations defining admissible discontinuities be designated $\mathbb{Q}_{ij} = (\mathbf{Q}_{ij}, \mathbf{q}_{ij})$; we then have, after re-expressing the inverse of the black operation in the white frame,

$$\mathbb{Q}_{ij} = \mathbb{W}(\lambda)_j \, \mathbb{P} \, \mathbb{W}(\mu)_i{}^{-1} \, \mathbb{P}^{-1} \qquad\qquad (9)$$

We note that $\mathbb{Q}_{ij}$ must be proper, but the component operations, $\mathbb{W}(\lambda)_j$, and $\mathbb{W}(\mu)_i$, can be both proper or improper, but not mixed.

There are two special cases where substitution of crystal symmetry operations into expression (9) leads to discontinuities which have already been discussed in this article. The first special case is when one of the two crystal operations is the identity; for example, if $\mathbb{W}(\mu)_i$ = $(\mathbf{I},0)$, then $\mathbb{Q}_{ij}$ = $\mathbb{W}(\lambda)_j$. In other words, defects which are admissible in the bulk of either of the adjacent crystals are also admissible in any interface. The second special case is when substitution into expression (9) leads to the identity, i.e. $\mathbb{Q}_{ij}$ = $(\mathbf{I},0)$, and hence the associated discontinuities do not exhibit dislocation or disclination character. This arises when the two symmetry operations involved are pairs of coincident operations, i.e. $\mathbb{W}(\lambda)_j = \mathbb{W}(\mu)_i = \mathbb{W}(c)_k$, or antioperations, $\mathbf{W}'$, and the discontinuities correspond to surface like features as described in the previous section. However, in the general case, expression (9) leads to a broad range of defects within the three basic categories of interfacial dislocations, disclinations and dispirations, as summarised below.

(i) Interfacial dislocations; these correspond to the case where $\mathbb{Q}_{ij}$ is a translation, $(\mathbf{I},\mathbf{q}_{ij})$, and the Burgers vector, $\mathbf{b}_{ij}$, is equal to $\mathbf{q}_{ij}$. There are several distinct ways in which this can arise. When the component crystal operations are both translations, the resulting Burgers vector is given by $\mathbf{b}_{ij} = \mathbf{t}(\lambda)_j - \mathbf{P}\mathbf{t}(\mu)_i$, and there is an interfacial "step" associated with the dislocation with height given approximately by h_{ij} = $\mathbf{n}.(\mathbf{t}(\lambda)_j +\mathbf{P}\,\mathbf{t}(\mu)_i)/2$. Many examples of dislocations in this class have been reported in the literature[14]; their crystallographic origin can be regarded as due to the breaking of translation symmetry.

Dislocations can also arise if coincident operations are orientationally aligned, i.e. $\mathbf{W}(\lambda)_j = \mathbf{P}\mathbf{W}(\mu)_i \mathbf{P}^{-1}$, but where the translational alignment is broken due to the relative position of the crystals, $\mathbf{p}$. In this case substitution into expression (9) leads to $\mathbf{b}_{ij}$ = $(\mathbf{W}(c)_k - \mathbf{I})\mathbf{p}$, and h_{ij} = $\mathbf{n}.\mathbf{w}(\lambda)_j$. Experimental observations of such defects in the interface between Al and GaAs have been reported [18]. In this case the maximum possible symmetry of the bicrystal is p2mm, but the actual space group is only p1 due to the relative position of the crystals. The interfacial structures on either side of the defects are related by one of the broken (coincident) (110) mirror planes. No step arises at the core in this case because $\mathbf{n}.\mathbf{w}(c)_k = 0$ in this case.

A third way in which dislocations can arise is when nonsymmorphic operations exist in the two crystals such that $\mathbf{W}(\lambda)_j = \mathbf{P}\mathbf{W}(\mu)_i \mathbf{P}^{-1}$, but where $\mathbf{w}(\lambda)_j$ is not equal to $\mathbf{P}\mathbf{w}(\mu)_i$. In other words the orthogonal parts are coincident but the glide-translations are not.

According to expression (9), these dislocations have $\mathbf{b}_{ij} = \mathbf{w}(\lambda)_j - \mathbf{Pw}(\mu)_i + (\mathbf{W}(\lambda)_j - \mathbf{I})\mathbf{p}$, and $h_{ij} = \mathbf{n}.(\mathbf{w}(\lambda)_j + \mathbf{Pw}(\mu)_i)/2$. An example of such a defect in the (001) interface between $NiSi_2$ and Si has been studied in detail[19]. The white and black operations in this case can be taken as the four-fold rotation and four-fold screw-rotation axes parallel to [001] in the $NiSi_2$ and Si respectively.

The final way in which interfacial dislocations can arise is due to the breaking of antisymmetry, $\mathbb{W'}_k$. Defects in this class do not arise because of incompatible surface features, as envisaged in the derivation of expression (9), but are characterised by a very similar formulation wherein $\mathbb{W}(\lambda)_j$ and $\mathbb{W}(\mu)_i$ are replaced by $\mathbb{W'}_k$. The resulting expression for the Burgers vector is $\mathbf{b} = (\mathbf{W'}_k - \mathbf{I})\mathbf{p}$, and $h = 0$. An example of such a defect in a grain boundary in Al has been studied experimentally[20].

(ii) Interfacial disclinations; for this case $\mathbb{Q}_{ij}$ is a pure rotation. These defects can arise if point symmetry operators are orientationally misaligned in the two crystals. Small angular misalignments can occur, for example, during epitaxial growth on vicinal or rough substrates, and this has been described in more detail elsewhere[21].

(iii) Interfacial dispirations; these are characterised by operations, $\mathbb{Q}_{ij}$, which comprise both rotation, $\mathbf{Q}_{ij}$, and translation, $\mathbf{q}_{ij}$. For a discussion of these defects and an example of an experimental observation, the reader is referred to other work[14].

Finally, we mention that the nature of the line-discontinuity which delineates the intersection of a domain boundary (in a nonholosymmetric crystal) with an interface can be readily included in the present scheme. In this case, the exchange operator characterising the domain boundary is substituted into the expression for $\mathbb{Q}_{ij}$ instead of $\mathbb{W}(\lambda)_j$ or $\mathbb{W}(\mu)_i$, as appropriate.

CONNECTIVITY AND CONSERVATION OF TOPOLOGICAL PROPERTIES

In the foregoing sections we have shown how discontinuities on surfaces, in the bulk, and at interfaces can be characterised in terms of symmetry operations. The object of the present section is to summarise the crystallographic origin of this diversity of discontinuities, and to discuss reactions among the different types. The topological notion of "connectivity" is helpful in regard to the former topic.

Connectivity

In earlier sections we have described a systematic method for identifying the character of surface, bulk, and interfacial discontinuities. This involves the determination of all the variant configurations, and the understanding that discontinuities arise when variants coexist. (Line-defects in bulk crystals can also be regarded in this way, although the "variant" forms are actually identical in this case). The number of variants possibly arising at each stage of the creation of a crystal, surface, or interface, depends on the extent to which symmetry exhibited at the previous stage is broken. Thus, to find all the admissible discontinuities, it is necessary to begin with the most symmetric configuration imaginable, and to identify the decompositions involved at each stage of dissymmetrisation. This is summarised in fig. 17(a), for surface and bulk discontinuities, and in fig.17(b) for surface-like and bulk-like features arising at interfaces. The extensive analogy between the two cases is evident from the diagrams. In the case of grain boundaries, where antisymmetry may be present, additional discontinuities can arise.

The decomposition diagrams illustrated in fig.17 emphasise the crystallographic origin of admissible discontinuities. For example, the set of interfacial defects characterised by $\mathcal{Q}_{ij}$ arises when the symmetry elements of the black and white crystals do not coincide. We refer to this as the extent to which the adjacent crystals are disconnected, a useful concept for analysing the number and nature of the discontinuities in the set $\mathcal{Q}_{ij}$ for any particular interface. It is helpful in the present discussion to consider briefly the two extremes of this condition. Completely connected crystals exhibit the same space group (but not necessarily the same structure) and the correspondence between their coordinate frames, $\mathcal{P}$, is the identity, $(\mathbf{I}, \mathbf{0})$, i.e. the two lattices have identical parameters and orientation. The opposite extreme of completely disconnected crystals means that no symmetry operations at all are coincident in the two crystals. In the former case, i.e., when expression (3) is satisfied for all the black and white symmetry operations, expression (9) becomes

$$\mathcal{Q}_{ij} = \mathcal{W}(\lambda)_j \; \mathcal{P} \; \mathcal{W}(\mu)_i^{-1} \; \mathcal{P}^{-1} = \mathcal{W}(c)_j \; \mathcal{W}(c)_i^{-1} = \mathcal{W}(c)_l \qquad (10)$$

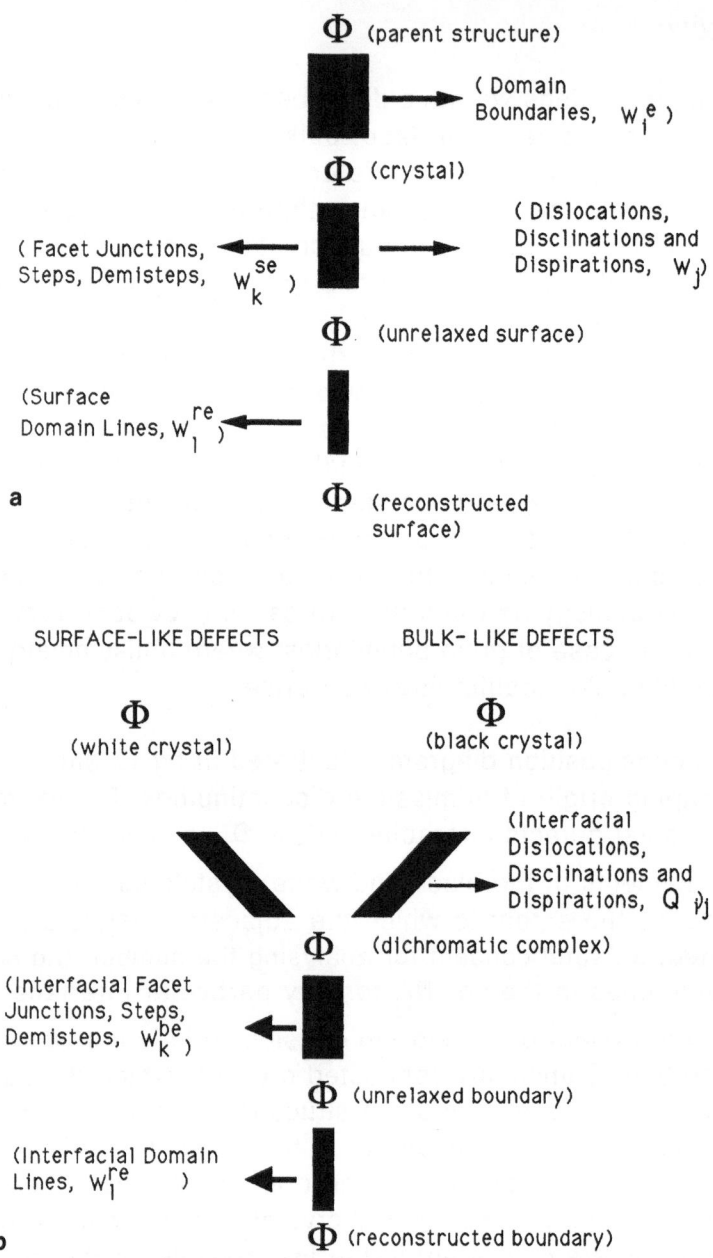

Fig. 17. Decomposition diagrams illustrating the crystallographic origin of various types of defects; (a) shows the dissymmetrisation steps for defects in single crystals and on surfaces, and (b) shows the analogous stages for interfacial defects.

In other words, the discontinuities which can exist in the bicrystal, including those at the interface, are, from the topological point of view, the same as those that can arise in either of the single crystals, irrespective of the interfacial orientation. Now consider departures from complete connection; the greater the extent to which the symmetry of the crystals is no longer coincident, the greater is the extent to which the sets of crystal discontinuities differ, and, according to expression (9), the wider becomes the range of distinct admissible interfacial features. Departures from complete connection arise either if the crystals have distinct space groups, or the correspondence between the two coordinate frames is not the identity (due to differences of lattice parameters, relative orientations or positions).

Conservation of topological properties

The topological properties associated with discontinuities in contiguous materials are conserved. This can be proved formally by constructing closed circuits, and mapping them into perfect crystal, as is outlined in the present section. The first stage is to construct a closed circuit linking the discontinuities of interest. The elementary steps in this circuit must be in "good" material, i.e. they must correspond unambiguously to operations in perfect material. The second stage is to map this circuit onto perfect crystal in order to determine the closure failure, if any. This can be established readily by mathematical means as follows. Each elementary translation and discontinuity in the initial circuit is represented by its characterising operation; if this sequence of operations is equal to the identity, then the circuit is closed in perfect material. Closure implies that the overall content of discontinuities linking the original circuit is zero. For example, consider the surface of Si as depicted in fig.11 where a step SS' is shown decomposing into two demisteps, D and D'. The circuit UVWXU is closed, and encircles the nodal reaction between the step and the demisteps, there being no other discontinuities linked by the circuit. Summing the component operations, taking their line-directions into account, and after cancellation of opposing translation operations, the sequence reduces to the operations characterising the demisteps and the step, i.e.

$$(\mathbf{I}, 1/2[01\bar{1}])\ (4^+, 1/4[1\bar{1}1])\ (4^-, 1/4[1\bar{1}1]) \qquad\qquad (11)$$

which is equal to the identity. In other words, the topological properties of the discontinuities are conserved at the nodal reaction.

Similarly, we can investigate the situation where a surface step, like SS', is connected to a bulk dislocation; the circuit now appears as shown in fig. 18(a). Summing the elementary translations around the circuit in this case leads to $(\mathbf{l},\mathbf{t})$, the operation characterising the step. The fact that this sequence does not equal the identity means that the circuit links a further discontinuity characterised by the inverse of $(\mathbf{l},\mathbf{t})$. In other words, taking the line-senses of the discontinuities into account, the step and dislocation are characterised by the same operation.

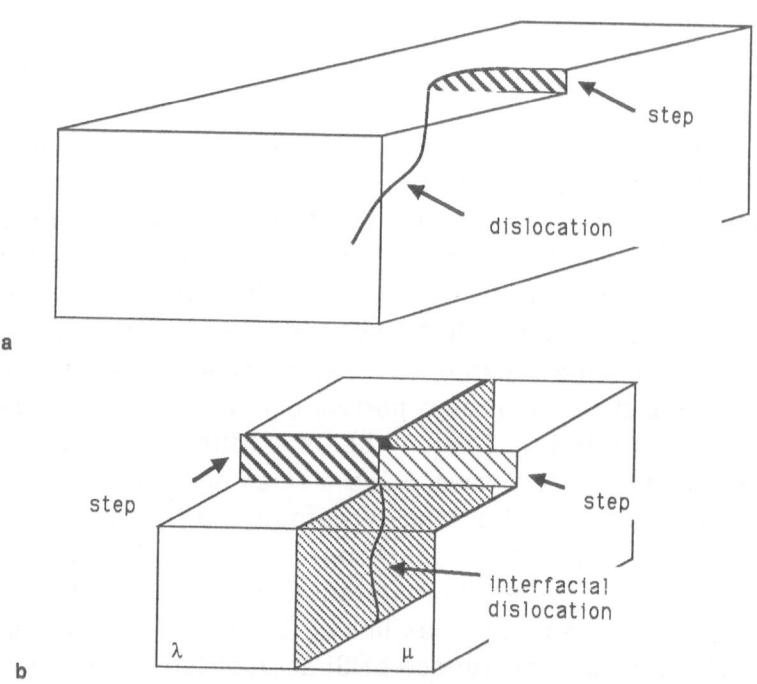

Fig. 18. Schematic illustrations showing, (a) the emergence of a dislocation from the bulk to become a surface step, and (b) the emergence of an interfacial dislocation leading to two surface steps.

The above discussion of a crystal dislocation emerging onto a surface illustrates the more general point that discontinuities which pass from the bulk onto a surface or interface preserve their intrinsic topological character despite the fact that there may be dramatic changes of structure. Thus, dislocations emerging onto surfaces become steps, or possibly surface domain lines on reconstructed surfaces.

There are further possibilities when a crystal dislocation impinges on an interface: (i) it may be transmitted through the interface if it is characterised by a coincident translation, thereby introducing an interfacial step, (ii) it may decompose into interfacial dislocations, or (iii) in a reconstructed interface, it may coincide with a translation-domain line.

The analysis presented above can be extended to more complicated situations such as an interfacial defect emerging onto a surface, as shown in fig. 18(b). Similar reasoning to that above leads to the conclusion that the topological properties are conserved at the node, i.e. the surface features, characterised by $\mathbb{W}(\lambda)_j$ and $\mathbb{W}(\mu)_i^{-1}$, are connected to an interfacial defect characterised by

$$\mathbb{Q}_{ij} = \mathbb{W}(\lambda)_j \; \mathbb{P} \; \mathbb{W}(\mu)_i^{-1} \; \mathbb{P}^{-1}.$$

REFERENCES

1. A. Zangwill, "Physics at Surfaces," Cambridge Univ. Press, Cambridge (1988).
2. T. Hahn, ed., "International Tables for Crystallography," Reidel, Dordrecht (1983).
3. J. Friedel, "Dislocations," Pergamon Press, Oxford (1967).
4. J. P. Hirth and J. Lothe, "Theory of Dislocations," McGraw-Hill, New York (1968).
5. F. R. N. Nabarro, "Theory of Dislocations," Clarendon Press, Oxford (1967).
6. A. V. Shubnikov and V.A. Koptsik, "Symmetry in Science and Art," Plenum Press, NewYork (1977).
7. R. C. Pond and W. Bollmann, The symmetry and interfacial structure of bicrystals, Phil. Trans. Roy. Soc. Lond. 292: 449 (1979).
8. R. C. Pond and D. S. Vlachavas, Bicrystallography, Proc. R.Soc. Lond. 386A: 95 (1983).
9. J. Knall, J. B. Pethica, J. D. Todd, and J. H. Wilson, Strusture of Si(113) determined by scanning tunneling microscopy, Phys. Rev. Lett. 66: 1733 (1991).
10. O. Haase, R. Koch, M. Borbonus, and K.H. Rieder, Role of regular steps on the formation of missing-row reconstructions: oxygen chemisorption on Ni(771), Phys. Rev. Lett. 66: 1725 (1991).
11. L. M. Peng and J. T. Czernuszka, Studies on the etching and annealing behaviour of α-Al_2O_3 ($\overline{1}012$) surfaces by reflection electron microscopy, Surface Science 243: 210 (1991)
12. R. M. Feenstra, A. J. Slavin, G. A. Held, and M. A. Lutz, Surface diffusion and phase transition on the Ge(111) surface studied by scanning tunneling microscopy, Phys. Rev. Lett. 66: 3257 (1991).
13. B. S. Swartzentruber, Y. W. Mo, and M. G. Lagally, Domain boundary control of edge roughness in vicinal Si(001), Appl. Phys. Lett. 58: 822 (1991).

14. R. C. Pond, Interfacial dislocations, in "Dislocations in Solids, 8" F.R.N. Nabarro, ed., North Holland, Amsterdam (1972).

15. A. Bourret, L. Billard and M. Petit, HREM determination of the structure of {211} Σ = 3 in germanium, Inst. Phys. Conf. Ser. 76: 23 (1985)

16. A. Bourret and J. J. Bacmann, Defect structure in the Σ = 27, 11 and (211) Σ = 3 symmetrical grain boundaries in germanium, Proc. J.I.M.I.S. 4, Minakami Spa (1985).

17. D. Loretto, J. M. Gibson, and M. Yalisove, Reconstruction of hetero-interfaces in MBE: $CoSi_2(001)$ on Si(001), Thin Solid Films 184: 309 (1990).

18. C. J. Kiely and D. Cherns, On the atomic structure of the Al-GaAs(100) interface, Phil. Mag. A 59:1 (1989)..

19. R. C. Pond and D. Cherns, Surface structure and the origin of 1/4<111> interfacial dislocations in $NiSi_2$:Si epitaxial films, Surf. Sci. 152: 1197 (1985).

20. R. C. Pond, Periodic grain boundary structures in aluminium. II. A geometrical method for analysing periodic grain boundary structure and some related transmission electron microscope observations, Proc. Roy. Soc. Lond. A 357: 471 (1977).

21. R. C. Pond, Interfacial Defects and Epitaxy, Mat. Res. Soc. Symp. Proc. 56, (1986).

THEORY OF ELECTRON STATES AT SURFACES AND INTERFACES

M. Schlüter

AT&T Bell Laboratories, Murray Hill, NJ 07974

This paper summarizes a series of lectures given on the theory of electron states at surfaces and interfaces. It emphasizes the modern Density Functional approach to ground state electronic structure properties and the quasi-particle approach to electronic excitations. Examples are given to illustrate the state of the art in both areas.

1. Fundamentals of Density Functional Theory

The density functional theory (DFT) represents an attempt to integrate the lattice ion potential V_{ext} with the mutual interaction of electrons. The whole theory is based on the remarkable theorem by Hohenberg and Kohn [1] which states that all properties of the many-particle ground state are given by the ground-state electron *density* distribution. The idea is to replace the system of interacting electrons by a system of *noninteracting* particles which move in a fictitious additional field such that their density remains the same and then solve the noninteracting system.

The energy of the system of *interacting* electrons in an external nuclear potential, written as a functional of density, is then

$$E[\rho] = T_s[\rho] + E_{Coul}[\rho] + \int V_{ext}(r)\rho(r)\,dr \qquad (1)$$
$$+ E_{xc}[\rho] ,$$

where the first three terms describe the energy of the noninteracting system and where E_{xc} accounts for the rest. $T_s[\rho]$ is the kinetic energy of the noninteracting electrons,

$$T_s[\rho] = \sum_i \int \psi_i(r) \left[-\frac{1}{2} \nabla^2 \right] \psi_i(r)\,dr$$

$$\tag{2}$$

and

$$E_{Coul}[\rho] = \frac{1}{2} \int \frac{\rho(r)\rho(r')}{|r-r'|}\,drdr'$$

is the usual Coulomb energy (where the electrons are treated as fixed and uncorrelated,

Equilibrium Structure and Properties of Surfaces and Interfaces
Edited by A. Gonis and G.M. Stocks, Plenum Press, New York, 1992

175

V_{ext} the ionic pseudopotential and $E_{xc}[\rho]$ the exchange correlation energy, which globally contains all interactions due to the Pauli exclusion principle and to the otherwise correlated motion of electrons. E_{xc} also includes the difference between the kinetic energy of an interacting system and $T_s[\rho]$ defined above.

As shown by Kohn and Sham [2], a variational solution of Eq. (1) can be obtained by solving a set of one-electron Schrödinger-type equations self-consistently:

$$[T + V(r)]\,\psi_i(r) = \varepsilon_i\psi_i(r)$$

(3)

with

$$T = -\frac{1}{2}\,\nabla^2$$

and

$$V(r) = \int \frac{\rho(r')}{|r - r'|}\,dr' + \frac{\delta E_{xc}[r]}{\delta\rho(r)} + V_{ext}(r)$$

and

$$\rho(r) = \sum_{occupied\ states} |\psi_i(r)|^2\ .$$

The occupation of the states $\psi_i(r)$ is done according to Fermi statistics with the eigenvalue ε_i serving as the energy for the purpose of determining occupancy of a state. The solution of the many-body problem is thus reduced to the solution of a set of one-particle equations, not more difficult to solve than the Hartree problem. The theoretical difficulty lies in determining the unknown functionals $E_{xc}[\rho]$ and $V_{xc}[\rho] = \delta E_{xc}[\rho]/\delta\rho(r)$. Here, a second important concept yields the framework for practical approximations. In analogy to the Thomas-Fermi method, an approximation (local-density functional or LDA) to E_{xc} and V_{xc} is given by regarding a small neighborhood of the electron system as behaving like jellium, i.e. $\rho = const.$, at the value of the local density. Then $E_{xc}[\rho] \approx \int \rho(r)\,E_{xc}^h[\rho(r)]\,dr$ and well-known approximations for the homogeneous electron gas exchange correlation energies E_{xc}^h can be used. These also include expressions for spin-polarized systems. Discussions on various aspects of density-functional theory can be found in several recent reviews [3-9] and electron gas data can be obtained from Refs. 10-14.

While there is a large body of successful applications of local-density or local-spin-density theory, there are also failures, notably in two areas. First, some ground-state properties of open-shell systems, such as 3d row transition metal atoms, are rather sensitive to the approximate form of exchange and correlation functionals. The presently used functionals sometimes yield wrong ground states. Similarly, some surface properties such as the existence of a long-range image potential are not given by simple LDA theories. The inability of the LDA to describe the stability of negative ions falls into this category as well. Improved functionals, going beyond the local approximation, seem to be necessary and in some of these cases a correct description of the ground state can be obtained by going beyond LDA [15-23].

The second point refers to the Kohn-Sham Eqs. (2). It is general practice, although there is no formal justification, to interpret the eigenvalues of (2) as quasi-particle energies, especially for extended systems where relaxation effects are negligible. While the overall features of experimentally measured quasiparticle spectra (photoemission or inverse photoemission) are quite well reproduced by the ε_i of Eqn. (2), sizeable quantitative errors occur. These questions generally lie outside the framework of Density Functional Theory and shall be discussed later below.

Before ending this chapter on Density Functional Theory, we will briefly quote a few "special" functionals which have been considered for less common circumstances. For finite temperatures, the original Hohenberg, Kohn, Sham theory has been extended formally by Mermin [24]. Relativistic functionals have been discussed by McDonald and Vosco [25] and a formal extension to include current densities and magnetic fields has been given by Vignale and Rasolt [26]. The special situation of electron-positron correlation has been discussed in a two-fluid model by Boronski and Nieminen [27], and superconducting functionals have been given by Oliveira, Gross and Kohn [28]. Finally, an extension to excited states in terms of a frequency dependent functional has been derived by Gross and Kohn [29].

2. Norm Conserving Pseudopotentials

Pseudopotentials were originally introduced to simplify electronic-structure calculations by eliminating the need to include atomic core states and the strong potentials responsible for binding them. Considerable success was achieved in describing the band structure and optical properties of semiconductors and simple metals with the use of empirical pseudopotentials [30]. In this approach, the total effective potential $V(r)$ in Eq. (2) was replaced by just a few adjustable terms in a Fourier expansion.

In a later, more sophisticated approach, a parametrized smooth model potential was used to replace the *ionic* part of $V_{ext}(r)$ in Eq. (2) which originates from the valence electrons only. The ionic parameters were adjusted to fit experimental band energies in a fully self-consistent calculation. The charge density given by the nodeless valence wave functions had no formal connection to the true charge (see Fig. 1) but was treated as the real valence charge in Eq. (3). These semi-empirical model ion potentials were then assumed to be transferable from one chemical environment to another [31,32].

The formal justification for the use of pseudopotentials was initially based upon the orthogonalized-plane-wave (OPW) method [33]. The approach yields nonempirical pseudopotentials [34,35] which are nonlocal in the sense that each angular-momentum component of the valence pseudowavefunction about an atomic center feels a different potential. The wave functions of OPW-type pseudopotentials have, however, a certain problem. The normalized pseudowavefunctions have the correct shape outside core region, but incorrect amplitudes (see Fig. 1). This is because neglect of the so-called orthogonality hole put too much charge inside the core region [30]. The problem is serious in self-consistent calculations, since it will cause errors in the Coulomb potential. The problem can, in principle, be overcome by going back to the original OPW scheme, but the procedure is cumbersome and obviates most of the advantages of using a pseudopotential.

The problem of the orthogonality hole is *not* a necessary consequence of replacing an all-electron potential by a pseudopotential. Below we shall outline a specific and practical (but not unique) approach to formally constructed pseudopotentials with many desirable properties [36].

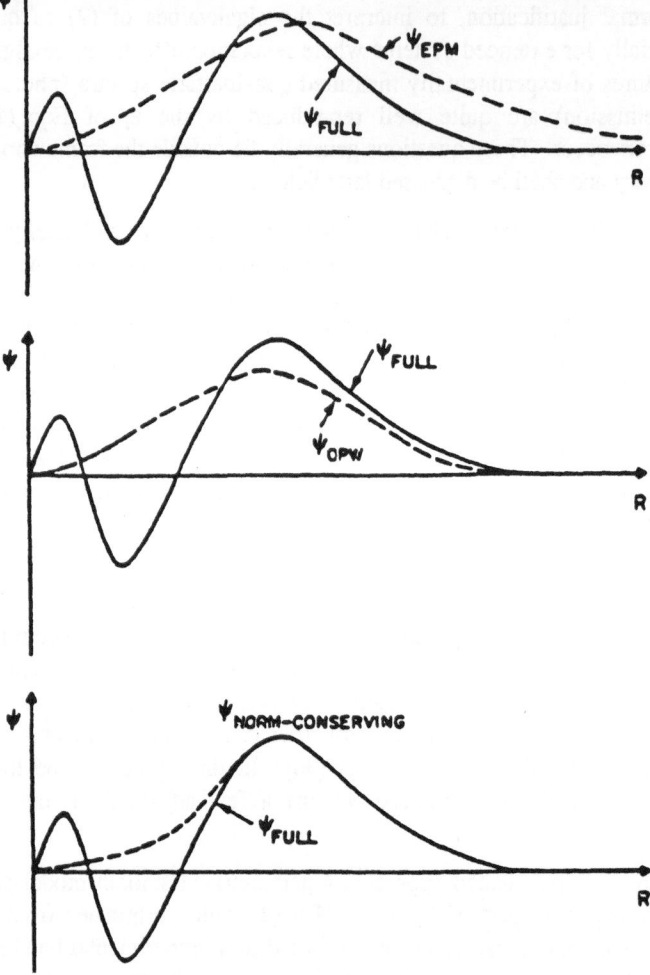

Fig. 1. Schematic view of different types of atomic pseudowavefunctions. An empirical potential wavefunctions (top) is compared to an OPW wavefunction (middle) and a norm-conserving wavefunction (bottom).

We start with a reference atom for which we have a self-consistent local-density calculation, including core states (i.e. solutions for Eqs. (2), (3)), at hand. A pseudowavefunction for any valence state of the atom need have just *two* properties to be consistent with our intentions: it should be nodeless, and it should, *when normalized*, become identical to the true valence wave function beyond some core radius R_c (see Fig. 1). Such a function can be constructed in arbitrarily many ways. For any one pseudowavefunction, the radial Schrödinger equation can be inverted to yield a pseudopotential which has this function as its eigenfunction at the correct eigenvalue. By this construction, it is clear that the pseudopotential and full potential are identical beyond R_c and that inside R_c the pseudopotential correctly simulates the scattering properties of the full potential at the eigenvalue energy and for the particular angular momentum under discussion. To be useful, the ionic (i.e. unscreened) portion of such norm-conserving pseudopotentials must be *transferable* to situations other than the atomic reference state. Such situations are, e.g., molecules, solids or excited atomic configurations. An identity, related to Friedel's sum rule, can be used to show that any *norm-conserving* pseudopotential satisfies an important transferability criterion. The identity is

$$4\pi \int_0^R r^2 \psi^2 \, dr = 2\pi \left[(r\psi)^2 \, \frac{d}{d\varepsilon} \, \frac{d}{dr} \ln \psi \right]_R , \qquad (4)$$

where $\psi(r)$ is the solution of the radial Schrödinger equation of energy ε (which is not necessarily an eigenvalue). The consequence of this identity is that, if two potentials v_1 and v_2, e.g. the fully potential and the pseudopotential, yield solutions ψ_1 and ψ_2 which have the same integrated charge inside R, the linear energy variation around ε of their logarithmic derivatives and thus scattering phase shifts are identical. The requirement on the atomic pseudowavefunction is that it agrees identically with the true wave function for $R > R_c$, when both are normalized, guarantees that the integrated charge is identical for $R > R_c$, and thus that the scattering properties of the pseudopotential and full potential have the same energy variation to linear order when transferred to other systems. A number of procedures have been proposed to construct norm-conserving pseudopotentials [36-42]. We refer the reader to Ref. 37 for more details.

The construction of any pseudopotential is naturally based on the assumption of a frozen core charge. It has been shown that the error in the total energy associated with this approximation is of second order in valence charge density changes [43]. Combined with the norm-conserving property, the pseudopotentials thus do not lead to any first-order transferability error. We shall give some examples below to illustrate this property. Second order effects can still be detrimental, however, and shall be discussed below.

For treating heavier atoms, relativistic effects can be included. Noting that, for valence states in the valence region, the relativistic Dirac equation reduces to the nonrelativistic Schrödinger equation, all relativistic effects on the valence electrons can be lumped into the form of the core pseudopotential [44]. This in turn enables one to treat heavy atoms in a nonrelativistic formalism. Two different pseudopotentials are found for each angular momentum l (i.e. $j = l + \frac{1}{2}$ and $j = l - \frac{1}{2}$). It is useful to define a j-average potential for scalar-relativistic calculations and a j-difference potential if spin-orbit splittings are of interest [45]. A set of such norm-conserving potentials for Au is shown in Fig. 2.

In Table 1 test calculations for a variety of excited atom configurations show the amount of transferability error. The errors are generally small and acceptable. Larger and partially unacceptable second order errors can, however, occur in particular situations. One reason for this is that the exchange and correlation potential $V_{xc}[r]$ used in self-consistent calculations is an inherently nonlinear function of the charge density. The partition into core and valence electrons and the construction of a pseudopotential for a reference atom implicitly linearizes the ρ-dependence of V_{xc} which leads to errors of second order in $\Delta\rho$. However, for e.g., highly spin-polarized situations the nonlinearities can be strong and the second-order errors can grow unacceptably large. A simple trick, first proposed by Louie *et al.*, alleviates the situation [46]. Transferability is essentially perfectly restored if one carries a frozen atomic-core charge ρ_{core} along and adds it to the actual valence charge ρ_{val}, each time V_{xc} is calculated. Table 2 illustrates the small transferability errors obtained this way for spin-polarized Ni [47].

Norm-conserving pseudopotentials have been extensively used in electronic structure calculations. There are several reviews [48-50] of applications to bulk semiconductors, metals, surfaces, defects, molecules and clusters. Undoubtedly, there are many variations in considering and solving the problem and further refinements are evolving. Below we will refer to some recent developments which have been of wide-spread utility.

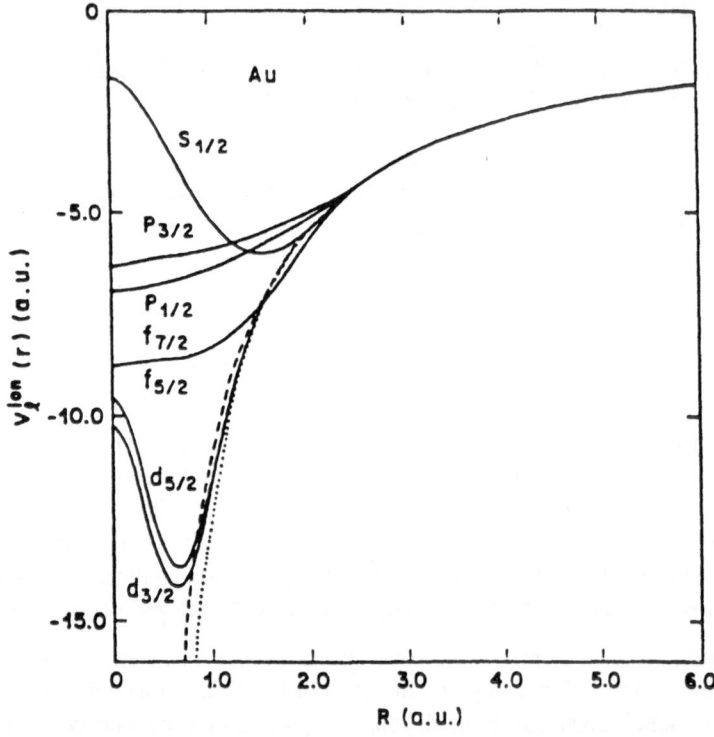

Fig. 2. Norm-conserving ion core pseudopotentials for gold (from ref. 37).

As mentioned, the introduction of a different pseudopotential for each ℓ-value (or j-value) generates a "partial" nonlocality. This potential operating on a wavefunction has to project out its particular ℓ-component. It is not a full projection in three dimensions since it does not act on the radial dependence of the wavefunction. If taken, e.g. between two plane waves (k, k') such partial projection creates a matrix $V(k, k')$ which is not separable into products. It has been pointed out by Kleinman and Bylander [51] that a full 3D projection factorizes $V(k, k')$ and therefore leads to sizeable computational savings. This modification can easily be added to the original norm conserving concept and is widely used today. Attempts to go beyond first-order norm conservation have been proposed by Shirley et al [52], and the use of multiple reference energies has been introduced by Vanderbilt [53]. Smoothness and therefore plane wave convergence of norm conserving pseudopotentials is a major concern and has been discussed in many forms. Most recent attempts to create "optimally" smooth potentials [54,55] yield dramatic convergence improvements. The concept of smoothness has been generalized by Vanderbilt [53] and an extension, formally similar to the augmented plane wave (APW) approach has been given. All these recent developments are based on the flexibility of the pseudopotential approach and have dramatically increased its computational range of usefulness.

Table 1. Results of test calculations of norm-conserving pseudopotentials for a variety of representative atoms. State 1 is reference state for pseudopotential construction, state 2 represent the (de-) excited state. Differences in total energy are given in columns 4 (full-core atom), and 5 (pseudopotential atom). All energies are given in eV. (From Ref. 37.)

| | | | Excitation Energies | |
| | | | Full Core | Pseudo |
Element	State 1	State 2	ΔE_{tot} (eV)	ΔE_{tot} (eV)
C	$s^2 p^2$	sp^3	8.23	8.22
Si	$s^2 p^2$	sp^3	6.79	6.79
Ge	$s^2 p^2$	sp^3	8.01	8.00
Sn	$s^2 p^2$	sp^3	7.02	7.01
Ni	$d^8 s^2$	$d^9 s^1$	−1.66	−1.57
Pd	$d^9 s^1$	$d^{10} s^0$	−1.47	−1.43
Pt	$d^9 s^1$	$d^{10} s^0$	−0.03	−0.003
Sm^{2+}	$f^6 d^0$	$f^5 d^1$	4.32	4.56

Table 2. Transferability test for norm-conserving pseudopotentials for spin-polarized Ni. The effect of including the frozen core charge ρ_c into the nonlinear exchange and correlation potential is evident. All energies in a.u. (From Ref. 47.)

Configuration		All Electron	ρ_c Exact Pseudo	ρ_c Gaussian Model Pseudo	$\rho_c = 0$ Pseudo
$d^8 s^2$ unpolarized	$E =$	−1516.832	−88.711	−48.766	−38.322
$d^8 s^2$, 100% polarized	$\Delta E =$	−0.034	−0.034	−0.035	−0.64
$d^7 s^{0.75} p^{0.25}$ unpolarized	$\Delta E =$	1.393	1.398	1.397	1.395
$d^7 s^{0.75} p^{0.24}$, 100% polarized	$\Delta E =$	1.285	1.290	1.283	1.204

3. Implementation

So far we have discussed the concept of Density Functional Theory to describe an interacting electron system and the concept of pseudopotentials to eliminate the core electron degrees of freedom. With these concepts in place, eqns. 1-3 have to be solved numerically for the surface or interface system under consideration. This involves a series of rather difficult numerical tasks and many attempts (bandstructure methods) have been made to handle these.

We here briefly describe a particular approach to the problem, i.e. the plane wave approach. Originally thought of as a perturbative expansion away from the free-electron limit [56], the plane wave approach today is capable of tackleing many complex situations. It is at the cutting edge of modern developments in the art of quantitative description of electronic properties of real materials. There is a long list of advantages of the plane wave approach, among them chiefly the simplicity for calculating all matrix elements. These are simply Fourier transforms or combinations of such and modern FFT (fast Fourier transform) computer algorithms can be employed. A further advantage lies in the fact that plane waves represent a complete orthogonal set; therefore systematic improvements are easy. There are no symmetry constraints and no spurious effects that can occur as for atom centered basis sets (e.g. so-called Pulay [57] forces). And finally, plane wave expanded wavefunctions lend themselves very conveniently for further use in the calculation of matrix elements of observables. There are, of course, a number of disadvantages associated with this approach as well. First of all, many plane waves are needed to describe the Kohn-Sham wave functions, charge densities and potentials (eqns. 2, 3) for realistic systems. To date, over 100,000 plane waves have been used to describe systems like C_{60} clusters [58]. At times, especially for finite size objects like clusters large numbers of plane waves are needed to describe the correct exponential decay of wavefunctions. Plane waves are nonlocal in real space and an improvement, needed e.g. near a particular atom, translates into an improved description everywhere. The use of a discrete number of plane waves requires a basic periodicity. For a-periodic systems, such as clusters, defects, dislocations, surfaces, etc. this is done by using large supercells [32] into which the object is placed, artificially periodically repeated, yet well separated to avoid spurious interactions. The reference of a plane wave set to a basic periodicity can introduce further spurious effects if this periodicity is changed (so-called Pulay stresses) and convergence effects have to be observed carefully. Nevertheless, because of the aforementioned advantages plane wave basis sets are increasingly popular for electronic structure calculations.

We now briefly discuss some important aspects of solving eqns. 1-3 once the choice of Hamiltonian and basis set has been made. For a system of N_{at} atoms with $N_{el} = N_{el} \cdot N_{at}$ electrons there are $3 N_{at} - 6$ structural and N_{pw} electronic degrees of freedom where N_{pw} plane waves are needed to describe the Kohn-Sham wavefunctions of N_{el} electrons. Traditionally, these degrees of freedom were considered separately in nested self-consistency loops as shown in Fig. 3.

The structural and the electronic optimizations were done in series. If traditional matrix diagonalization procedures were used the asymptotic cost in computer time scaled like $\sim N_{pw}^3$. This diagonalization was done for every structural arrangement, for every stage of charge self-consistency and for every k-point in the Brillouin zone needed to integrate the correct change density. The total energy was evaluated explicitly [59]. Much thought has been given to optimizing [60,61] the k-space integration procedure initiated by the key paper by Baldereschi [62]. Similarity, elaborate acceleration schemes [63-64] have

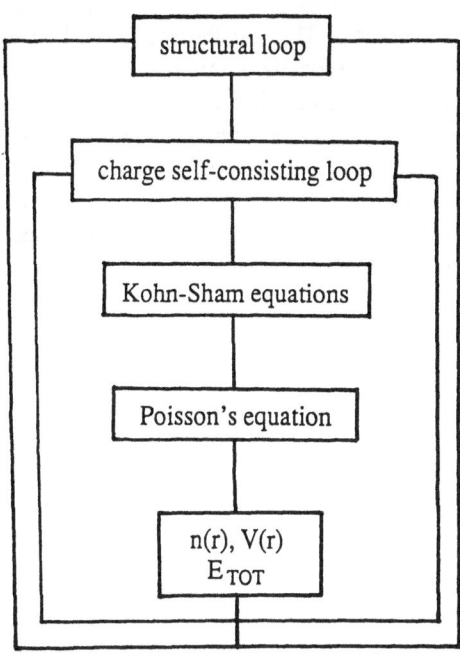

Fig. 3. Schematic diagram for a self-consistent calculation.

been devised to stabilize and speed-up the charge self-consistency feedback ("charge sloshing"). Such instabilities are amplified for situations with large unit cells and/or high density of states near the Fermi level. Nevertheless, systems with $N_{at} \approx 10$ using up to $N_{pw} \approx 2000$ can routinely be studied using the traditional approach. These systems are typically limited to elements with "soft" pseudopotentials, i.e. they exclude first row atoms and third row transition metal atoms.

There have been numerous attempts [65,66] to avoid the N_{pw}^3 computing bottle-neck of completely diagonalizing the plane wave Hamiltonian matrix. Typically, only a few times $N_{el} \ll N_{pw}$ Kohn-Sham wavefunctions are only needed and iterative diagonalization procedures should be used. Moreover, it was argued that full charge self-consistency is not needed for each atomic configuration except for the fully relaxed groundstate. The break-through of combining the optimization of structural and electronic degrees of freedom came with the seminal paper by Car and Parrinello [68] in 1985.

In the quantum molecular dynamics (QMD) method of Car and Parrinello structural and electronic degrees of freedom one treated simultaneously. Basically, the set of unknown plane wave coefficients $a_n(G)$ describing the wavefunctions of N_{el} electrons is treated as classical "particles" with the additional constraint that the resulting N_{el} one-electron wavefunctions be kept orthogonal. This constraint enters the QMD equations in Lagrange multiplier form, together with a fictitious classical electron mass and kinetic energy. At the end of the simulation this classical kinetic has to be zero placing the electrons onto the Born-Oppenheimer surface of atomic motion. The advantages of this type of approach are manifold. Most importantly, real finite (atom)-temperature simulations can be done more efficiently. Secondly, because the limiting orthogonalization steps scale ~ $N_{el}^2 N_{pw}$ (rather than N_{pw}^3) larger numbers of plane waves can be used to describe complex systems. This increases the number of atoms (N_{at}) that can realistically be treated by about one order of magnitude beyond what is possible with traditional diagonalization methods. The Car-Parrinello method employs full electron dynamics. Therefore trapping in secondary minima in *electron* phase space can in principle be avoided. If the fictitious classical electron kinetic energy is neglected, then the method becomes akin to relaxation-type iterative methods. Here steepest descent or conjugate gradient type methods are most powerful [69] and are often used in place of the original Car-Parrinello method. Pros and cons of a variety of these methods are discussed in recent review articles [70,71]. Summarizing these developments it is probably fair to say that in the last five years a "quantum step" has taken place in our technical ability to describe the electronic properties of real materials.

4. Quasiparticle Theory

The DFT, described in chapter 1 gives a framework for the interaction effects in an inhomogeneous system, which in the simple LDA has proven to be quantitatively rather good for many ground state properties. The Kohn-Sham equation, the effective one-particle Schrödinger equation yields wavefunctions which form a quite accurate ground state density. The eigenvalues from the same equation do not, however, yield one-particle excitation energies with the same accuracy. Sham and Kohn already pointed out that the eigenvalues of the DFT one-particle Schrödinger equation (even with the exact DFT potential in place of the LDA potential) do not correspond to the quasi-particle energies, except at the Fermi level. The quasiparticles in a solid are described by a Schrödinger type equation [10]:

$$(T + V_{ext} + V_H) \, \psi_{nk}(\mathbf{r})$$

$$+ \int d\mathbf{r}' \Sigma(\mathbf{r},\mathbf{r}';E_{nk}) \, \psi_{nk}(\mathbf{r}') = E_{nk} \psi_{nk}(\mathbf{r}) \,,$$

(5)

including the kinetic energy, the external potential due to the ion cores, the average electrostatic (Hartree) potential, and the nonlocal, energy dependent and in general complex self-energy operator. The self-energy operator Σ describes the exchange and correlations among the electrons. The real part of the eigenvalue gives the energy of the quasiparticle while the imaginary part corresponds to the lifetime.

Approximations to the self-energy in the context of the LDA have been given already by Sham and Kohn [72]. Later Pickett and Wang [73] have improved upon this idea and calculated quasi-particle energies for realistic semiconductors. Since then, much progress has been made to construct the self-energy with the LDA orbitals but without the LDA on the self-energy itself [74-80].

The physically important approximation is for the self-energy operator, here described by the so-called *GW* approximation [81]:

$$\Sigma(\mathbf{r},\mathbf{r}';E) =$$

$$i \int \frac{dE'}{2\pi} \, e^{-\delta E'} \, \mathbf{G}(\mathbf{r},\mathbf{r}';E-E') W(\mathbf{r},\mathbf{r}';E') \,,$$

(6)

where $\delta = 0^+$. The full interacting Green's function G as well as the dynamically screened Coulomb interaction W enter explicitly. The latter is fully described by the dielectric matrix ε^{-1} which screens v, the bare Coulomb interaction:

$$W(\mathbf{r},\mathbf{r}';E) = \int d\mathbf{r}'' \varepsilon^{-1}(\mathbf{r},\mathbf{r}'';E) v(\mathbf{r}''-\mathbf{r}') \,.$$

(7)

For the Green's function, a quasiparticle approximation is used:

$$G(\mathbf{r},\mathbf{r}';E) = \sum_{n,k} \frac{\psi_{nk}(\mathbf{r}) \psi_{nk}^*(\mathbf{r}')}{E - E_{nk} - i\delta_{nk}} \,.$$

(8)

Here ψ_{nk}, E_{nk} are quasiparticle wavefunctions and energies implicitly defined by Eq. (5), $\delta_{nk} = 0^+$ for occupied states and $\delta_{nk} = 0^-$ for empty states.

The full dynamical dielectric matrix is required. This represents a major difficulty for evaluating the self-energy operator for a semiconductor. Due to the charge inhomogeneity (covalent bonding), the so-called local fields in the screening response are physically very important. These local fields yield the variation in the screened Coulomb interaction around an added electron as it moves from regions of high density, e.g. bonds, to regions of low density, e.g. interstitials. The local fields are described by the off-diagonal elements of the dielectric matrix in a reciprocal space representation, q,G. Various approaches have been applied to obtain the full dynamical response. Hybertsen and Louie [74] have taken the problem in two steps. First, the static dielectric matrix is calculated as a ground state property from the LDA. Second, the dielectric matrix is extended to finite frequency using a generalized plasmon pole model which represents each component q,G,G' of the imaginary part of dielectric matrix by a delta function. No adjustable parameters enter. A variation on this approach has been introduced by von der Linden and Horsch [82] with similar results. Godby, Schlüter and Sham [75] directly calculated the dynamical dielectric matrix without the plasmon pole approximation but only for imaginary frequencies which is appropriate for integrated quantities and allows quasiparticle energies to be extracted only in the near gap region.

Clearly proper treatment of the dielectric response of semiconductors is a difficult and important component in the self-energy approach. Some progress has been made in finding a model for the screening which contains the essential role of local fields. Since the plasmon pole model is found to be adequate (except for lifetimes, of course), attention can be focussed on the static response function. The important part of the local fields for the self-energy problem is the variation in the local screening potential near the site of the added electron as the site is varied. This approximately follows the local charge density, as is intuitively clear: more electrons nearby lead to more complete screening locally. This is exploited in a model introduced by Hybertsen and Louie [83]. The model consists in taking the screening potential around in added electron at r' to be the same as for a homogeneous medium of the local density at r'. The model for the homogeneous medium must include the finite dielectric constant. Thus ε_0 is the single parameter entering the model.

A somewhat different model has been proposed by Gygi and Baldereschi [84]. The model consists in separating the self-energy in the semiconductor into a long-range and short-range part. The latter is taken to be the corresponding expression for the homogeneous electron gas at the local density in the crystal. For the long-range part, the static screened exchange plus Coulomb hole approximation is made using a model (diagonal) dielectric function. In this fashion nonlocality is preserved in the long-range part (also at intermediate length scales) and local field effects are included via the LDA framework. The accuracy of this approximation is about 0.2 eV for a variety of group IV and III-V semiconductors.

Once the screening response has been established, the self-energy must be evaluated and Eq. (5) solved. In the Green's function Eq. 8, the quasiparticle wavefunctions and spectrum are initially approximated by the solutions of the LDA Kohn-Sham equations. Then the spectrum is updated. The LDA wavefunctions are usually adequate. Equation (5) can be solved perturbatively to first order in the difference $\Sigma - V_{xc}$ (V_{xc} is the LDA exchange-correlation potential.) For bulk semiconductors, numerical precision of 0.1 eV can be typically achieved for band gaps and dispersions. The diverse approaches for handling the screening response generally give final quasiparticle energies in good agreement with each other.

When electron-hole interactions (excitons) are neglected, optical properties can be interpreted in terms of an independent quasiparticle picture. Then the quasiparticle band structure calculated as described above can be directly compared to experiment.

Consider first the minimum absorption edge. The LDA typically yields a value too small in comparison to experiment. The quasiparticle gap calculated using the self-energy approach, however, is in excellent agreement with experiment for a wide range of semiconductors and insulators, as illustrated in Table III. Higher energy optical transitions at critical points in the Brillouin zone can also be calculated and be compared to reflectivity data. Results for Si and Ge are shown in Table IV.

Table III. Comparison of the minimum gap in the LDA spectrum to experiment and results of the self-energy approach (QP).

E_g	LDA	QP	Exp^c
diamond	3.9	5.6^a	5.48
Si	0.52	$1.29^{a,b}$	1.17
Ge	<0	0.75^a	0.744
LiCl	6.0	9.1^a	9.4
AlAs	1.18	1.99^b	2.14
GaAs	0.56	1.48^b	1.52

^{a}Ref. 74 ^{b}Refs. 75,76
^{c}Ref. 85

The energies of quasiparticles as probed by photoemission or inverse photoemission are of course directly related to the energies calculated here. For semiconductors, generally very good agreement is found with the GW calculations [47-76].

The essential physics behind the *GW* self-energy approach to quasiparticle energies in semiconductors is three-fold. First, the self-energy operator $\Sigma(\mathbf{r},\mathbf{r}';E)$ is nonlocal, i.e. the "potential" seen by a quasiparticle at site $\mathbf{r}$ depends on its own shape at $\mathbf{r}'$. Most importantly, the range of this nonlocality encompasses typical variations in crystal Bloch functions of order interatomic distances or longer. The self-energy operator thus, for example, strongly discriminates between bonding (valence band) and anti-bonding (conduction band) states. Second, the screened Coulomb interaction which enters Σ includes the strong local field effects described above. Third, dynamical renormalization effects in Σ are retained. The picture of quasi particles renormalized by interaction with plasmons implicit in the plasmon pole model is accurate provided local field effects are

Table IV. Comparison of the calculated direct transition energies from Ref. 74 to critical point energies derived from optical spectroscopy. The results for Ge include the spin-orbit interaction.

	Theory	Expt.[a]	
Si			
$\Gamma_{25'v} \to \Gamma_{2c}$	4.08	4.2	E_0
$\Gamma_{25'v} \to \Gamma_{15c}$	3.35	3.4	E_0'
$L_{3'v} \to L_{1c}$	3.54	3.45	E_1
$L_{3'v} \to L_{3c}$	5.51	5.50	E_1'
$X_{4v} \to X_{1c}$	4.43	4.44, 4.60	E_2
Ge			
$\Gamma_{8v} \to \Gamma_{7c}$	0.71	0.887	E_0
$\Gamma_{7v} \to \Gamma_{7c}$	1.01	1.184	$E_0 + \Delta_0$
$\Gamma_{8v} \to \Gamma_{6c}$	3.04	3.006	E_0'
$\Gamma_{8v} \to \Gamma_{8c}$	3.26	3.206	$E_0' + \Delta_0'$
$L_{4,5v} \to L_{6c}$	2.18	2.25	E_1
$L_{4,5v} \to L_{4,5c}$	5.76	5.576	E_1'
$X_{5v} \to X_{5c}$	4.45	4.501	E_2

[a]Ref. 85

taken into account. Including all these effects quantitatively in the *GW* approximation yields excellent results for semiconductors and insulators. Higher order diagrams (vertex corrections) seem to be less important. However, direct calculations of these terms with similar accuracy have yet to be done.

The next step in an *ab initio* theory of the optical response of semiconductors is to examine the role of excitonic interactions in the dynamical optical response. Other topics related to optical properties which are of interest include the behavior near the metal-insulator transition.

There is a formal resemblance of Dyson's equation, which gives the quasiparticle energies, to the Schrödinger equation for the effective-one-electron eigenvalues of DFT: each is a Schrödinger equation for fictitious, noninteracting electrons moving in a effective potential (although in one case this "potential", the self-energy, is nonlocal and energy dependent). This does not mean in general that there is a relation between the quasiparticle energies E_i and the DFT eigenvalues $E_{i,DFT}$.

However, the highest occupied density functional eigenenergy does represent the chemical potential in the conductor case and the valence band edge in the insulator or semiconductor case. The band gap of an insulator or a semiconductor can be defined precisely in terms of the ground state energy as a function $E_{tot}^{(M)}$ of the number of particles M. If the insulating ground state has N particles, the conduction band edge is the change of the total ground state energy when an electron is added and the valence band edge is given by the change when an electron is removed:

$$E_c \equiv E_{tot}^{(N+1)} - E_{tot}^{(N)} \qquad (9)$$

$$E_v \equiv E_{tot}^{(N)} - E_{tot}^{(N-1)} .$$

The band gap is naturally the difference:

$$E_g = E_c - E_v . \qquad (10)$$

One can show [86] with the help of the variational theorem that the highest DFT eigenvalue is given by the energy to remove an electron:

$$E_{M,DFT}^{(M)} = E_{tot}^{(M)} - E_{tot}^{(M-1)} . \qquad (11)$$

The Hohenberg-Kohn theorem is implicitly valid for a fixed number of electrons, M. The density function equation and the exchange-correlation potential in it, V_{xc}, are implicitly defined as functions of M. Since $M = N$ gives the insulating ground state, when an electron is removed, the highest occupied state is not changed much (only to $O(1/N)$) and $V_{xc}^{(N-1)}$ is the same as $V_{xc}^{(N)}$ but that when an electron is added across the gap, the $(N + 1)th$ state is very different and there is a discontinuity [87,88] in $V_{xc}^{(N)}$ as N is changes to $N + 1$:

$$V_{xc}^{(N+1)} = V_{xc}^{(N)} + \Delta , \qquad (12)$$

where Δ is independent of position since the electron density is changed only to $O(1/N)$. It follows that the difference between the true gap, Eq. (10) and the density functional gap given by

$$E_{g,DFT} = E_{N+1,DFT}^{(N)} - E_{N,DFT}^{(N)} \qquad (13)$$

is just the potential discontinuity, Δ:

$$E_g = E_{g,DFT} + \Delta . \qquad (14)$$

The situation is depicted diagrammatically in Fig. 4.

Δ has been shown to be a significant fraction of the energy gap for a one-dimensional model semiconductor within a two-plane wave basis set [86,89], and also for real semiconductors [75]. This is important. If Δ is small, exact DFT will give accurate band structures, and attention should be paid to going beyond the LDA in density-functional theory; if Δ is large, no attempt to add nonlocal-density corrections to the LDA will improve the calculated DFT band structures, and one must either go outside DFT, or take the discontinuity in V_{xc} into account explicitly as a correction to the DFT band structure when quasiparticle energies are needed. Thus Δ lies at the heart of the relationship between the quasiparticles and DFT for semiconductors and insulators. In a periodic solid, where the momentum $\mathbf{k}$ is a good quantum number for both the DFT and quasiparticle wave functions, the rigorous extension $\Delta(\mathbf{k})$ is possible to all the minimum direct band gaps throughout the Billouin zone with energies below the Auger threshold.

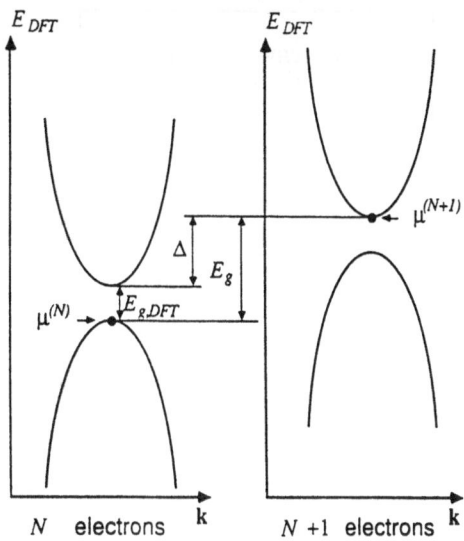

Figure 4. Illustration of the significance of Δ, the discontinuity in V_{xc}. The exact DFT Kohn-Sham one-electron energies are shown in the form of a band structure for the N and $(N+1)$ particle system. The two differ in a uniform increase of the eigenvalues by Δ, as explained in the text. The quasiparticle gap E_g is the difference between the two eigenvalues indicated: $E_g = E_{N+1,DFT}^{(N+1)} - E_{N,DFT}^{(N)}$. It is evident that $E_g = E_{g,DFT} + \Delta$.

A practical method to calculate an (approximate, but non-LDA) exchange correlation potential $V_{xc}(\mathbf{r})$ and therefore Δ is by way of solving Dyson's equation first for a given (approximate e.g. GW) self-energy $\Sigma(r,r',\omega)$ and then by requesting that both the DFT potential V_{xc} and the self-energy Σ produce the same total electron density. This leads to the integral equation, implicitly defining V_{xc} for a known Σ:

$$\text{Im} \int_{-\infty}^{\mu} [G_{DFT}(\Sigma - V_{xc})\, G]_{r,r}\, d\omega = 0 , \qquad (15)$$

where the one particle Green's function G_{DFT} contains V_{xc} and G contains Σ. The different states of such calculations are illustrated in a flow diagram in Fig. 5.

The discontinuity Δ in V_{xc} upon addition of an electron to the N-electron system may be obtained by subtracting the calculated DET minimum gap from the calculated GW minimum quasiparticle gap. Δ is shown in Table V and graphically in Fig. 6. Although Δ increases with the band gap, it is a remarkably constant proportion of the error in the LDA band gap, showing that even diamond is sufficiently locally free-electron-like for the LDA to be a good approximation to the true DFT potential. The fact that Δ is in each case a substantial proportion of the band gap shows that attempts to calculate improved quasiparticle energies within DFT by going beyond the LDA would bring no significant improvement.

It is interesting to compare this V_{xc} with previous attempts to go beyond the LDA. The most successful such potentials that have stayed within the general framework of

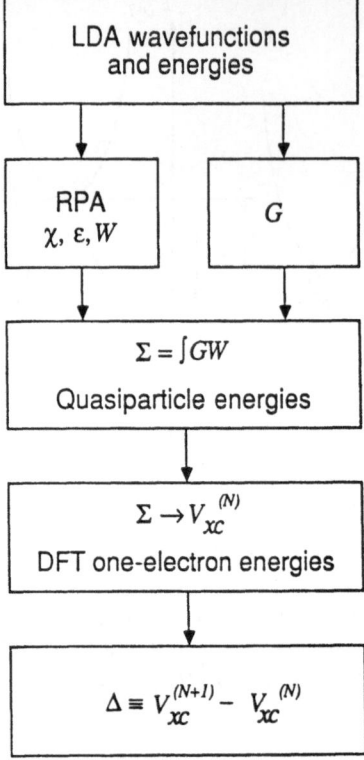

Fig. 5. Flow diagram of the calculation, starting from the self-consistent LDA pseudopotential calculation, forming the screened Coulomb interaction W and the Green's function G, combining them to form Σ and then using Σ to obtain the quasiparticle energies E_i and also the true exchange-correlation potential V_{xc}.

DFT have been the weighted-density approximation [15] (WDA), the average-density approximation [15] (ADA), and gradient-expansion techniques [16] (but only when explicitly designed to obey the important sum rule on the exchange-correlation hole). In each case the Kohn-Sham eigenvalues changed. The general trend is the same with each of the three non-LDA V_{xc}'s: the band gaps are increased by between 0.05 and 0.34 eV, so that the 55% error in the LDA minimum gap in silicon, for example, is reduced to 26-43%. A substantial correction therefore remains.

Table V. Δ, the discontinuity in the exchange-correlation potential; E_g, the calculated minimum quasiparticle band gap; and the LDA gap error $E_g - E_{g,LDA}$ for the four materials (from Ref. 75).

Material	Δ (eV)	E_g (eV)	LDA gap error (eV)	Δ(LDA gap error) (%)
Si	0.58	1.24	0.72	81
GaAs	0.67	1.58	0.91	74
AlAs	0.65	2.18	0.81	80
Diamond	1.12	5.33	1.43	78

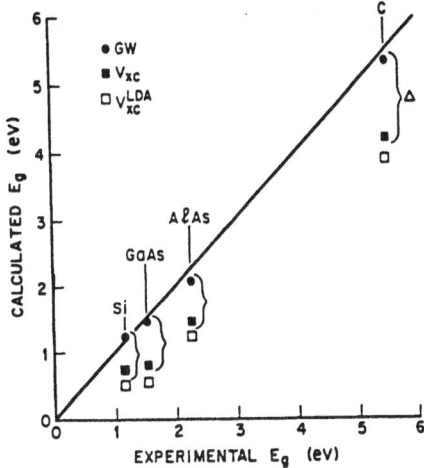

Fig. 6. The calculated minimum band gap (i) the *GW* approximation, (ii) DFT, and (iii) the LDA plotted against the experimental band gap. The 45° line is a guide to the eye. Δ, the discontinuity in the exchange-correlation potential, is indicated.

5. Ordered Buried Boron (2×1) Layers in Silicon(100)

We now turn to some recently discovered semiconductor surface/interface problems which has been studied using some of the techniques described in the previous chapters. It has been found in semiconductors that the symmetry of certain ordered adatom reconstructions can be preserved when epitaxial overlayers are grown at low temperatures [90]. Consequently, the overgrowth results in a "δ-doped" buried layer of impurity atoms with a high degree of lateral translational order. We discuss here the system of boron on Si(100) which has recently been prepared in the laboratory. Low-T ($\sim 300\,°C$) Si overgrowth over B on a Si(100), in contrast to Si(111) [91], surface preserves not only the original boron order, but also results in a crystalline instead of an amorphous Si overlayer. This system is thus a good candidate for producing electrically-active ordered doping layers in crystalline silicon with the potential for high carrier mobility [91,92]. *Ab initio* total energy calculations [93] have been performed of substitutional boron both in the Si(100)-(2 × 1) surface and in a (100)-(2 × 1) buried layer using methods described in previous chapters and in Ref. 94.

We first discuss the results of calculations on the Si(100) surface, which forms a (2 × 1) "dimer" reconstruction [95]. When boron is deposited on this surface, a new reconstruction occurs with a similar (2 × 1) periodicity [90]. This pattern is found to be stable with 1/2 monolayer boron coverage and becomes increasingly disordered at higher coverages. Using this information, one can conduct a limited search, in order to understand how an ordered, buried δ-doped layer might be formed, for the equilibrium substitutional configuration of the boron (2 × 1) covered Si(100) surface. It is found that dimer formation remains favorable. However, some sites are much lower in energy than others. Figure 7 shows a side view of the Si(100) surface with the sites of a substitutional boron atom labelled A-D. The sub-surface position B was found to be 0.11, 0.65, and 0.40 eV lower in energy than sites, A, C, and D, respectively. Boron thus occupies a subsurface position on Si(100), in analogy to Si(111) [96]. No experimental data are available yet for the boron position on the Si(100) surface. Since position A is only slightly higher in energy, it is likely that an experimentally prepared surface (30 sec anneal at 450 °C) contains some disordered mixture of A and B positions. Nevertheless, high quality δ-doped buried layers are more likely to be grown because the lowest energy position for the substitutional boron atom is subsurface in contrast to a site on the surface.

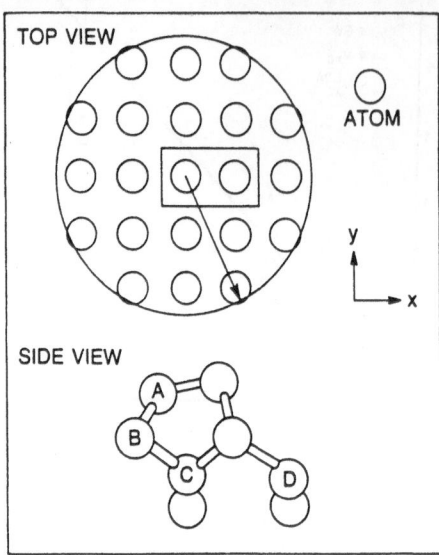

Fig. 7. Sketch of the structural arrangement on Si(100) (2 × 1). The top figure shows a (2 × 1) unit cell with a boron acceptor Bohr-radius superimposed. The bottom figure shows a side view of the surface with individual atomic sites for boron substitution labelled A-D.

We now discuss the buried boron layer, beginning with the geometric structure. For the calculations, a perfect substitutional (2 × 1) pattern was assumed for a half monolayer of boron buried in crystalline silicon, with both atomic and volumetric relaxation. Consequently, there is one boron atom per unit cell. This model of the buried layer is reasonable for the following reasons. It has been found from transmission electron diffraction and from grazing incidence x-ray diffraction that an epitaxial layer of Si can be grown at 300°C on this surface without qualitative change of the (2 × 1) pattern [92]. Segregation studies using Auger spectroscopy indicate that at least 50% of the boron remains in the original ordered layer. Finally, channeling studies suggest substitutional boron sites [92]. The most important geometric result is a large contraction of the interlayer spacing in the z-anis direction. Because boron has a covalent radius 0.29 angstroms smaller than that of silicon, one expects a sizable reduction in volume/atom near boron layer. However, the (2 × 1) symmetry constrains the x-axis and y-axis interlayer spacings to remain at their bulk values but allows the z-axis spacings to change. The DFT calculations give a rigid body displacement of the Si layers, upon crossing the boron layer, of $\Delta z = -0.35$ Å, with the distortion confined to two to three layers on each side. Beating patterns seen in grazing incidence x-ray scattering [97] can be fit with $\Delta z = -0.45 \pm 0.1$ Å, with the distortion extending over fewer than 4 layers on either side which is in excellent agreement with the theoretical predictions.

Finally, we discuss the electronic structure of the boron buried layers. Each boron atom in the δ-doped layer adds one hole to the valence band. A top view of this layer is shown in Figure 7. The large circle shows the extent of the Bohr radius of the hole wavefunction in the effective mass approximation. Because of the large overlap between neighboring boron atoms, one expects a strong interaction between their electronic states. Figure 8 shows the z-axis profile $(x - y$ integrated) of the hole distribution which extends over $\sim 7\text{Å}$ (FWHM). This result is in rough agreement with the (9.4 Å) width of an isolated boron effective mass impurity.

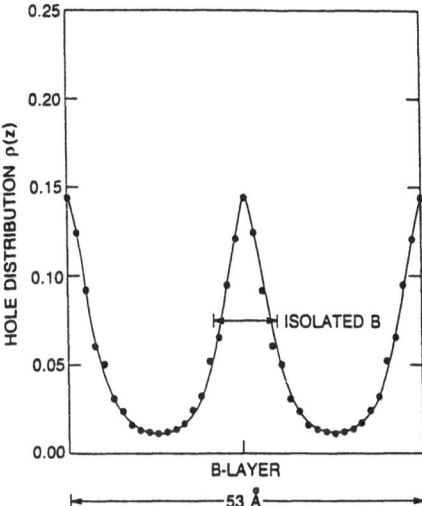

Fig. 8. Calculated hole distribution profile along the (100) direction, integrated over the (2×1) unit cell. The width at half maximum is comparable to that of an isolated boron impurity.

Because the wavefunctions of the (2×1) ordered boron atoms overlap strongly in the $x - y$ plane, a deep level 2-D impurity bandstructure results instead of a shallow impurity level. Furthermore, the anisotropic crystal field $(x \neq y \neq z)$ splits the p-like states into three bands with different masses. The bandstructure is shown in Figure 9, where the Fermi-energy is the zero of energy. As usual, the projected Si bulk bandstructure is superimposed onto the impurity bandstructure. The Fermi-level lies slightly above the valence band maximum, consistent with the experimental finding that 100% of the boron dopants are electrically active [92]. It is instructive to further analyze the impurity bandstructure. Figure 10 shows charge density contour plots of p holes in the individual three impurity bands (at $k = 0$). The contours indicate p-like functions, strongly localized at the boron atoms. At $k = 0$ the crystal field places the p_z orbital above the p_x orbital (along the short (2×1) direction) and the p_y orbital, which are nearly degenerate. As expected, the dispersion of the planar p orbitals is stronger along the orbital direction (σ) then perpendicular to it (π). Because of the strong orbital overlap this dispersion difference is approximately independent of short or long (2×1) direction. The bandstructure in Figure 9 is coded to reflect the orbital character and band crossings. Keeping in mind that this is an LDA bandstructure, there is some uncertainty in the position of the gap states with respect to the band edges, as we discussed at length in chapter 4.

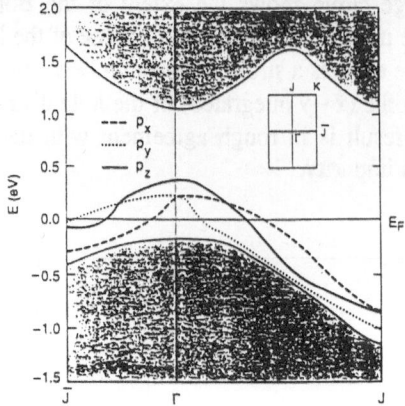

Fig. 9. Two-dimensional bandstructure of Si(100) (2 × 1) B. The boron induced impurity bands are coded according to their p_x (dashed line), p_y (dotted line), or p_z (full line) character. Overlayered is the projected Si bold band-structure, with the bulk conduction band rigidly shifted so the gap matches the experimental value (see chapter 4).

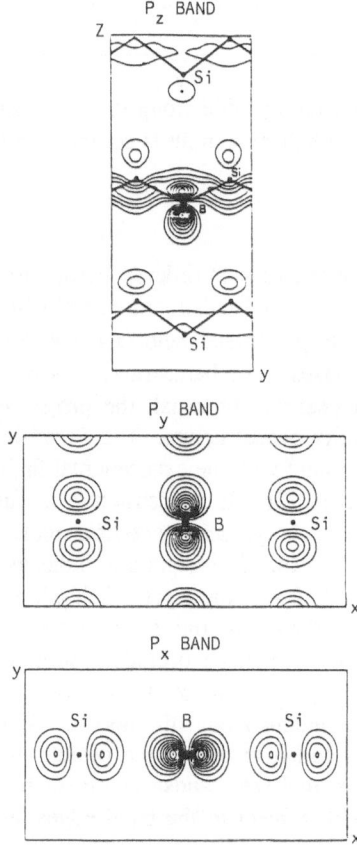

Fig. 10. Charge density contour plots of hole carrier distributions arising from the three impurity bands at $k = 0$ in Figure 9.

As seen from Figure 9 the Fermi-level crossed all three impurity bands giving rise to three separate sheets of the Fermi surface with varying orbital character. There exists the possibility for a series of optical interband transitions in the 0.5-1.0 eV range. Carriers are all hole like and strongly confined (within ~ 4Å) to the 2D dopant layer. Their effective masses in the x,y plane are very anisotropic. Along the dispersive σ-directions the p_x, p_y masses are comparable to the Si light hole mass, while along the π-direction they are very heavy. The p_z mass is more isotropic and similar to the Si heavy hole mass.

The next step is to investigate how much residual boron disorder will limit electron mobilities and ultimately how this disorder can be controlled. The confined 2D nature of the hole wavefunction strongly samples this disorder. More knowledge about the disorder will help to understand details of the impurity scattering, and hopefully suggest avenues to increase mobilities that could lead to new improved semiconductor devices.

6. Surface State Spectroscopy of Si(111)(2 × 1)

We shall discuss in this chapter a case for which both accurate Density Functional calculations and Quasiparticle calculations were needed to understand the correlation between structure and electronic excitations. The system is the Si(111) (2 × 1) surface whose reconstruction pattern is obtained upon cleavage; it is metastable and converts into the stable (7 × 7) pattern after heating above 600°C. Much detailed work has been done to understand the "simple" (2 × 1) reconstruction pattern. The atomic structure has been studied by low energy electron diffraction [98], medium energy ion backscattering [99] and scanning tunneling spectroscopy [100]. Most studies address the electronic structure, among those are direct [101-103] and inverse photoemission [104], differential reflectivity [105], electron energy loss spectroscopy [106,107], photothermal deflection spectroscopy [108], excite and probe photoemission [109], and scanning tunneling spectroscopy [110]. All these studied basically support the "buckled chain model", proposed in 1981 by K. Pandey [110]. A perspective view of this reconstruction is shown in Fig. 11 where the ideal, unreconstructed surface (top) is compared to the (2 × 1) chain geometry (bottom).

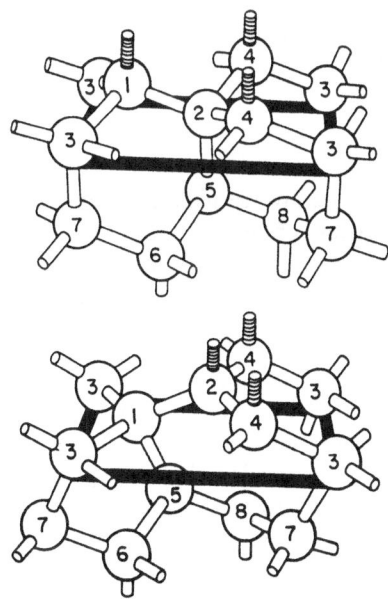

Fig. 11. Perspective view of the atomic structure of the ideal (top) and reconstructed Si(111) (2 × 1) surface (bottom).

The reconstruction can be formed simply by depressing atom (1) and moving the bond between atom pairs (2,5) to atom pairs (1,5). Several DFT total energy studies have been done [112-114] to obtain the atomic equilibrium structure of this reconstruction. More recent calculations [113,114] agree well with the experimentally derived structural parameters [98,99]. In particular it is found that the surface dimer (atoms 2,4 in Fig. 11) buckles with a large (~0.5Å) amplitude. This buckling is driven by a stabilizing net electron charge transfer from the "down" atoms to the "up" atoms within the chain. An accurate knowledge of this buckling distortion is of importance since it influences the electronic surface structure. The electronic structure of Si(111) (2 × 1) is dominated by a pair of highly dispersive π (bonding) and π^* (antibonding) bands arising from the dangling bonds of the chain atoms (2,4 in Fig. 11). As can be seen from Fig. 12, the bands disperse along the chains ($\bar{\Gamma} - \bar{J}$) while they are relatively flat perpendicular to them ($\bar{J} - \bar{K}$).

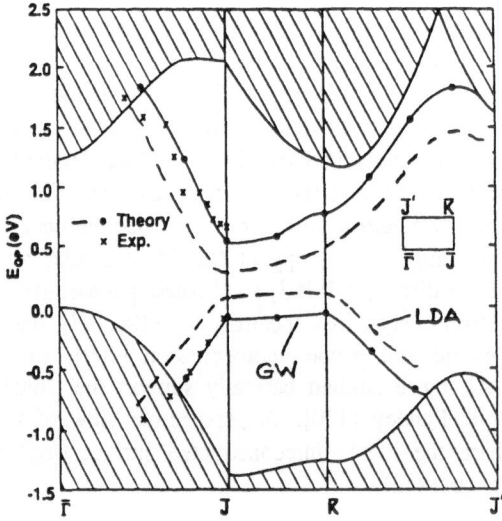

Fig. 12. Calculated surface bandstructure for the (2 × 1) chain model of Si(111). The dashed line is the LDA result while the solid line is the GW quasiparticle result of Ref. 113.

From photoemission [103] and inverse photoemission [104] it is possible to infer a band gap between these states of $E_g^{exp} \approx 0.65 \pm 0.1\ eV$ (at room temperature). Converged LDA calculations give a π/π^* "bandgap" at $\bar{J}$ of $\varepsilon_g \approx 0.25\ eV$; another example of the DFT bandgap under estimate, discussed in Chapter 4.

Northrup, Hybertsen and Louie [113] have recently done a quantitative GW quasiparticle calculation for these electron states. As expected they find a large increase in the gap value to $E_g^{QP} \approx 0.62\ eV$ which is in quantitative agreement with the photoemission data. Interestingly, this case of a surface bandstructure with a subgap energy scale fits beautifully into the trends of bulk energy level corrections. This can be seen in Fig. 13 where the energy correction $\Delta E = E^{QP} - E^{LDA}$ is plotted vs E^{LDA}. The surface states, having both valence and conduction band admixtures need corrections which interpolate the bulk values.

Photoemission experiments measure single quasiparticle distributions. In an optical excitation experiment, quasiparticle interactions (or excitonic effects) can take place. This is indeed found here: low T optical absorption [105] shows an onset for quasiparticle pair creation near 0.47 eV which can be interpreted as $E_g^{exp} - \Delta E$ where $\Delta E \approx 0.2\ eV$ is an exciton binding energy. Large excitonic effects had indeed been proposed [115] for surface states. Northrup, Hybertsen and Louie [113] have taken this idea a step further and actually calculated the optical absorption cross section from the quasiparticle band structure including a parametrized excitonic interaction. Figure 14 shows their results in comparison to the experimental differential reflectivity [105]. The main points to note are the shift in oscillator strength to lower energies and the sharpening up of the spectrum, a clear signature of excitonic effects.

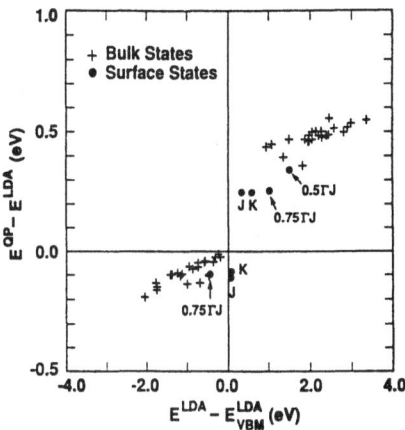

Fig. 13. Quasiparticle self-energy corrections (with respect to LDA energies) of bandstates in silicon. The results for the Si(111) (2 × 1) surface states are included (Ref. 113).

An interesting feature of this excitation spectrum is its temperature dependence. The specific changes, measured by optical reflectivity [105] and electron energy loss [106] are a red-shift and a broadening of the peak with increasing temperature. This behavior is consistent with weak coupling, "large" polarons, as predicted by Chen $et\ al.$ [116] for the π-bonded chain model. A detailed analysis requires the coupling of a rather low energy (~ 10 meV) dispersionless phonon. A mode of this kind has been observed in inelastic He-atom scattering [117], but is not predicted by simple vibrational models for the π-bonded surface [118]. To explain the existence of such mode one has precisely to invoke coupling to the electronic excitation degree's of freedom. This has been done in a model calculation by Alerhand and Mele [119]. In addition to the usual short range forces between surface atoms, a long range polarization term is added to the dynamic phonon matrix which represents the renormalization of phonons by coupling to electronic transitions from π to π^*. For the specific case it has been found that this coupling results in a significant mode-softening effect for vibrations with wave vectors Q along the zone edge, giving rise to the rather dispersionless mode around 10 meV. This type of mode

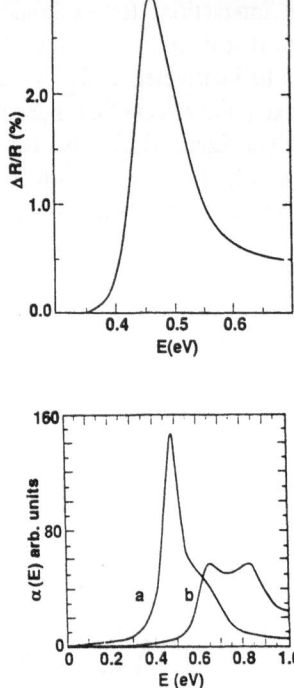

Fig. 14. Comparison between calculated optical cross sections for non-interacting quasiparticles (b) and pairs including excitonic interactions (bottom). The experimental reflectivity spectrum is shown in the upper figure (Ref. 105).

softening is known [120] for transition metals and compounds where this type of electron phonon coupling is also seen as the main driving force for superconductivity. The existence of this mode has also recently been confirmed by direct Car-Parrinello type molecular dynamics simulations [114] of the (2 × 1) surface.

In conclusion, the Si(111) (2 × 1) surface is a system of considerable complexity that has been studied both experimentally and theoretically in much detail. In particular, a variety of modern quantitative state-of-the-art approaches, described in this lecture, have been used successfully to explain rather subtle physical phenomena.

Acknowledgements

I like to thank my collaborators and colleagues at Bell Laboratories and elsewhere for many stimulating interactions. Much of the work described in these lectures is due to them.

References

1. P. Hohenberg and W. Kohn, Phys. Rev. B136, 864 (1964).

2. W. Kohn and L. J. Sham, Phys. Rev. A140, 1133 (1965).

3. *Theory of the Inhomogeneous Electron Gas*, eds. S. Lundvist and N. H. March, Plenum press, New York, 1983.

4. *Local Density Approximations in Quantum Chemistry and Solid State Physics*, eds. J. P. Dahl and J. Avery, Plenum Press, New York, 1984.

5. *Density Functional Methods in Physics*, eds. R. M. Dreizler and J. Da Providencia, Plenum press, New York, 1985.

6. O. Gunnarsson and R. O. Jones, Physica Scripta, **21**, 394 (1980).

7. M. Schlüter and L. J. Sham, Physics Today **2**, 36 (1982).

8. "Density Functional Theory of Atoms and Molecules", by R. G. Parr and W. Yang, Oxford University Press, New York, 1989.

9. R. O. Jones and O. Gunnarsson, Rev. Mod. Phys. **61**, 689 (1989).

10. K. Singwi, A. Sjolander, M. P. Tosi and R. H. Land: Phys. **B1**, 1044 (1970).

11. O. Gunnarsson and B. I. Lundquist, Phys. Rev. **B14**, 4274 (1976).

12. U. von Barth and L. Hedin, J. Phys. **C5**, 1629 (1972).

13. E. P. Wigner, Phys. Rev. **42**, 1002 (1934).

14. D. M. Ceperley and B. J. Alder, Phys. Rev. Lett. **45**, 566 (1980).

15. O. Gunnarsson, M. Jonson and B. I. Lundquist, Phys. Rev. **B20**, 3136 (1979).

16. D. C. Langreth and M. J. Mehl, Phys. Rev. Lett. **47**, 446 (1981).

17. L. Fritsche and H. Gollisch, Z. Phys. **B48**, 209 (1982).

18. M. S. Hybertsen and S. G. Louie, Phys. Rev. **B30**, 5777 (1984).

19. W. Kohn, L. J. Sham, Phys. Rev. **140**, A1133 (1965).

20. J. P. Perdew and Y. Wang, Phys. Rev. **B33**, 8800 (1986).

21. A. D. Becke, Phys. Rev. **A38**, 3098 (1988).

22. J. P. Perdew, A. Zunger, Phys. Rev. **B23**, 5075 (1981).

23. A. Svane, O. Gunnarsson, Phys. Rev. **B37**, 9919 (1988).

24. D. Mermin, Phys. Rev. **137**, A1441 (1965).

25. A. McDonald, S. Vosko, J. Phys. **C12**, 2977 (1979).

26. G. Vignale, M. Rasolt, Phys. Rev. **B37**, 10685 (1988).

27. E. Boronski, R. M. Nieminen, Phys. Rev. **B34**, 3820 (1986).

28. L. Oliveira, H. Gross, W. Kohn, Phys. Rev. Lett. **60**, 2430 (1988).

29. H. Gross, W. Kohn, Phys. Rev. Lett. **55**, 2850 (1985).

30. For a review, see M. L. Cohen and V. Heine, *Solid State Phys.* **24**, 38; (1970) V. Heine and D. Weaire, *Solid State Phys.* **24**, 249 (1970).

31. J. A. Appelbaum and D. R. Hamann, Phys. Rev. **B8**, 777 (1973).

32. M. Schlüter, J. R. Chelikowsky, S. G. Louie and M. L. Cohen, Phys. Rev. Lett. **34**, 1385 (1975).

33. C. Herring, Phys. Rev. **52**, 1169 (1940).

34. J. C. Phillips and L. Kleinman, Phys. Rev. **116**, 287 (1959).

35. F. Bassani and V. Celli, J. Phys. Chem. Solids **20**, 64 (1961).

36. D. R. Hamann, M. Schlüter and C. Chiang, Phys. Rev. Lett. **34**, 1494 (1979).

37. G. B. Bachelet, D. R. Hamann and M. Schlüter, Phys. Rev. **B26**, 4199 (1982).

38. C. Topp and J. J. Hopfield, Phys. Rev. **B7**, 1295 (1974).

39. A. Redondo, W. A. Goddard III and T. C. McGill, Phys. Rev. **B15**, 5038 (1977).

40. A. Zunger, J. Vac. Sci. Technol. **16**, 1337 (1979).

41. P. A. Christiansen, Y. S. Lee and K. S. Pitzer, J. Chem. Phys. **71**, 4445 (1979).

42. G. P. Kerker, J. Phys. **C13**, L189 (1980).

43. U. VonBarth and C. D. Gelatt, Phys. Rev. **B21**, 2222 (1980).

44. L. Kleinman, Phys. Rev. **B21**, 2630 (1980).

45. G. B. Bachelet and M. Schlüter, Phys. Rev. **B25**, 2103 (1982).

46. S. G. Louie, S. Froyen and M. L. Cohen, Phys. Rev. **B26**, 1738 (1982).

47. H. S. Greenside and M. Schlüter, Phys. Rev. **B28**, 535 (1983).

48. M. L. Cohen in "Highlights of Condensed Matter Theory", Proc. 89th Intl. School of Physics, "Enrico Fermi", North Holland, New York, 1985, p. 16.

49. S. G. Louie in *Electronic Structure, Dynamics and Quantum Structure Properties of Condensed Matter*, eds. J. Devreese and P. E. vanCamp, Plenum Press, New York, 1985, p. 335.

50. J. R. Chelikowsky, M. L. Cohen, "Handbook of Semiconductors" ed. Landsberg, Elsevier, New York, 1992, in print.

51. L. Kleinman, D. M. Bylander, Phys. Rev. Lett. **48**, 1425 (1982).

52. E. Shirley, D. Allen, R. Martin, J. D. Joannopoulous, Phys. Rev. **B40**, 3652 (1989).

53. D. Vanderbilt, Phys. Rev. **B41**, 7892 (1990).

54. A. M. Rappe, K. Rabe, E. Kaxiras, J. D. Joannopoulous, Phys. Rev. **B41**, 1227 (1990).

55. N. Troullier, J. L. Martins, Phys. Rev. **B43**, 1993 (1991).

56. See ref. 30.

57. P. Pulay, Mol. Phys. **17**, 197 (1969).

58. N. Troullier, J. L. Martins, to be published.

59. J. Ihm, A. Zunger, M. L. Cohen, J. Phys. **C12**, 4409 (1979), Erratum, J. Phys. **C13**, 3095 (1979).

60. D. J. Chadi, M. L. Cohen, Phys. Rev. **B8**, 5747 (1973).

61. H. Monkhorst, J. D. Pak, Phys. Rev. **B13**, 5189 (1976).

62. A. Baldereschi, Phys. Rev. **B7**, 5212 (1973).

63. C. G. Broyden, Math. Comp. **19**, 577 (1965).

64. D. G. Anderson, J. Assoc. Comp. Mach. **12**, 547 (1965).

65. E. R. Davidson, J. Comput. Phys. **17**, 87 (1975).

66. D. M. Wood, A. Zunger, J. Phys. A**18**, 1343 (1985).

67. J. L. Martins, M. L. Cohen, Phys. Rev. B**37**, 6134 (1988).

68. R. Car, M. Parrinello, Phys. Rev. Lett. **55**, 2471 (1985).

69. M. Teter, M. Payne, D. Allen, Phys. Rev. B**40**, 12255 (1989).

70. M. C. Payne, M. P. Teter, D. C. Allen, J. D. Joannopoulos, to be published.

71. D. K. Remler, P. A. Madden, Mol. Phys. **70**, 921 (1990).

72. L. J. Sham, W. Kohn, Phys. Rev. **145**, 561 (1966).

73. W. E. Pickett, C. S. Wang, Phys. Rev. B**30**, 4719 (1984).

74. M. S. Hybertsen and S. G. Louie, Phys. Rev. Lett. **55**, 1418 (1985); Phys. Rev. B**34**, 5390 (1986).

75. R. W. Godby, M. Schlüter and L. J. Sham, Phys. Rev. Lett. **56**, 2415 (1986); Phys. Rev. B**37**, 10159 (1988).

76. R. W. Godby, M. Schlüter and L. J. Sham, Phys. Rev. B**35**, 4170 (1987).

77. M. S. Hybertsen and S. G. Louie, Phys. Rev. Lett. **58**, 1551 (1987); Phys. Rev. B**38**, 4033 (1988).

78. M. S. Hybertsen and M. Schlüter, Phys. Rev. B**36**, 9683 (1987).

79. S. B. Zhang, et al., Solid State Comm. **66**, 585 (1988).

80. J. Northrup, M. S. Hybertsen, S. G. Louie, Phys. Rev. Lett. **66**, 500 (1991).

81. L. Hedin and S. Lundquist, Solid State Phys. **23**, 1 (1969).

82. W. von der Linden and P. Horsch, Phys. Rev. B**37**, 8351 (1988).

83. M. S. Hybertsen and S. G. Louie, Phys. Rev. B**37**, 2733 (1988).

84. F. Gygi and A. Baldereschi, Phys. Rev. Lett. **63**, 2160 (1989).

85. Landolt-Bornstein, Vol. III, Springer, New York, p. 17 (1982).

86. L. J. Sham, M. Schlüter, Phys. Rev. B**32**, 3883 (1985).

87. J. P. Perdew, M. Levy, Phys. Rev. Lett. **51**, 1884 (1983).

88. L. J. Sham, M. Schlüter, Phys. Rev. Lett. **51**, 1888 (1983).

89. M. Lannoo, M. Schlüter, L. J. Sham, Phys. Rev. B**32**, 3890 (1985)

90. R. L. Headrick, L. C. Feldman, and I. K. Robinson, Appl. Phys. Lett. **55**, 442 (1989).

91. R. L. Headrick, B. E. Weir, A. F. J. Levi, B. Freer, J. Bevk, and L. C. Feldman, J. Vac. Sci. Tech. A**9**, 2269 (1991).

92. R. L. Headrick, B. E. Weir, A. F. J. Levi, D. J. Eaglesham, and L. C. Feldman Appl. Phys. Lett. **47**, 2779 (1990).

93. M. Needels, M. S. Hybertsen, M. Schlüter, to be published, Proc. Intl. Conf. Defects Semicoind. Bethlehem, 1991.

94. M. Needels, J. D. Joannopoulous, Y. Bar-Yam, and S. T. Pantelides, Phys. Rev. B **43**, 4208 (1991).

95. R. E. Schlier and H. E. Farnsworth, J. Chem. Phys. **30**, 917 (1959).

96. R. L. Headrick, I. K. Robinson, E. Vleig, and L. C. Feldman, Phys. Rev. Lett. **63** 1253 (1989); P. Bedrossian, R. D. Meade, K. Mortensen, D. M. Chen, J. A. Golovchenko, and D. Vanderbilt, *ibid.*, 1257 (1989); I. W. Lyo, E. Kaxiras, and Ph. Avouris, *ibid.*, 1261 (1989).

97. B. E. Weir, R. L. Headrick, and L. C. Feldman, to be published.

98. F. J. Himpsel, P. M. Marcus, R. M. Tromp, I. P. Batra, M. Cook, F. Jona, and H. Liu, Phys. Rev. **B30**, 2257 (1984).

99. R. M. Tromp, L. Smit, and J. F. van der Veen, Phys. Rev. Lett. **51**, 1672 (1983).

100. R. M. Feenstra, W. A. Thompson, A. P. Fein, Phys. Rev. Lett. **56**, 608 (1986).

101. F. J. Himpsel, P. Heimann, and D. E. Eastman, Phys. Rev. **B24**, 2003 (1981).

102. F. Houzay, G. M. Guichar, R. Pinchaux, and Y. Petroff, J. Vac. Sci. Technol. **18**, 860 (1981).

103. R. I. G. Uhrberg, G. V. Hansson, J. M. Nicholls, and S. A. Flodstrom, Phys. Rev. Lett. **48**, 1031 (1982).

104. P. Perfetti, J. M. Nicholls, and B. Reihl, Phys. Rev. **B36**, 6160 (1987); A. Cricenti, S. Selci, K. O. Magnusson, and B. Reihl, Phys. Rev. **B41**, 12908 (1990).

105. F. Ciccacci, S. Selci, G. Chiarotti, and P. Chiaradia, Phys. Rev. Lett. **56**, 2411 (1986).

106. N. J. DiNardo, J. E. Demuth, W. A. Thompson, and Ph. Avouris, Phys. Rev. **B31**, 4077 (1985).

107. R. Matz, H. Luth, and A. Ritz, Solid State Commun. **46**, 343 (1983).

108. M. A. Olmstead and N. M. Amer, Phys. Rev. Lett. **52**, 1148 (1984).

109. J. Bokor, R. Storz, R. R. Freeman, and P. H. Bucksbaum, Phys. Rev. Lett. **57**, 881 (1986).

110. J. A. Stroscio, R. M. Feenstra, and A. P. Fein, Phys. Rev. Lett., **57**, 2579 (1986).

111. K. C. Pandey, Phys. Rev. Lett. **47**, 1913 (1981); **49**, 223 (1982).

112. J. E. Northrup and M. L. Cohen, Phys. Rev. Lett. **49**, 1349 (1982); J. Vac. Sci. Technol. **21**, 333 (1982); Phys. Rev. **B27**, 6553 (1983).

113. J. E. Northrup, M. S. Hybertsen and S. G. Louie, Phys. Rev. Lett. **66**, 500 (1991).

114. J. Ancilotto, W. Andreoni, A. Selloni, R. Car, and M. Parrinello, Phys. Rev. Lett. **65**, 3148 (1990).

115. R. DelSole and A. Selloni, Phys. Rev. **B30**, 883 (1984).

116. C. D. Chen, A. Selloni and E. Tosatti, Phys. Rev. **B30**, 7067 (1984).

117. J. P. Toennies, N. Harten and C. R. Woell, Phys. Rev. Lett., **57**, 2947 (1986).

118. O. L. Alerhand, D. C. Allan and E. J. Mele, Phys. Rev. Lett. **55**, 2700 (1985).

119. O. L. Alerhand and E. J. Mele, Phys. Rev. Lett. **59**, 657 (1987).

120. W. Weber and C. M. Varma, Phys. Rev. **B19**, 6142 (1979).

EMBEDDING FOR SURFACES AND INTERFACES

J.E. Inglesfield

Theoretical Physics I, Catholic University of Nijmegen
Toernooiveld, NL-6525 ED Nijmegen

1 Introduction

Surfaces and interfaces are an important and interesting feature of the physical world: as
they are the two-dimensional junction between three-dimensional systems, the electronic
and atomic structure of surfaces and interfaces shows novel properties such as localized
states. Surfaces are of great technological importance in corrosion and catalysis, and
metal-semiconductor and semiconductor-semiconductor interfaces are important in electronics; friction occurs when two surfaces rub - is this a surface or an interface property?
We need no further justification for trying to solve the Schrödinger equation for these
systems, for the electronic properties determine all the other properties. In this article
I shall discuss the general problem of solving the Schrödinger equation in this type of
problem, characterized by an "interesting" surface or interface region joined on to the
bulk, and I shall try to illustrate some of the physics of these systems by my examples.

2 Basics

2.1 The many-electron problem in one-electron form

The electrons in any material interact with the nuclei, which are much heavier and
can be assumed to be at fixed positions, and with each other. The success of modern
electronic structure calculations is based on a viable method for treating the electron-
electron interaction, called *density functional theory* [1]. This is based on the fact that
the ground state charge density of the electrons is unique to the particular external
potential in which these electrons move (*i.e.* the nuclear potential). Following Kohn
and Sham [2], the density $\rho_0(\mathbf{r})$ of the real, interacting electrons is reproduced by the
charge density of non-interacting electrons moving in the nuclear potential $V_{\mathrm{nuc}}(\mathbf{r})$ plus
an effective one-electron potential which simulates the effect of the electron-electron
interaction. This consists of the Hartree potential $V_H(\mathbf{r})$, the electrostatic potential due
to the charge density of all the electrons in the system, plus the exchange-correlation
potential $V_{xc}(\mathbf{r})$, which accounts for the dynamically correlated motion of electrons -

Equilibrium Structure and Properties of Surfaces and Interfaces
Edited by A. Gonis and G.M. Stocks, Plenum Press, New York, 1992

when the electron under consideration moves, the other electrons move out of its way, lowering the potential energy. The resulting one-electron Schrödinger equation is [1]:

$$\left[-\frac{1}{2}\nabla^2 + V_{\mathrm{nuc}}(\mathbf{r}) + V_H(\mathbf{r}) + V_{xc}(\mathbf{r})\right]\psi_i(\mathbf{r}) = \epsilon_i\psi_i(\mathbf{r}), \tag{1}$$

giving wave-functions from which the ground state charge density can be constructed:

$$\rho_0(\mathbf{r}) = \sum_{\mathrm{occupied}\,i} |\psi_i(\mathbf{r})|^2, \tag{2}$$

and eigenvalues from which the ground state energy can be found:

$$E_0 = \sum_{\mathrm{occupied}\,i} \epsilon_i - \frac{1}{2}\int d^3\mathbf{r}\,V_H(\mathbf{r})\rho_0(\mathbf{r}) - \int d^3\mathbf{r}\,V_{xc}(\mathbf{r})\rho_0(\mathbf{r}) + E_{xc}. \tag{3}$$

In (3), the exchange-correlation energy E_{xc} is the difference in energy between the actual, interacting system and the effective one-electron system defined by (1) with the same density. It is a functional (*i.e.* a function of a function) of the charge density, and the exchange-correlation potential is related to it by functional differentiation:

$$V_{xc}(\mathbf{r}) = \frac{\delta E_{xc}[\rho_0(\mathbf{r})]}{\delta \rho_0(\mathbf{r})}. \tag{4}$$

All the complexity of the many-body problem for the ground state now sits in the determination of E_{xc}.

What makes density functional theory useful is the fact that a very good approximation exists for the exchange-correlation energy - the local density approximation [1, 2, 3] in which E_{xc} is written as:

$$E_{xc} \simeq \int d^3\mathbf{r}\,\rho_0(\mathbf{r})\epsilon_{xc}(\rho_0(\mathbf{r})), \tag{5}$$

where $\epsilon_{xc}(\rho_0)$ is the exchange-correlation energy per electron of a uniform electron gas with the local density $\rho_0(\mathbf{r})$. This is saying that we can divide space up into little regions, in each of which the electrons interact as if they were part of an infinite, extended electron gas with the same density. Although *a priori* we would only expect (5) to be valid when the electron density varies very slowly, over a length scale greater than the screening length, say, in fact it works extremely well even in systems like atoms, molecules and surfaces in which the density varies rapidly [4]. With this prescription for E_{xc}, V_{xc} is given by:

$$V_{xc}(\mathbf{r}) = \frac{d}{d\rho}[\rho\epsilon_{xc}(\rho)]\bigg|_{\rho=\rho_0(\mathbf{r})}. \tag{6}$$

Both the Hartree and exchange-correlation potentials depend on the charge density, which in turn depends on the potentials - the system of equations must be solved self-consistently:

---

[1] Hartree atomic units are used, with $e = \hbar = m = 1$

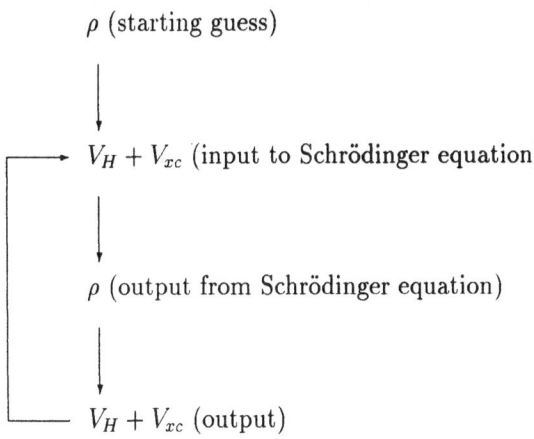

ρ (starting guess)

$V_H + V_{xc}$ (input to Schrödinger equation)

ρ (output from Schrödinger equation)

$V_H + V_{xc}$ (output)

Unfortunately this process is always unstable, and the simplest way to handle this is to mix only a small fraction of the output potential from the n'th iteration in to the input potential for the $n + 1$'th iteration:

$$V_{\text{input}}^{(n+1)} = (1 - \alpha)V_{\text{input}}^{(n)} + \alpha V_{\text{output}}^{(n)}. \tag{7}$$

Surface calculations can be wildly unstable [5], and more sophisticated schemes based on the Newton-Raphson method [6] for solving non-linear equations generally speed up the determination of the self-consistent charge density and potential.

We should note here that density functional theory is designed only to give the ground state properties of the system (in fact, anything which can be written as a functional of the ground state charge density). The individual wave-functions $\psi_i(\mathbf{r})$ do not represent the wave-functions of real electrons (more accurately, quasiparticles), and only appear as an aid to calculating ρ_0; similarly the individual eigenvalues ϵ_i do not represent real excitation energies. This is unlike Hartree-Fock theory, where (within the rather severe limitations of the theory) the one-electron energies ϵ_i represent the energy needed to remove an electron from an unoccupied state, or the energy gained when an electron goes into an unoccupied state (Koopman's theorem). Nevertheless, the density functional eigenvalues are frequently interpreted as excitation energies for the processes of adding or removing an electron, and compared as such with peaks in photoemission and inverse photoemission experiments. The wave-functions $\psi_i(\mathbf{r})$ are also used to construct the transition matrix elements.

2.2 The potential at the surface

Having got an effective single particle Schrödinger equation, we now turn to the form of the potential $V = V_{\text{nuc}} + V_H + V_{xc}$ felt by the electrons at the surface [7]. In a bulk crystal, the total potential V has three-dimensional periodicity, but when the crystal is chopped in two to make a surface, the periodicity in the perpendicular direction is destroyed. However, the resulting semi-infinite crystal still has two-dimensional periodicity parallel to the surface - as long as there are no surface defects. This means that the potential at a point $(\mathbf{R}, z)$ ($\mathbf{R}$ is the vector parallel to the surface, and z measures distance perpendicular) is the same as at $(\mathbf{R} + \mathbf{R}_I, z)$, where $\mathbf{R}_I$ is a vector in the two-dimensional surface lattice or mesh:

$$V(\mathbf{R}, z) = V(\mathbf{R} + \mathbf{R}_I, z). \tag{8}$$

This is illustrated in figure 1, which shows equipotential contours on a plane through the Ag(001) surface. We see that the potential is periodic parallel to the surface, but in

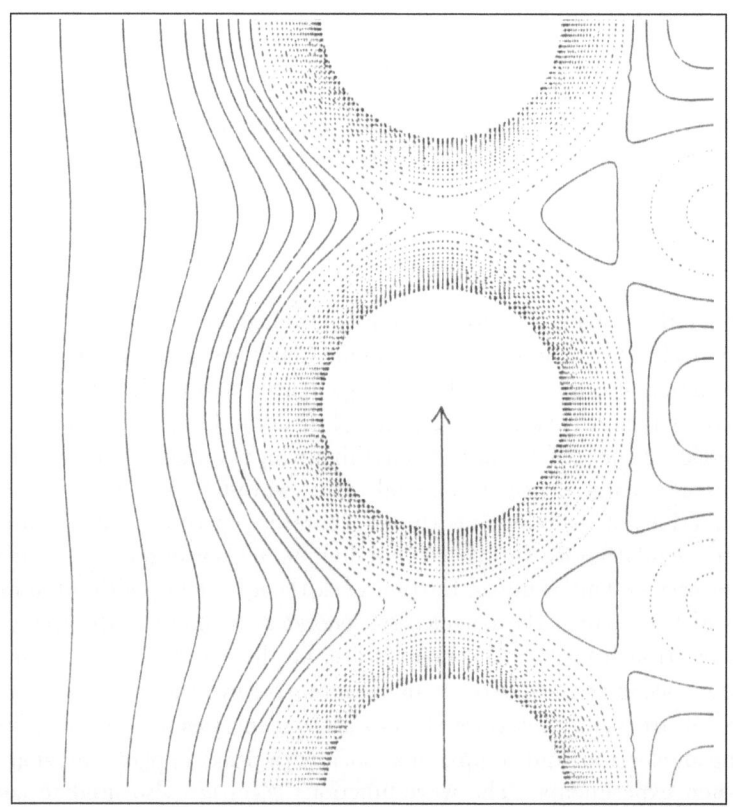

Figure 1. Potential felt by an electron at the Ag(001) surface. The figure shows contours of constant potential on a plane passing through the surface layer of atoms. Solid contours represent a positive potential relative to the average potential between the atoms in the bulk, and dashed contours show a negative potential. The holes in the middle of the atoms are simply where the potential is too deep for my plotting program! The arrow shows the vector of the 2-dimensional surface mesh.

the z-direction it goes from the constant potential in the vacuum to the bulk potential, and there is no overall periodicity in this direction.

As an electron goes into the solid through the surface, the potential goes smoothly from the vacuum potential a long way from the surface, through the image potential, to the bulk potential. At a metal surface, an electron feels the bulk potential beyond the first or second layer of atoms. There are relatively small changes in potential in the top layer or two, and this region is called the *selvedge*. The changes in the static properties due to the surface, such as the charge density and the bonding, are largely confined to the selvedge. The reason why the potential goes so rapidly to the bulk is perfect screening (in the case of metals), which means that any potential perturbation - such as a surface - is screened away within a Thomas-Fermi screening length or so. Note that even though the potential *does* go to the bulk value, overall the whole system has lost periodicity in the perpendicular direction. It is this form of potential which is the basis of our embedding ideas.

The two-dimensional periodicity of the surface means that the electronic wave-functions can be labelled by a two-dimensional Bloch wave-vector. Displacing $\mathbf{R}$ in (1) by a surface lattice vector $\mathbf{R}_I$, and using (8), we see that the displaced wave-function satisfies just the same Schrödinger equation:

$$\left[-\frac{1}{2}\nabla^2 + V(\mathbf{R}, z)\right]\psi(\mathbf{R} + \mathbf{R}_I, z) = E\psi(\mathbf{R} + \mathbf{R}_I, z). \tag{9}$$

Setting aside any degeneracy of ψ, this means that the displaced wave-function must be the same as the original wave-function, apart from a phase factor of modulus unity, which can be written as $\exp i\mathbf{K}.\mathbf{R}_I$:

$$\psi(\mathbf{R} + \mathbf{R}_I, z) = \exp i\mathbf{K}.\mathbf{R}_I \psi(\mathbf{R}, z). \tag{10}$$

If ψ is degenerate it is also possible to choose the wave-functions with this Bloch property. The two-dimensional Bloch wave-vector $\mathbf{K}$ is thus a good quantum number which we can use to label the wave-functions at the surface. Just as in the case of a bulk crystal, where the wave-functions are labelled by the three-dimensional Bloch wave-vector, $\mathbf{K}$ is not defined to within a *surface* reciprocal lattice vector $\mathbf{G}$. We can add $\mathbf{G}$ to $\mathbf{K}$ without changing the phase factor in (10), because the definition of the surface recips is:

$$\mathbf{G}.\mathbf{R}_I = 2\pi \times \text{integer}. \tag{11}$$

This lack of uniqueness in the Bloch wave-vector means that we can choose $\mathbf{K}$ to lie in the surface Brillouin zone.

The fact that we can add $\mathbf{G}$ to $\mathbf{K}$ without affecting the two-dimensional Bloch property of the wave-function is intimately connected with the diffraction properties of surfaces [8, 9, 10]. A low energy beam of electrons fired at the surface in a particular direction has a free electron wave-function away from the surface, with a well-defined wave-vector $\mathbf{K}$:

$$\psi_{incident}(\mathbf{R}, z) = \exp i\mathbf{K}.\mathbf{R} \exp ik_z z. \tag{12}$$

This is scattered by the surface into reflected waves travelling back from the surface, giving a total wave-function in the vacuum of the form:

$$\psi(\mathbf{R}, z) = \exp i\mathbf{K}.\mathbf{R} \exp ik_z z + \sum_{\mathbf{G}} A_{\mathbf{G}} \exp i(\mathbf{K} + \mathbf{G}).\mathbf{R} \exp -ik_{\mathbf{G}} z. \tag{13}$$

The two-dimensional periodicity of the surface potential diffracts the electrons through the surface recips $\mathbf{G}$, without affecting the Bloch property of ψ, which still satisfies (10).

209

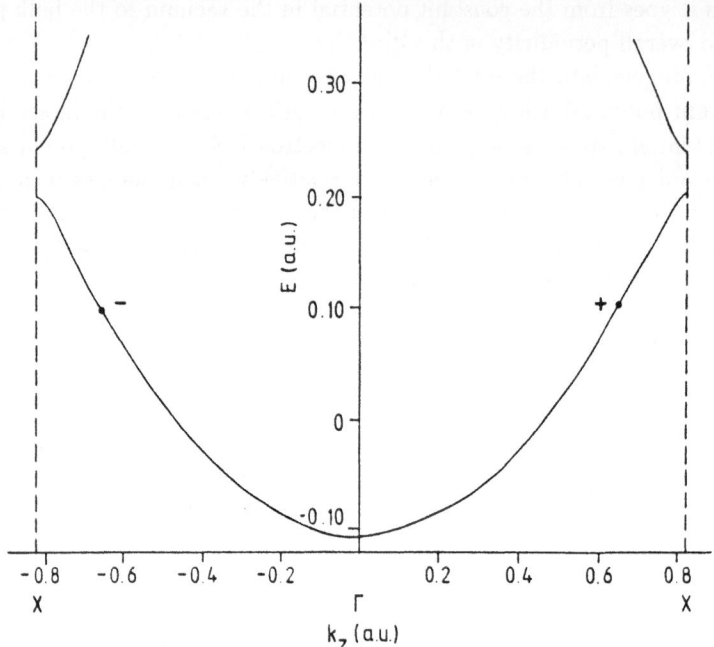

Figure 2. Bulk band structure of Al along $X\Gamma X$, corresponding to $\mathbf{K} = 0$ in the (001) surface Brillouin zone. The minus sign indicates a wave travelling towards the surface, and the plus sign represents a wave travelling away from the surface.

As the reflected waves travel in directions determined by $\mathbf{K} + \mathbf{G}$, imaging the electrons on a fluorescent screen gives a map of the surface reciprocal mesh. The amplitudes $A_{\mathbf{G}}$ in (13) contain information about the location of the atoms themselves with respect to the corresponding real space surface mesh. This is LEED - low energy electron diffraction, but exactly the same principles go over to other surface diffraction techniques, such as helium atom diffraction.

2.3   Bulk states at the surface

There are two types of electronic states with energy below the vacuum zero, that is, confined to the semi-infinite solid by the surface potential barrier: bulk states which hit the surface and are reflected, which we consider in this section, and surface states which are localized at the surface which we consider in the following section.

Away from the surface we know that an electron feels the same potential as in the bulk, so we can build up a solution of the Schrödinger equation in this region out of bulk solutions. Fixing the wave-vector $\mathbf{K}$ parallel to the surface, let us consider the bulk energy bands as a function of k_z, the wave-vector component perpendicular to the surface. If we are working at $\mathbf{K} = 0$ ($\bar{\Gamma}$), for example, in the *surface* Brillouin zone for fcc(001), the bulk bands projecting onto this wave-vector lie along $X\Gamma X$ in the *bulk* Brillouin zone, as shown in figure 2. The state indicated by the $-$ in figure 2 represents a bulk wave-function with wave-vector component $\mathbf{K} = 0$, energy E, travelling towards the surface. In the same way that the LEED electron is scattered externally by the surface, this state is reflected back into the solid by the surface, into states with wave-vector $\mathbf{K} + \mathbf{G}$ and energy E. All the intensity is reflected, because the energy lies below the vacuum zero and the electron cannot escape, and energy is conserved in the elastic scattering by the surface potential. There is a finite probability of finding the electron *just* outside the surface, because the electron wave-function decays exponentially into the vacuum.

In the case shown in figure 2, there is only one travelling wave reflected by the surface (indicated by + on the figure) - the bands with surface wave-vector $\mathbf{K} + \mathbf{G}$ are just the same, and do not introduce any new states. But as well as the one *travelling* wave, we must also consider evanescent waves [11], which in a nearly-free electron system (like figure 2) have the form:

$$\phi^+_{\mathbf{K},\mathbf{G};E} \sim \exp i(\mathbf{K} + \mathbf{G}).\mathbf{R} \exp -\gamma_{\mathbf{G}} z, \tag{14}$$

with:

$$|\mathbf{K} + \mathbf{G}|^2 - \gamma^2_{\mathbf{G}} = 2E. \tag{15}$$

This wave-function is not allowed in an infinite crystal, of course, because it blows up in the $-z$ direction. But it is an allowed solution of the Schrödinger equation in the semi-infinite system because the surface stops it blowing up. Not only is it allowed, these evanescent solutions of the bulk Schrödinger equation *must* be considered when building up solutions in the presence of a surface. As it has a reduced wave-vector $\mathbf{K}$, and energy E, the surface scatters some of the incident wave into this state, though because it is evanescent it does not carry any flux away from the surface - this is all carried by the travelling wave (+ on figure 2). There is a one-to-one relationship between reflected travelling + evanescent waves like (15), and the surface reciprocal lattice vectors. That is, for each $\mathbf{G}$ there is either a wave travelling away from the surface with the form:

$$\phi^+_{\mathbf{K},\mathbf{G};E} \sim \exp i(\mathbf{K} + \mathbf{G}).\mathbf{R} \exp i k_{\mathbf{G}} z, \tag{16}$$

with:

$$|\mathbf{K} + \mathbf{G}|^2 + k^2_{\mathbf{G}} = 2E, \tag{17}$$

or an evanescent wave like (15). This one-to-one property, which is fairly obvious in the nearly-free electron case, persists even when we turn on a strong crystal potential to distort the bands from the nfe energy bands [11].

With the full potential, the wave-function away from the surface can now be written down as:

$$\psi = \phi^- + \sum_{\mathbf{G}} r_{\mathbf{G}} \phi^+_{\mathbf{K},\mathbf{G};E} \tag{18}$$

- the incident travelling wave + reflected travelling and evanescent waves labelled by $\mathbf{G}$. In the vacuum, assuming a step potential (a common starting point for the simplest surface calculations), the solution has the form:

$$\psi = \sum_{\mathbf{G}} t_{\mathbf{G}} \exp i(\mathbf{K} + \mathbf{G}).\mathbf{R} \exp \hat{\gamma}_{\mathbf{G}} z \tag{19}$$

- waves decaying into the vacuum. To find the unknown coefficients $r_{\mathbf{G}}$, $t_{\mathbf{G}}$ in (18, 19) we match amplitude and derivative across the surface at $z = 0$, to the right of which we have (approximately) the perfect bulk potential and a solution given by (18), and to the left of which we have the (approximately) constant vacuum potential and a solution given by (19). This can be done by expanding the functions at $z = 0$ in two-dimensional Fourier series made up of plane waves $\exp i(\mathbf{K} + \mathbf{G}).\mathbf{R}$. If there are $\mathcal{N}$ $\mathbf{G}$'s in the Fourier series, this gives $2\mathcal{N}$ matching conditions from which we can find $\mathcal{N}$ $r_{\mathbf{G}}$'s and $\mathcal{N}$ $t_{\mathbf{G}}$'s. We have just the right number of conditions to find the unknown coefficients, and we can thus find the wave-function uniquely. For every bulk state travelling towards the surface we can thus find a solution for the semi-infinite solid with a surface, and at each value of $\mathbf{K}$ the bulk bands give continua of states at the surface, with energy gaps.

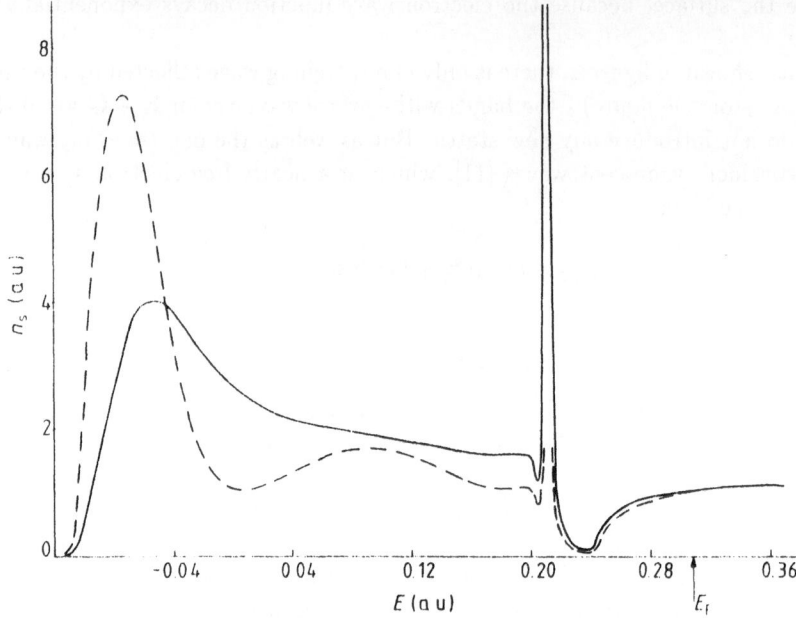

Figure 3. Solid line: surface density of states at Al(001), with $\mathbf{K} = 0$; dashed line: density of states in the second layer.

The matching method for finding bulk states at the surface is rather cumbersome and impractical [12], but it does show how the wave-functions can really be explicitly constructed. It can also be used to find the LEED wave-function, with the solution in vacuum given by (13) and the solution in the crystal given by (18), without ϕ^- travelling towards the surface [13].

2.4   Surface states

If E lies in a bulk energy gap at a particular wave-vector $\mathbf{K}$, there is no state ϕ^- representing a bulk wave incident on the surface. However, we can still try to match the $\mathcal{N}$ decaying exponentials:

$$\psi = \sum_{\mathbf{G}} r_{\mathbf{G}} \phi^+_{\mathbf{K},\mathbf{G};E} \tag{20}$$

onto (19), giving us $2\mathcal{N}$ homogeneous equations in $r_{\mathbf{G}}$, $t_{\mathbf{G}}$, which may have a non-trivial solution at discrete energies in the gap. When this happens, we have a surface state - a wave-function decaying exponentially both into the vacuum and into the crystal [14].

Rather than studying the individual wave-functions at the surface - impractical when we are dealing with bulk states infinitely close together in energy, but each of infinitesimal weight - it is convenient to work with the *local density of states*. This is defined by the sum of states:

$$\sigma(\mathbf{r}, E) = \sum_i \delta(E - E_i)|\psi_i(\mathbf{r})|^2, \tag{21}$$

and is the charge density of states with energy E. We can also define a local density of states at a fixed wave-vector $\mathbf{K}$, in which case the sum runs over those states with wave-vector $\mathbf{K}$ parallel to the surface. Figure 3 shows the local density of states with zero $\mathbf{K}$ at the Al(001) surface, and this shows very clearly the bulk states at the surface, with an energy gap in which lies a surface state.

3 Slab calculations

Most surface calculations do not actually consider the properties of a semi-infinite solid, with the two distinct types of wave-functions which we have just described - bulk states hitting the surface and bouncing back, and localized, discrete surface states. Most calculations in fact treat a slab of material, typically 5 or so layers thick, with two surfaces [15].

The reason why slab calculations are ubiquitous is that they use the same technology as bulk electronic structure calculations. In a bulk crystal the three-dimensional periodicity, and the use of Bloch's theorem, means that it is only necessary to solve the Schrödinger equation in one unit cell. A periodic array of slabs, separated by empty space, is simply a three-dimensional crystal with a large unit cell in the direction perpendicular to the surface and can be treated by the same methods as bulk crystals, though with a large basis set to cope with the large unit cell. This slab superlattice provides one way of calculating the electronic structure of surfaces. An alternative is to use a single slab, with the boundary condition that the wave-functions decay exponentially into the vacuum on either side of the slab (actually, they are taken to vanish on some plane in the vacuum). Apart from the fact that the basis functions are then standing waves in the perpendicular direction, like $\sin kz$ or $\cos kz$ rather than $\exp ikz$, the technology is again much the same as in the bulk crystal.

Highly accurate slab methods have been developed, so that it is now possible to calculate the energy of the W(001) surface reconstruction, for example, involving energy differences of the order of 1 mRyd per surface atom [16]. Slab calculations can describe surface properties like the charge density, potential and surface energy accurately because these are rather *local* properties, and are not much affected by the presence of a second surface separated from the first by only a few layers of bulk-like material. The work-function of W(001), for example, is shown in table 1 as a function of slab thickness [17], and we see that it rapidly approaches a constant value (the experimental value is 4.63 eV). There are two main reasons for the locality of the charge density and energy.

Table 1. Work-function of W(001) in slab geometries.

layers	ϕ eV
1	5.47
3	4.82
5	4.62
7	4.64

First, these properties involve sums over occupied states; although individual states are affected by the presence of a surface - for example the wave-length of a standing wave is affected - sums over states are much less affected. This is analogous to the black-body cavity radiation theorem, which says that the local spectral density of radiation is independent of the nature of the walls of the cavity more than a wave-length or so away. In the case of electronic states this becomes a statement that the local density of states (21) is rather independent of boundary conditions, and integrating over energy gives a still more local property. The second reason why these are local properties is that the short Thomas-Fermi screening length ensures that the potential perturbation due to a surface is rapidly screened away (in metals, at any rate).

Slab calculations have a major drawback, however, namely that the *individual* wave-functions are different from those at the surface of a semi-infinite solid. Fixing the wave-vector **K** parallel to the surface, a slab has a discrete spectrum of states below the

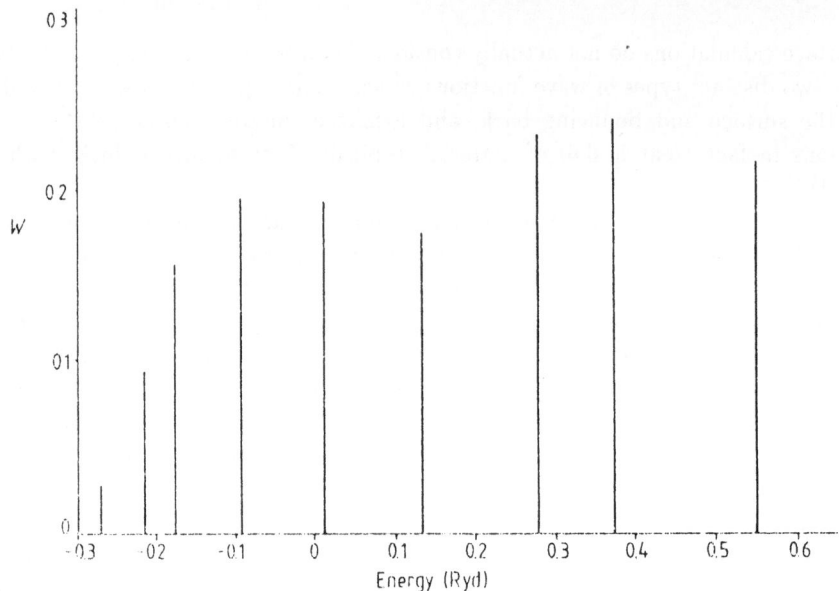

Figure 4. Charge density in top layer of atoms in a 7-layer slab calculation for Al(001), for states with $\mathbf{K} = 0$, as a function of their energy.

vacuum zero, because it is a finite potential well. Figure 4 shows the surface density of states for Al(001) in a 7-layer slab calculation, at $\mathbf{K} = 0$, and we see that it bears little resemblance to figure 3, showing the actual surface density of states in semi-infinite geometry. There is no distinction in slab geometry between the discrete surface states and the continua of bulk states hitting the surface, and surface states can only be identified by some rather arbitrary criterion of localization of charge density in the surface layers. In the case of the Al(001) slab, the surface states on the two surfaces interact, resulting in the two states at E=0.28 and 0.38 Ryd (figure 4). Clearly, we must do better if we wish to compare our results with experimental measurements of surface density of states, using photoemission for example (figure 5) [18].

4 Green functions

Green functions are the mathematical tool which enables changes in potential, or changes in boundary conditions to be included in the solution of Schrödinger's equation [19], and they form the basis for understanding the embedding method, and other methods of treating the surface problem.

The Green function describes the response of a system at point $\mathbf{r}$ to a perturbation at point $\mathbf{r}'$. The Green function for the Schrödinger equation satisfies:

$$\left(-\frac{1}{2}\nabla^2 + V(\mathbf{r}) - E\right) G(\mathbf{r}, \mathbf{r}'; E) = \delta(\mathbf{r} - \mathbf{r}') \tag{22}$$

- the ordinary Schrödinger equation with a δ-function on the right-hand-side. The Green function depends on $\mathbf{r}$, $\mathbf{r}'$ and energy E; it is not an eigenfunction of the Schrödinger equation, though it can be built up from eigenfunctions:

$$G(\mathbf{r}, \mathbf{r}'; E) = \sum_i \frac{\psi_i(\mathbf{r})\psi_i^*(\mathbf{r}')}{E_i - E}. \tag{23}$$

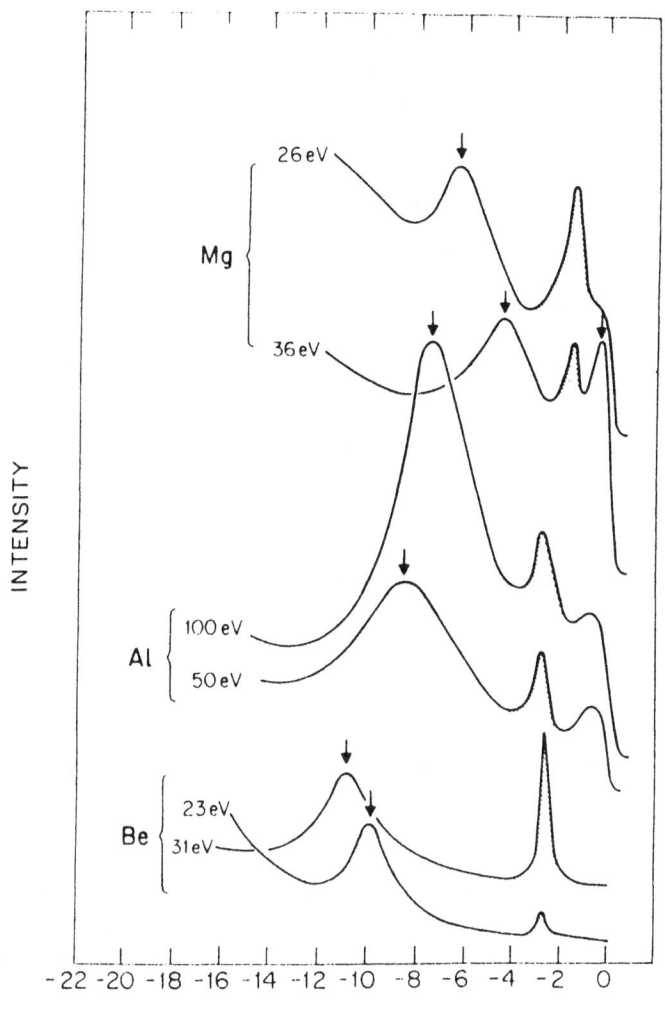

Figure 5. Normal photoemission from the surfaces of simple metals, showing the sharp peak (shaded) of the surface state [18].

Making use of the completeness properties of the ψ_i's, *i.e.*:

$$\sum_i \psi_i(\mathbf{r})\psi_i^*(\mathbf{r}') = .\delta(\mathbf{r} - \mathbf{r}'), \tag{24}$$

we see that (23) satisfies (22). A simple way of understanding (24) is to write the inhomogeneous Schrödinger equation (22) in operator (matrix, if you prefer) form;

$$(H - E)G = I, \tag{25}$$

where I is the identity operator, so that:

$$G = (H - E)^{-1}. \tag{26}$$

The matrix element of this between eigenfunctions ψ_i, ψ_j is:

$$\langle i|G|j\rangle = \frac{\delta_{i,j}}{E_i - E}, \tag{27}$$

corresponding to (23). Note that the Green function has poles at the eigenfunctions of the system.

The local density of states (21) can be found immediately from the Green function. From (23) we see that σ is related to G by:

$$\sigma(\mathbf{r}, E) = \frac{1}{\pi}\mathcal{I}m\, G(\mathbf{r}, \mathbf{r}; E + i\epsilon). \tag{28}$$

Green functions enable us to find the change in the system due to a perturbation δV. Let us write the Green function which satisfies the unperturbed equation as G_0, and the Green function for the perturbed system as G:

$$(H(\mathbf{r}'') - E)G_0(\mathbf{r}'', \mathbf{r}) = \delta(\mathbf{r}'' - \mathbf{r}) \tag{29}$$

$$(H(\mathbf{r}'') + \delta V(\mathbf{r}'') - E)G(\mathbf{r}'', \mathbf{r}') = \delta(\mathbf{r}'' - \mathbf{r}'). \tag{30}$$

Multiplying (29) by $G(\mathbf{r}', \mathbf{r}'')$, (30) by $G_0(\mathbf{r}, \mathbf{r}'')$, integrating over $\mathbf{r}''$ and subtracting we obtain:

$$G(\mathbf{r}, \mathbf{r}') = G_0(\mathbf{r}, \mathbf{r}') - \int d\mathbf{r}'' G_0(\mathbf{r}, \mathbf{r}'')\delta V(\mathbf{r}'')G(\mathbf{r}'', \mathbf{r}') \tag{31}$$

(we have made use of the fact that H is Hermitian, and that $G(\mathbf{r}, \mathbf{r}') = G(\mathbf{r}', \mathbf{r})$). This is Dyson's equation, and is the most important equation in Green function theory. Solving this equation by iteration, that is, by starting off with the approximation that $G \approx G_0$ and iterating gives the perturbation series:

$$G = G_0 - G_0\delta V G_0 + G_0\delta V G_0\delta V G_0 + \cdots. \tag{32}$$

Alternatively, we can expand the Green functions as matrices and use matrix inversion to give us the exact result:

$$G = (1 + G_0\delta V)^{-1}G_0. \tag{33}$$

Dyson's equation methods are particularly suitable for treating the spatially localized perturbation δV due to an impurity in a solid [20]. The Green function for the perfect crystal can be found from (23), and then using (33) the Green function for the defective crystal can be determined. There is a problem in using Dyson's equation at surfaces, because the perturbation in this case extends over a (semi-) infinite region of space, where half the crystal is cut away and replaced by vacuum. However, it *can* be used with localized basis functions, in the tight-binding or linear combination of atomic orbitals

(LCAO) methods. In this case the perturbation due to the surface just corresponds to removing the hopping integrals across the surface - cutting the bonds - and it is then very easy to use Dyson's equation to find G for the half-crystal. This was used in the early days to find the surface electronic structure in model tight-binding systems [21], and more recently it has been used within the framework of an empirical tight-binding approximation for the bulk electronic structure to study the electronic structure of ideal and reconstructed semiconductor surfaces [22]. One can go much further than this, and Dyson's equation methods have been developed into accurate, self-consistent schemes for finding surface electronic structure within the LCAO framework [23].

4.1 Green functions and boundary conditions

Green functions allow the boundary conditions in the solution of a differential equation to be straightforwardly included. An example of this is in electrostatics, where the solution of Laplace's equation can be found in an enclosed region of space, given either the potential or the electric field on the boundary. The relevant Green function in this case is the Coulomb interaction, $1/|\mathbf{r} - \mathbf{r}'|$. In the case of Schrödinger's equation, the important result which we shall use in developing the embedding method is that knowing the normal derivative of a wave-function on the boundary of some region, and the Green function for that region, we can find the wave-function everywhere. This provides a way of extending the solution of the Schrödinger equation from the surface region into the substrate, where we know the perfect crystal Green function.

The wave-function which we wish to find in a particular region R satisfies the homogeneous Schrödinger equation at energy ε:

$$\left(-\frac{1}{2}\nabla^2 + V(\mathbf{r}) - \varepsilon\right)\psi(\mathbf{r}) = 0, \tag{34}$$

and we assume that we know the Green function G_0 for this region, satisfying:

$$\left(-\frac{1}{2}\nabla^2 + V(\mathbf{r}) - \varepsilon\right)G(\mathbf{r}, \mathbf{r}'; \varepsilon) = \delta(\mathbf{r} - \mathbf{r}'). \tag{35}$$

Multiplying (34) by G_0, (35) by ψ, subtracting the equations and integrating throughout region R we obtain:

$$\psi(\mathbf{r}) = \frac{1}{2}\int_R d\mathbf{r}'[G_0(\mathbf{r}, \mathbf{r}')\nabla^2\psi(\mathbf{r}') - \psi(\mathbf{r}')\nabla^2 G_0(\mathbf{r}, \mathbf{r}')], \tag{36}$$

and then from Green's theorem:

$$\psi(\mathbf{r}) = -\frac{1}{2}\int_S d\mathbf{r}_S \left(G_0(\mathbf{r}, \mathbf{r}_S)\frac{\partial\psi(\mathbf{r}_S)}{\partial n_S} - \psi(\mathbf{r}_S)\frac{\partial G_0(\mathbf{r}, \mathbf{r}_S)}{\partial n_S}\right), \tag{37}$$

where S is the boundary of the region. We can now construct G_0 to have zero derivative on S, and then (37) gives us an equation relating the amplitude of the wave-function in R and its normal derivative on S:

$$\psi(\mathbf{r}) = -\frac{1}{2}\int_S d\mathbf{r}_S G_0(\mathbf{r}, \mathbf{r}_S)\frac{\partial\psi(\mathbf{r}_S)}{\partial n_S}. \tag{38}$$

This is the analogy of the electrostatics method for finding the potential inside R given the field on the boundary.

In scattering theory we often come across the notion of logarithmic derivative, relating the derivative and the amplitude of the wave-function at the surface of a muffin tin potential for example. Equation (38) gives us the generalization of a logarithmic derivative. Putting $\mathbf{r}$ on S:

$$\psi(\mathbf{r}_S) = -\frac{1}{2}\int_S d\mathbf{r}'_S G_0(\mathbf{r}_S, \mathbf{r}'_S)\frac{\partial\psi(\mathbf{r}'_S)}{\partial n_S}, \tag{39}$$

and taking the inverse we obtain:

$$\frac{\partial\psi(\mathbf{r}_S)}{\partial n_S} = -2\int_S d\mathbf{r}'_S G_0^{-1}(\mathbf{r}_S, \mathbf{r}'_S)\psi(\mathbf{r}'_S). \tag{40}$$

By the inverse in (40) we mean:

$$\int_S d\mathbf{r}'_S G_0^{-1}(\mathbf{r}_S, \mathbf{r}'_S)G_0(\mathbf{r}'_S, \mathbf{r}''_S) = \delta(\mathbf{r}_S, \mathbf{r}''_S). \tag{41}$$

G_0^{-1} is the generalized logarithmic derivative. We shall see in the next section that it gives us the embedding potential.

5 Embedding

The idea of embedding [24, 25] is that the perturbation in potential produced by the surface is confined only to the top few layers of atoms, and we can divide the system up into two regions - region I which consists of the surface layers and the vacuum, and region II which is the unperturbed substrate. In region I we treat the Hamiltonian explicitly, obtaining the full solution of the Schrödinger equation, and region II is replaced by an *embedding potential* which is added on to the surface Hamiltonian. The embedding potential scatters the surface electrons in the same way as the bulk crystal, in other words it is a pseudopotential which forces the electron wave-functions in the surface region to have the correct logarithmic derivative to match onto the substrate wave-functions (hence the connection with (40)). The advantage of embedding is that we have only to deal with a system of finite thickness - the surface region - and just as in a slab calculation we can use any convenient basis set to expand the wave-functions. The difference with ordinary slab calculations is that the embedding potential takes care of the substrate.

We start off from a variational principle, building up a trial wave-function from an arbitrary function ϕ in region I. This has to be extended through the whole of space, and we (notionally) extend it into the substrate (region II) with the solution of the bulk Schrödinger equation at some trial energy ϵ, ψ, which matches in amplitude onto ϕ over the interface S between the two regions:

$$\phi(\mathbf{r}_S) = \psi(\mathbf{r}_S). \tag{42}$$

In general we cannot match the normal derivative as well as the amplitude if ϕ is an arbitrary function, though of course both match in the full solution of Schrödinger's equation. Nevertheless we have a perfectly acceptable trial function, for which the expectation value of the energy is:

$$E = \frac{\int_I d\mathbf{r}\,\phi H\phi + \epsilon\int_{II} d\mathbf{r}\,\psi^2 + \frac{1}{2}\int_S d\mathbf{r}_S\,\phi\partial\phi/\partial n_S - \frac{1}{2}\int_S d\mathbf{r}_S\,\phi\partial\psi/\partial n_S}{\int_I d\mathbf{r}\,\phi^2 + \int_{II} d\mathbf{r}\,\psi^2}. \tag{43}$$

The surface integrals come from the effect of the kinetic energy operator on the discontinuity in the normal derivatives $\partial\phi/\partial n_S$ and $\partial\psi/\partial n_S$ across the interface (remember: a kink in a wave-function costs kinetic energy).

The fundamental principle of embedding is that the terms in (43) involving ψ can be eliminated, in terms of ϕ. From our Green function result (40), we can write the normal derivative term in the numerator of (43) as:

$$-\frac{1}{2}\int_S d\mathbf{r}_S \phi \frac{\partial \psi}{\partial n_S} = \int_S d\mathbf{r}_S \int_S d\mathbf{r}'_S \phi(\mathbf{r}_S) G_0^{-1}(\mathbf{r}_S, \mathbf{r}'_S) \phi(\mathbf{r}'_S), \tag{44}$$

where we have made use of (42). The volume integrals are more difficult, but they can also be eliminated in a related way. We differentiate the Schrödinger equation satisfied by ψ with respect to energy:

$$H\psi = \epsilon\psi$$
$$H\frac{\partial \psi}{\partial \epsilon} = \psi + \epsilon\frac{\partial \psi}{\partial \epsilon}. \tag{45}$$

We now multiply the top equation by $\partial\psi/\partial\epsilon$, the bottom equation by ψ, integrate through region II and subtract, giving us:

$$\int_{II} d\mathbf{r}\psi^2 = -\frac{1}{2}\int_{II} d\mathbf{r}\left(\psi\nabla^2\frac{\partial\psi}{\partial\epsilon} - \frac{\partial\psi}{\partial\epsilon}\nabla^2\psi\right). \tag{46}$$

The use (once again) of Green's theorem converts the volume integral into a surface integral over the interface S:

$$\int_{II} d\mathbf{r}\psi^2 = \frac{1}{2}\int_S d\mathbf{r}_S\left(\psi\frac{\partial^2\psi}{\partial n_S \partial\epsilon} - \frac{\partial\psi}{\partial\epsilon}\frac{\partial\psi}{\partial n_S}\right). \tag{47}$$

But the second term in (47) vanishes, because we are keeping $\psi(\mathbf{r}_S)$ constant, so we finish up with the expression for the normalization integral in II:

$$\int_{II} d\mathbf{r}\psi^2 = \frac{1}{2}\int_S d\mathbf{r}_S\psi\frac{\partial^2\psi}{\partial n_S \partial\epsilon}$$
$$= -\int_S d\mathbf{r}_S \int_S d\mathbf{r}'_S \phi(\mathbf{r}_S)\frac{\partial G_0^{-1}}{\partial\epsilon}\phi(\mathbf{r}'_S). \tag{48}$$

We can then write the expectation value of the energy as:

$$E = \left(\int_I d\mathbf{r}\phi H\phi + \frac{1}{2}\int_S d\mathbf{r}_S\phi\frac{\partial\phi}{\partial n_S}\right.$$
$$+ \left.\int_S d\mathbf{r}_S \int_S d\mathbf{r}'_S \phi(\mathbf{r}_S)G_0^{-1}(\mathbf{r}_S,\mathbf{r}'_S)\phi(\mathbf{r}'_S) - \epsilon\int_S d\mathbf{r}_S \int_S d\mathbf{r}'_S \phi(\mathbf{r}_S)\frac{\partial G_0^{-1}}{\partial\epsilon}\phi(\mathbf{r}'_S)\right)$$
$$\left/ \left(\int_I d\mathbf{r}\phi^2 - \int_S d\mathbf{r}_S \int_S d\mathbf{r}'_S \phi(\mathbf{r}_S)\frac{\partial G_0^{-1}}{\partial\epsilon}\phi(\mathbf{r}'_S)\right). \right. \tag{49}$$

This is a variational expression for ϕ in which G_0^{-1} - the embedding potential - ensures that ϕ matches onto the solution in the substrate.

To find the actual wave-function in the surface region we expand ϕ in terms of a set of basis functions:

$$\phi(\mathbf{r}) = \sum_i a_i\chi_i(\mathbf{r}), \tag{50}$$

and minimizing E gives a matrix equation for the coefficients:

$$\sum_j [H_{ij} + (G_0^{-1})_{ij} + (E - \epsilon)\partial(G_0^{-1})_{ij}/\partial\epsilon]a_j = E\sum_j S_{ij}a_j. \tag{51}$$

Here,

$$H_{ij} = \int_I dr\chi_i(\mathbf{r})H\chi_j(\mathbf{r}) + \frac{1}{2}\int_S dr_S\chi_i(\mathbf{r}_S)\frac{\partial\chi_j(\mathbf{r}_S)}{\partial n_S}$$

$$(G_0^{-1})_{ij} = \int_S dr_S \int_S dr_S'\chi_i(\mathbf{r}_S)G_0^{-1}(\mathbf{r}_S,\mathbf{r}_S')\chi_j(\mathbf{r}_S')$$

$$S_{ij} = \int_I dr\chi_i(\mathbf{r})\chi_j(\mathbf{r}) \tag{52}$$

- H_{ij} is the matrix element of the Hamiltonian in the surface plus vacuum region, with an additional interface integral which ensures Hermiticity; S_{ij} is the overlap matrix element in this region. $(G_0^{-1})_{ij}$ is the matrix element of the embedding potential, converting what would otherwise be a slab calculation into a calculation including the substrate. The energy-derivative term, entering (49) from the normalization of the wave-function in the substrate, reappears here as a first-order energy correction so that G_0^{-1} is evaluated at the right energy. To find the Green function in the surface region, which we need in the continuum of bulk states, we expand it as follows in terms of our basis functions:

$$G(\mathbf{r},\mathbf{r}';E) = \sum_{ij} g_{ij}(E)\chi_i(\mathbf{r})\chi_j(\mathbf{r}'). \tag{53}$$

Then g_{ij} satisfies:

$$\sum_j [H_{ij} + (G_0^{-1})_{ij} - ES_{ij}]g_{jk} = \delta_{ik}. \tag{54}$$

By evaluating the embedding potential at the energy E at which we are working, the energy derivative in (51) disappears.

5.1   Properties of the embedding potential

The embedding potential is complex at energies which lie within the allowed energy bands of the substrate. This is easy to understand, because a real embedding potential (or no embedding potential) results in discrete states, as region I has finite thickness. (This assumes that the wave-vector $\mathbf{K}$ parallel to the surface is fixed.) The complex character of the embedding potential broadens these discrete states into the true continuum of states of the system joined on to the substrate. The discrete states of slab or cluster calculations are often broadened to simulate the effect of the substrate, but this is usually done in an *ad hoc* way: only embedding gets it right, by adding the energy-dependent, complex embedding potential to the Hamiltonian.

The embedding potential has some of the properties of a pseudopotential [26]. The pseudopotential is normally a potential which replaces the actual potential inside the core of an atom, scattering valence electrons in the same way but eliminating the awkward core states and simplifying the form of the valence wave-functions in the core region. The embedding potential "pseudizes" away the whole of the substrate. A very interesting analogy with pseudopotential theory is in the relationship between the normalization of the wave-function in the substrate, and the energy-dependence of the embedding potential (48). In pseudopotential theory the so-called orthogonalization hole, which is the difference between the pseudo-charge and the actual charge inside the atomic core, is given by the energy-dependence of the pseudopotential [26]:

$$\int_{\text{core}} dr(\psi_{\text{ps}}^2 - \psi^2) = \int_{\text{core}} dr\psi_{\text{ps}}\frac{\partial V_{\text{ps}}}{\partial E}\psi_{\text{ps}}. \tag{55}$$

This result is very important for modern pseudopotentials [27], which by chosing to be norm conserving (*i.e.* the left side of (55) is zero) are automatically energy-independent to first order in E. Amusingly, the same relationship as (48) and (55) holds for the self-energy Σ in quasiparticle theory, where part of (phase) space is eliminated and replaced by the energy-dependent Σ [28].

6 Embedding at surfaces

6.1 LAPW basis set

With the embedding potential we only solve the Schrödinger equation explicitly in the true surface region, where the potential is different from the bulk, and we can use any convenient basis set. Linearized Augmented Plane Waves (LAPW's) provide a good choice [29]. They are constructed by dividing the surface region up into muffin tins around each atom, where the potential is approximately spherical, the interstitial region where the potential is approximately flat, and the vacuum region where the potential is mostly a function of distance from the surface. The LAPW is a plane wave in the interstitial region, and inside the muffin tins it consists of the linear combination of solutions of the atomic Schrödinger equation with the spherically averaged potential at some "pivot" energy, and their energy derivatives, which match on to the plane wave over the surface of the muffin tin in amplitude and derivative. The use of the solution at a fixed energy plus its energy derivative accounts for the "L" in LAPW - we are linearizing the solution with respect to energy dependence. In the vacuum region the solution is the linear combination of solutions of the vacuum Schrödinger equation with the planar-averaged (z-dependent) potential at a fixed energy, with the energy derivatives, which matches onto the plane wave on the edge of the slab. The Hamiltonian matrix elements in (52) are evaluated using the full potential, even though the LAPW's themselves are built up using the muffin tin potential. Expressions for the matrix elements are given by Inglesfield and Benesh [30].

6.2 Shifting the embedding surface

The boundary of region I, surface S, is conveniently taken to be a flat plane. However, this inevitably intersects muffin tins, and this might be expected to lead to complicated expressions for the matrix elements of the embedding potential. However, it is possible to take the matrix elements with only the plane wave part of the LAPW's. Let us consider one muffin tin intersected by the flat embedding plane, considering the various sub-regions $i \ldots iv$ and interfaces $S_a - S_d$ shown in figure 6. H is the Hamiltonian inside the muffin tin, in regions i and iii, and H_0 is the Hamiltonian in the interstitial region ii and iv, which we extend into region iii by any convenient method (*e.g.* by defining H_0 with zero potential). There are now *two* Hamiltonians for region iii - the actual Hamiltonian H, and the pseudo-Hamiltonian H_0.

The original variational principle consists of minimizing:

$$\langle \phi | H | \phi \rangle_{i+iii} + \langle \Phi | H_0 | \Phi \rangle_{ii} + \langle \psi | H_0 | \psi \rangle_{iv} \tag{56}$$

subject to normalization of the full wave-function, and the requirement that ϕ, Φ and ψ are continuous across all the interfaces in amplitude and derivative:

- $\phi = \Phi$ over S_a, with continuous derivatives

- $\Phi = \psi$ over S_b, with continuous derivatives

- $\phi = \psi$ over S_c, with continuous derivatives.

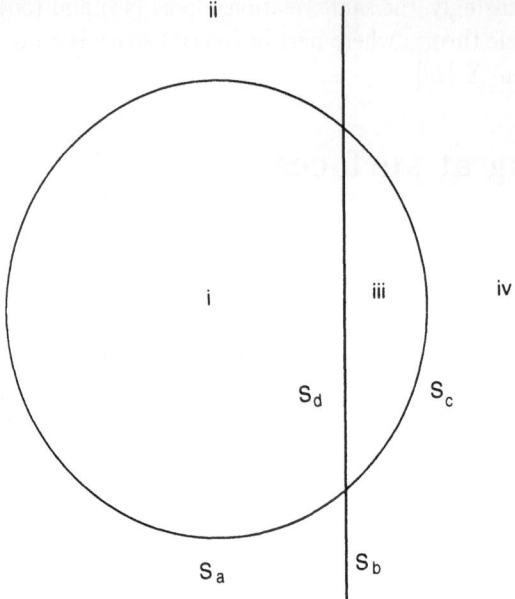

Figure 6. Muffin tin intersected by flat embedding plane.

This is equivalent to finding the stationary value of:

$$\langle\phi|H|\phi\rangle_{i+iii} + \langle\Phi|H_0|\Phi\rangle_{ii} - \langle\Phi|H_0|\Phi\rangle_{iii} + \langle\psi|H_0|\psi\rangle_{iii+iv} \tag{57}$$

subject to the continuity conditions:

- $\Phi = \phi$ over S_a, with continuous derivatives

- $\Phi = \psi$ over S_b and S_d, with continuous derivatives

- $\Phi = \phi$ over S_c, with continuous derivatives.

We actually have three trial functions defined in region iii, and figure 7 shows how these are related to each other by the boundary conditions. When (57) is stationary, these boundary conditions ensure that $\Phi = \psi$ inside region iii, giving us back (56) and the solution of the Schrödinger equation of the original problem.

We now go on to use embedding ideas, with our arbitrary trial function (equivalent to ϕ in section 5) defined as ϕ in region i, and region iii with Hamiltonian H, and Φ in regions ii and iii with Hamiltonian H_0. This is constructed with the first and third boundary conditions listed above. As in section 5 we extend this solution with an exact solution of the Schrödinger equation, this time satisfying:

$$H_0\psi = \epsilon\psi, \tag{58}$$

in regions iii and iv with the amplitude-only boundary conditions:

- $\psi = \Phi$ over S_b and S_d.

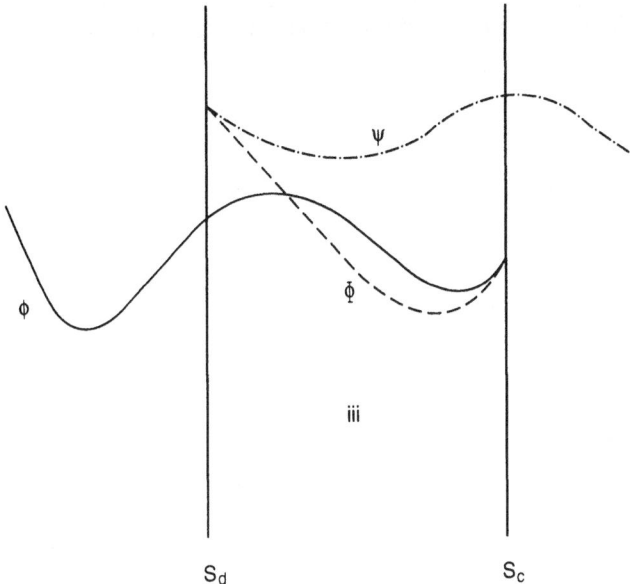

Figure 7. Boundary conditions within region *iii*.

The discontinuity in derivative between ψ and Φ over S_b and S_d means that (57) acquires extra terms, and we must look for the stationary value of:

$$\langle\phi|H|\phi\rangle_{i+iii} + \langle\Phi|H_0|\Phi\rangle_{ii} - \langle\Phi|H_0|\Phi\rangle_{iii} + \frac{1}{2}\int_{S_b+S_c} d\mathbf{r}_S \Phi(\partial\Phi/\partial n_S - \partial\psi/\partial n_S) + \epsilon\langle\psi|\psi\rangle_{iii+iv}.$$
$$(59)$$

Now the last two terms can be replaced by the embedding method, exactly as in section 5, but we note that even though the embedding plane intersects the muffin tin, the embedding potential replaces a region involving only the interstitial potential H_0. Moreover, ψ in the substrate has to be matched onto Φ over the embedding surface S_b and S_d, so the matrix elements of the embedding potential are between Φ only. This is a great simplification with the LAPW basis set, as we can expand Φ in the plane-wave part. The only unusual part of the procedure is that $\langle\Phi|H_0|\Phi\rangle_{iii}$ appears in the Hamiltonian expression with a minus sign.

6.3 Determining the embedding potential

The embedding potential G_0^{-1} is expanded in a Fourier series over the flat interface which we can now use between the surface and the substrate. A straightforward procedure for finding the embedding potential is to start from the reflection properties of the perfect crystal [25]. If we truncate the perfect crystal at S with a constant potential to the left, and shine a plane wave $\exp(i[\mathbf{K}+\mathbf{G}_m].\mathbf{R})\exp(ik_z z)$ onto it, this wave will be scattered, and the total wave-function to the left of S can be written as:

$$\psi(\mathbf{r}) = \exp(i[\mathbf{K}+\mathbf{G}_m].\mathbf{R})\exp(ik_z z) + \sum_{m'} R_{m',m}\exp(i[\mathbf{K}+\mathbf{G}_{m'}].\mathbf{R})\exp(\gamma_{m'}z), \quad (60)$$

where R is the reflection matrix, and:

$$|\mathbf{K}+\mathbf{G}_{m'}|^2 - \gamma_{m'}^2 = |\mathbf{K}+\mathbf{G}_m|^2 + k_z^2. \quad (61)$$

The Fourier coefficients of ψ and $\partial\psi/\partial n_s$ on S are then given by:

$$\begin{aligned} \psi_{m'} &= \delta_{m',m} + R_{m',m} \\ \psi'_{m'} &= \gamma_{m'}(-\delta_{m',m} + R_{m',m}). \end{aligned} \qquad (62)$$

But these are related by (40), so we see that the embedding potential is given by:

$$G_0^{-1} = \gamma(1 - R)(1 + R)^{-1}/2. \qquad (63)$$

The reflection matrix R can be conveniently found using the layer doubling method developed by Pendry [13] for LEED and photoemission calculations. First of all, the reflection and transmission matrices of a single atomic layer are calculated using a two-dimensional version of KKR, and then the reflection and transmission of two layers can easily be found. The procedure is repeated to give the reflection and transmission matrices of 4, 8, 16 ..., rapidly converging (when the energy has a small imaginary part) to give R for the semi-infinite crystal.

Recently a new method of calculating the embedding potential has been developed [31], based on the idea that a layer of bulk material embedded into the bulk has the Green function from which this bulk embedding potential can be found. This gives a non-linear equation for the embedding potential which can be solved by Newton-Raphson techniques. The advantage of this new method is that the solution of the the bulk layer problem can be carried out using the same basis set as the surface part of the problem, with the full potential, giving full compatibility between all parts of the embedding procedure. This idea is related to the " removal invariance principle" of Zhang and Gonis [32], based on multiple scattering theory.

6.4 Self-consistency at the embedded surface

In section 2.1 we saw that the charge density and the potential in the Schrödinger equation have to be determined self-consistently, and only minor modifications need to be made to to standard methods of solving Poisson's equation in the case of the embedded surface. The charge density (per spin) can be found immediately from the Green function (53), via the local density of states (28):

$$\rho(\mathbf{r}) = \frac{1}{\pi}\Im m \int^{E_F} dE G(\mathbf{r}, \mathbf{r}'; E + i\epsilon). \qquad (64)$$

The local density of states can have a lot of structure, in the d-bands of transition metals for example, so it is better to deform the contour of integration. This extends in (64) from below the bottom of the valence bands to the Fermi energy, just above the real axis, but because G is analytic in the upper half-plane this can be deformed into a semi-circle in the upper half-plane, again starting on the real axis below the bottom of the band and finishing on the real axis at E_F. Even if the density of states is highly structured along the real axis, G varies smoothly along the contour except where it approaches the real axis at E_F. Gauss-Chebychev provides a good numerical technique for carrying out the integration, particularly as it concentrates the sampling points near the ends of the range [30].

In addition to the integral over energy, we must sum over wave-vectors $\mathbf{K}$ in the surface Brillouin zone to obtain the total charge density. We sample at the special $\mathbf{K}$ points given by Cunningham [33] in the irreducible part of the surface BZ.

Knowing the charge density, we next solve Poisson's equation to find the Hartree contribution to the potential, for the next iteration cycle. The boundary conditions on the potential have to be included, and we stipulate that there is no electric field deep in the vacuum as $z \to -\infty$ (the total system is neutral at self-consistency), and the potential matches on to the bulk potential on the interface S with the substrate. These conditions fully determine the potential in the surface embedded region I. To solve Poisson's equation with these conditions we use a method due to Weinert [34], in which the charge density inside each atomic muffin tin is replaced by a pseudo-charge density with the same multipole moments whose Fourier transform can easily by found. This is added on to the Fourier expansion of the charge density in the interstitial region and from this the Fourier expansion of the potential can be found. This is only valid in the interstitial region, because of the use of the pseudo-charge density in the muffin tins, but the actual potential inside the muffin tins can be found just by integrating over the atomic charge density with the Coulomb interaction, with an extra surface contribution so that the potential joins on correctly to the interstitial potential. The exchange-correlation potential is almost trivially found, using the local density approximation (6).

Because of the notorious instability of direct iteration of the potential, we either use linear mixing à la (7), or the quasi-Newton-Raphson scheme due to Broyden [6] and developed by Srivastava [35]. Embedded calculations have a tendency to be more unstable than ordinary slab calculations, because the number of electrons in the embedded region fluctuates - the charge is determined by an integral up to the bulk E_F, whereas in slab calculations the total number of electrons is constant. Howver this leads to no real problems, and we just let the calculations proceed to convergence, at which surface charge neutrality automatically occurs to within a small fraction of an electron. The freedom in embedded calculations for charge to flow in and out is an advantage in calculations where an external electric field is applied, because this is automatically screened in our approach.

7 Results

7.1 Ag(001)

Figure 8 shows the charge density at the Ag(001) surface, determined using two atomic layers in the embedded region. The third and deeper layers of atoms have been replaced by the embedding potential; we can see that this is doing its job because the tops of the third layer of atoms are correctly simulated. This calculation gives a work-function for Ag(001) of 4.9 eV, compared with the experimental value of 4.6 eV.

7.2 Al(001)

The surface density of states for Al(001) at $\mathbf{K} = 0$, obtained with two embedded layers, is shown in figure 3. This is clearly much more useful for comparing with photoemission results than the 7-layer slab (figure 4). In particular, photoemission shows that the surface state lies very close to the bulk band edge, and this is apparent in the calculation. The charge density of the surface state at $\mathbf{K} = 0$ is shown in figure 9, and we can see the way that this piles up on the surface atoms - this is why the state shows up strongly in photoemission (figure 5). The surface state is decaying into the solid, but because it lies close to the bulk band edge it decays relatively slowly, and still has considerable amplitude beyond the second layer of atoms. Note that the charge density is essentially p-like with respect to the atoms. This is because the surface state lies near a band edge at which the bulk wave-functions are p-like (n.b. the inverted gap is necessary for surface states to occur).

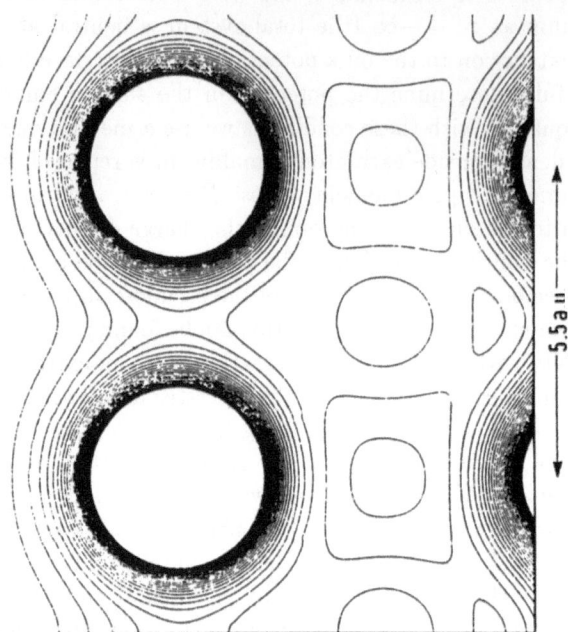

Figure 8. Contours of charge density at Ag(001), on a plane perpendicular to the surface passing through the top layer of atoms, and between the second layer.

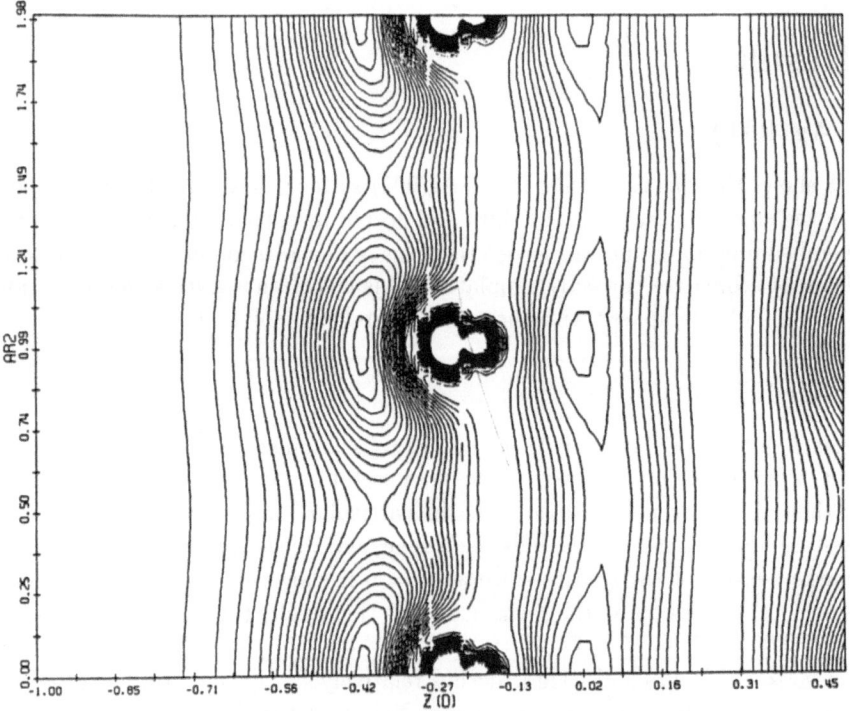

Figure 9. Charge density of surface state on Al(001) at $\mathbf{K} = 0$.

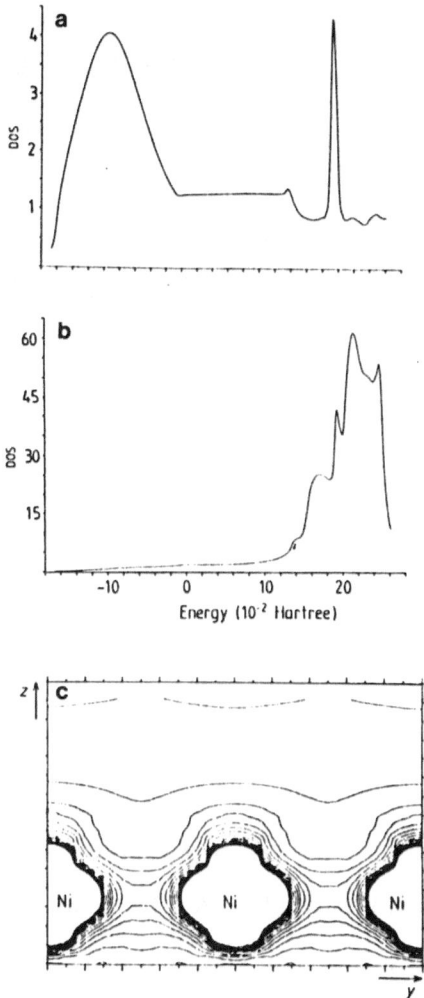

Figure 10. Electronic structure of Al-Ni interface. (a) Density of states with $\mathbf{K} = 0$ in Al muffin tin, (b) density of states in Ni muffin tin, (c) Charge density of interface state at $E = 0.192$ a.u.

7.3 Interfaces

Interfaces can be treated by embedding on both sides [36] - the interface has two substrates which are replaced by embedding potentials on both sides of the interface. As an example of this we look at an Al-Ni(001) junction, in which a common lattice constant was used for both materials. Figure 10 shows the density of states in the Al and Ni muffin tins at the interface, for states with zero wave-vector component parallel to the interface. The big peak in the Al density of states at E=0.192 a.u. occurs in the bulk Al band gap, as well as in a band gap of states with Δ_1 symmetry in bulk Ni. It is a localized *interface* state, and consists mainly of d_{z^2} orbitals on the Ni atoms (figure 10).

References

[1] P. Hohenberg and W. Kohn, Phys. Rev., **136**, B864 (1964)

[2] W. Kohn and L.J. Sham, Phys. Rev., **140**, A1133 (1965)

[3] N.D. Lang, Solid State Phys., **28**, 225 (1973)

[4] A.R. Williams and U. von Barth, in *Theory of the Inhomogeneous Electron Gas*, (eds., S. Lundqvist and N.H. March), Plenum, New York, 1983

[5] N.D. Lang and W. Kohn, Phys. Rev. B, **1**, 4555 (1970)

[6] G.C. Broyden, Math. Comput., **19**, 577 (1965)

[7] J.E. Inglesfield, in *Electronic Properties of Surfaces* (M. Prutton, ed.), Adam Hilger, Bristol (1984)

[8] M. Prutton, *Surface Physics*, Oxford University Press (1983)

[9] D.P. Woodruff and T.A. Delchar, *Modern Techniques of Surface Science*, Cambridge University Press (1986)

[10] A. Zangwill, *Physics at Surfaces*, Cambridge University Press (1988)

[11] V. Heine, Proc. Phys. Soc., **81**, 300 (1963)

[12] J.A. Appelbaum and D.R. Hamann, Phys. Rev. B, **6**, 2166 (1972)

[13] J.B. Pendry, *Low Energy Electron Diffraction*, Academic, New York (1974)

[14] S.J. Gurman and J.B. Pendry, Phys. Rev. Letts., **31**, 637 (1973)

[15] M. Posternak, H. Krakauer, A.J. Freeman and D.D. Koelling, Phys. Rev. B, **21**, 5601 (1980)

[16] C.L. Fu, S. Ohnishi, E. Wimmer and A.J. Freeman, Phys. Rev. Letts., **53**, 675 (1984)

[17] S. Ohnishi, A.J. Freeman and E. Wimmer, Phys. Rev. B, **29**, 5267 (1984)

[18] H.J. Levinson, F. Greuter and E.W. Plummer, Phys. Rev. B, **27**, 727 (1983)

[19] E. N. Economou, *Green's Functions in Quantum Physics*, Springer, Heidelberg (1979)

[20] F. Beeler, O.K. Anderson and M. Scheffler, Phys. Rev. B, **41**, 1603 (1990)

[21] D. Kalkstein and P. Soven, Surface Sci., **26**, 85 (1971)

[22] J. Pollmann, Fetskörperprobleme, **XX**, 117 (1980)

[23] J. Pollmann, R. Kalla, P. Krüger, A. Mazur and G. Wolfgarten, Applied Phys. A, **41**, 21 (1986)

[24] J.E. Inglesfield, J. Phys. C: Solid State Phys., **14**, 3795 (1981)

[25] G.A. Benesh and J.E. Inglesfield, J. Phys. C: Solid State Phys., **17**, 1595 (1984)

[26] V. Heine, Solid State Phys., **24**, 1 (1970)

[27] G.B. Bachelet, D.R. Hamann and M. Schlüter, Phys. Rev. B, **26**, 4199 (1982)

[28] L. Hedin and S. Lundqvist, Solid State Phys., **23**, 1 (1969)

[29] H. Krakauer, M. Posternak and A.J. Freeman, Phys. Rev. B, **19**, 1706 (1979)

[30] J.E. Inglesfield and G.A. Benesh, Phys. Rev. B, **37**, 6682 (1988)

[31] S. Crampin, J.B.A.N. van Hoof, M. Nekovee and J.E. Inglesfield, J. Phys.: Condensed Matter, *to be published*

[32] X.-G. Zhang and A. Gonis, Phys. Rev. Letts. **62**, 1161 (1989)

[33] S.L. Cunningham, Phys. Rev. B, **10**, 4988 (1974)

[34] M. Weinert, J. Math. Phys., **22**, 2433 (1981)

[35] G.P. Srivastava, J. Phys. A: Math. General Phys., **17**, L317 (1984)

[36] C.P. Farquhar and J.E. Inglesfield, J. Phys.: Condensed Matter, **1**, 599 (1989)

LOW-ENERGY ELECTRON DIFFRACTION AND ELECTRON HOLOGRAPHY: EXPERIMENT AND THEORY

M.A. Van Hove

Center for Advanced Materials
Materials Sciences Division
Lawrence Berkeley Laboratory
Berkeley, CA 94720, USA

1. INTRODUCTION

Over 600 surface structures have been determined during the 20 years before 1991 in terms of interlayer relaxations, bond lengths, bond angles and adsorption sites[1,2]. Over half of these were analyzed by low-energy electron diffraction (LEED), the oldest technique in this field. Compared to other techniques of surface structure determination, LEED gives the largest flexibility of application, being suitable for most materials (metals, semiconductors, intermetallic or ionic compounds, with or without atomic or molecular adsorbates of arbitrary kind, etc.) and most structural types (reconstructed or not, with adsorption as an overlayer or an underlayer, ordered or disordered, in submonolayer or multilayer amounts, etc.). This success results from the fact that the basic physics of the LEED process are well understood and have been harnessed to accurately reproduce experimental data[3-5].

At the other end of the age spectrum of techniques for surface crystallography is the newest: electron holography. This technique is closely related to LEED and several other electron diffraction methods (like photoelectron diffraction and Auger electron diffraction). We shall examine its prospects here, since it provides the great appeal of direct atomic-scale imaging. Compared to scanning tunneling microscopy (STM), electron holography should offer more detailed positional information (bond length and directions, in particular) to a depth of several atomic layers.

In recent years the technique of LEED has been substantially expanded and increasingly applied toward both more complex surfaces and more accurate results. Complexity arises in several ways. Larger unit cells, as in the (7×7) reconstruction of the Si(111) surface, contain more independent atoms whose positions are to be found. New techniques have permitted the fitting of over 30 independent structural parameters at once, a great improvement over conventional LEED methods. Higher structural accuracy

Equilibrium Structure and Properties of Surfaces and Interfaces
Edited by A. Gonis and G.M. Stocks, Plenum Press, New York, 1992

231

demands that more atomic coordinates be determined, since one atom's relaxation affects the determination of the positions of all other atoms. Today, accuracies on the order of 0.02 Å are becoming commonplace.

New types of surface structures are also increasingly being explored. Disordered overlayers have already received considerable theoretical and experimental attention. Other forms of disorder, particularly defects, must also be studied. Steps at surfaces, a very important type of defect, are increasingly being investigated. New theoretical methods have been proposed for such systems.

The higher accuracy expected nowadays, as well as the related increase in the number of coordinates to be fit to experiment, pose a serious problem in the structural search. The slow traditional process of testing one by one the more likely structural models[4] is no longer viable. Automated search algorithms (commonly used in x-ray crystallography) are now being applied to solve this logjam, despite the inherent complications due to multiple scattering. At the same time, a larger and more accurate experimental database is required, which can only be measured with faster, more automated instrumentation.

In the following, we shall first review the state of the art in surface crystallography by LEED. Experimental approaches and innovations will be addressed in section 2. The basic methods of conventional LEED theory will be briefly reviewed in section 3, since these underlie the more recent developments. Recent progress in theoretical methods of structural determination by LEED will be discussed in sections 4 through 8 for different types of complexity: surfaces with large unit cells, with incommensurate overlayers, with disordered overlayers, with complex relaxations and with steps. The various methods will be illustrated with examples of solved surface structures.

In section 9, we shall describe electron holography, which is only in its early development stage but is unusually elegant and exciting. It shares the same basic ingredients with LEED, and is quite similar to diffuse LEED, which has in fact given rise to one version of electron holography. Other forms of electron holography, based on photo- and Auger-electron diffraction, are better known and will also be discussed in section 9.

2. EXPERIMENTAL DEVELOPMENTS

In the LEED experiment[6,7] a well-collimated beam of electrons in the 10 - 300 eV range is back-scattered from the surface of a crystal. The elastically reflected electrons that carry the surface structural information are then separated from the inelastically scattered electrons by retarding grids and detected. Ordered surface structures produce diffracted beams in directions implied by the two-dimensional surface periodicity. Disordered surface structures yield diffuse diffraction into all directions. In either case, one needs to measure the diffraction probability into a collection of appropriate directions. The database of such measured "intensities" should be large enough to solve for all the unknown structural parameters of the surface.

Accuracy of diffraction measurement demands both good surface preparation and good control of angles (angle of incidence, angle of crystallographic cut and other diffractometer parameters). Regarding surface preparation, it should be stressed that impurities at low concentrations are often responsible for stabilizing otherwise metastable structures. For instance, alkali adatoms in amounts of 0.01 monolayers can induce the missing-row reconstruction of several fcc(110) surfaces.

Angular accuracy plays an increasingly important role when more structural parameters are to be determined. For example, it has been found that a 1° error in the incidence angle can induce errors of 0.1 Å in atomic positions parallel to the surface.

Early LEED intensity measurements used the Faraday cup, a tedious approach. The post-acceleration technique later allowed the diffraction pattern to be displayed on a fluorescent screen, from which intensities were measured by spot photometry or by photography. More recently, the video camera has become the favored detector of diffraction patterns displayed on a fluorescent screen[8,9]. This efficient and convenient mode of operation is called video LEED.

Developments in electron-gun designs permit one to obtain incident electron beams with smaller diameters (1lm instead of 1mm) or to obtain greater or variable coherence length. Also, lower incident electron beam currents (10^{-9} A instead of the more standard 10^{-3} A) are utilized in order to minimize radiation damage to the crystal or to the adsorbed monolayer. Surface charging can also be drastically reduced in this manner on insulating surfaces.

A lower incident beam current allows the use of modern position-sensitive detectors, instead of the fluorescent screen: resistive anodes or wedge-and-strip detectors are coupled with microchannel plates and make it possible to digitally record the complete angular distribution of diffracted intensities[10,11]. From such angular distributions one can also generate energy-dependent beam intensities (I-V curves) or other data sets in a computer. This type of diffraction experiment is called digital LEED. It permits the detection of diffraction beam intensities with much higher signal-to-noise ratio than video LEED. As a result, it is for example possible to carry out the diffuse LEED experiment, whereby intensity modulations at all collection angles can be detected to permit structure analysis from disordered monolayers.

The diffuse LEED intensities require special treatment before comparison with theory[12-15]. First, because the diffuse intensities are typically weak, there is a danger that they are affected by inelastic intensity contributions. The energy filtering that is normally applied in LEED does not exclude inelastic electrons which have suffered phonon losses or other energy losses of about 0.25 eV or less. This issue has been studied by Ibach and Lehwald[16]: they found that on the whole inelastic losses do not affect the measured diffuse intensities adversely. However, there are particular combinations of angles and energies where the elastic intensity is very low (due to destructive interference) while the inelastic losses are larger. One might apply a narrower energy-acceptance window, as in high-resolution electron energy loss spectroscopy, but this is difficult because of the small energy differences involved and the non-uniformities of the filtering grids. An other alternative, suggested by Ibach and Lehwald, is to subtract the diffuse intensities of the clean surface from those of the overlayer-covered surface. At least those phonon losses common to both surfaces will then approximately cancel out. This is the common procedure now for measuring diffuse intensities.

Another concern with diffuse intensities, at least when measured as a function of exit angle, relates to the fact that diffuse intensities can depend strongly on any long-range order present in the disordered layer. This manifests itself in the familiar streaking or tendency to form broad spots which is often observed with LEED. As we shall discuss more fully in section 6, the diffuse intensities can be viewed as the product of an intensity due to the long-range order (the "structure factor") and an intensity due to the short-range order (the "form factor"). If we are interested in the short-range order, i.e. the constant local bonding geometry, then we should eliminate the perturbing influence of the structure factor, because it is not normally included in the LEED theory: it could only be included if one knew the degree of long-range order in the first place. This elimination is relatively easily accomplished in the case of purely two-dimensional disorder (thus excluding step disorder, for example). One can then use the logarithmic derivative of the diffuse intensity with respect to energy, keeping the parallel momentum transfer constant[12-15]. Since the structure factor remains constant with energy for constant parallel momentum transfer, it disappears. One must of course then compare these data to the logarithmic derivative of the calculated diffuse intensity.

3. BASIC LEED THEORY

We here briefly review the theoretical ingredients that go into conventional LEED methods. These are also used in more modern versions of the theory, designed for the various applications which will be discussed later. We shall not develop the underlying mathematical formalisms, but stay with a physical description of the processes that are included in the theory. More detailed treatments are available in the literature[3,6,17].

The muffin-tin model is utilized to represent the scattering potential of the surface atomic lattice felt by LEED electrons, as in band-structure calculations (in fact, potentials developed for band-structure calculations are often excellent for substrates in LEED). The muffin-tin model consists of spherically-symmetrical ion-core potentials, surrounded by a constant muffin-tin level. However, this muffin-tin level may change from one atomic layer to another. This happens of course in particular at the interface between lattice and vacuum, and can also occur between a substrate and an overlayer.

A layered structure is often adopted in LEED to describe a surface. Atomic layers parallel to the surface are defined, in a way appropriate for each particular theoretical method. In many cases, a combined-space representation is chosen, which depends intimately on this layer approach. Namely, the wavefield is expanded in terms of spherical waves within those layers, while it is expanded in terms of plane waves in the gaps between those layers. These plane waves correspond directly to the diffracted beams.

In the spherical-wave representation it is most common to use free-space Green functions to describe wave propagation from one atom to another. These functions are partial (i.e. spherical) waves. The partial-wave expansion of the scattered wave is truncated at a finite value l_{max} of the angular momentum l. This yields $(l_{max}+1)^2$ partial waves at each atom, due to the allowed values of the magnetic quantum number m. The appropriate truncation l_{max} is determined by the scattering potential, whose effect on the spherical waves is described by phase shifts.

In the spherical-wave representation there are various ways of obtaining self-consistent solutions to the multiple-scattering problem. A "giant-matrix" inversion as proposed by Beeby does the job in a closed form. Then there are perturbation expansions, such as Reverse Scattering, which can converge to the same result, when multiple scattering is not too strong.

In the plane-wave representation many methods are available to treat multiple scattering between atomic layers (whose diffraction properties must have been calculated with spherical waves). One is the Bloch-wave method with wave-matching across the surface: this relies on periodicity perpendicular to the surface in the substrate, and allows any deviations from bulk structure at the surface in the wave-matching step. Another method is layer doubling, which requires sufficient damping of the wavefield into the bulk for convergence: two layers are joined into a pair, then two pairs are joined into a quartet of layers, etc., until convergence of the diffraction properties. A popular perturbation expansion of this problem is offered by renormalized forward scattering, in which the predominance of forward scattering is used to achieve rapid convergence (as long as multiple scattering is not too strong and damping not too weak). This last method is the most efficient for stacking layers from a computational point of view. However, it more frequently fails to converge. Both layer doubling and renormalized forward scattering break down for small interlayer spacings, because many exponentially evanescent waves are then able to carry electrons from layer to layer, resulting in large matrices and difficult numerical problems.

We have mentioned damping: this is the effect of inelastic energy losses, which reduce the elastically surviving flux of electrons. The most common way of taking this into account is through a small imaginary part of the scattering potential. It is usually assumed homogeneous and isotropic. Another equivalent approach is to introduce a mean free path length, that also exponentially dampens waves. This damping is responsible for the finite width of peaks in I-V curves (intensity vs. energy plots of diffracted beams).

Finally, all proper LEED theories include Debye-Waller factors: these represent the effect of thermal vibrations, namely the elastic or quasi-inelastic removal of electron flux into diffuse directions of diffraction. At each scattering in a multiple-scattering chain a Debye-Waller factor is allowed to remove a part of the electron flux.

The new methods introduced for complex and disordered surfaces use the same physical ingredients and many of the same calculational techniques described in this section. The difference often lies in a new packaging of multiple-scattering paths into units that are better adapted to the new tasks. Alternatively, some multiple-scattering paths are neglected. Or a linear expansion is carried out to rapidly compute diffraction intensities for slightly distorted geometries. The end result either is equivalent to that of the old way, or is a suitable approximation to it.

The central problem of LEED theory is to reduce the computational cost of the diffraction simulations. In Table 1, the cost of the conventional LEED computation is characterized in terms of its dependence on l_{max}, N (the number of atoms in the two-dimensional surface unit cell), A (the two-dimensional unit cell area, which affects the number of plane waves), d (the minimum spacing between layers, which also affects the number of plane waves) and the energy E (on which depend both the number of plane waves and the value of l_{max}). The high power with which this cost rises clearly indicates where the problem lies. From Table 1 it would appear that it is sufficient to lower the energy: however, this would reduce the possible experimental database below the minimum required, and it would lead to inaccuracies due to uncertainties in the scattering potentials for surface atoms at energies below 20 eV.

Table 1. Approximate scaling of the computational cost of conventional dynamical LEED with some critical parameters for methods using the spherical-wave vs. the plane-wave representation

computational cost proportional to:		parameter
in spherical-wave space	in plane-wave space	
$l_{max}^{4 \text{ to } 6}$		l_{max} = max. ang. mom.
$N^{2 \text{ to } 3}$		N = atoms / unit cell
	$A^{2 \text{ to } 3}$	A = unit cell area
	$d^{-4 \text{ to } -6}$	d = interlayer spacing
$E^{2 \text{ to } 3}$	$E^{2 \text{ to } 3}$	E = energy

A different kind of difficulty with LEED for complex and disordered surfaces is the issue of finding an efficient procedure to determine the many unknown structural parameters. This problem is not unique to LEED, of course. It exists equally well, for example, in standard x-ray crystallography. While an improvement in the computation of diffraction from a single structure is very helpful, an efficient procedure to determine many unknown structural parameters is even more essential for a reliable result. This is especially true because the unknown parameters are correlated: they cannot be determined one by one, but must be determined all at once. The conventional trial-and-error approach prevalent in LEED rapidly becomes prohibitively cumbersome.

In the next sections we describe several general strategies that have been proposed to overcome the computational problems of LEED discussed above. Examples of structures solved with those methods are given.

4. LARGE 2D UNIT CELLS

Surface reconstructions and large adsorbates (such as molecules) often give rise to large two-dimensional unit cells. This affects the computational cost as shown in Table 1 through the quantities A and N (N, the number of atoms in the unit cell, tends to increase in proportion to A).

4.1 Full Use of Symmetry

A very effective and obvious approach to performing complex structural analyses is to exploit any available symmetries to speed up the calculations. This has been done routinely for many years in that part of the LEED problem which uses the plane-wave representation[3,4], and less frequently wherever the spherical-wave representation is used[18].

In recent years, the latter approach has been applied very effectively[19,20]. Thus, complex reconstructions have been analyzed in considerable detail in this manner. For example, the Pt(110)-(1×2) and (1×3) missing-row reconstructions were studied down to fourth-layer relaxations, both perpendicular and parallel to the surface[19,21]. Even the highly complex Si(111)-(7×7) structure has been analyzed in terms of individual bond lengths, despite its 200 displaced atoms in the (7×7) unit cell[20]. Several complex structures of metal adsorption on Si and Ge(111) have similarly been studied[22-26], as well as the Ge(111)-c(2×8) reconstruction[27].

4.2 Beam Set Neglect (BSN)

Next we consider an approximate method that is plane-wave oriented. The intention is to identify and eliminate sets of plane waves that do not contribute significantly. The "beam set neglect" (BSN) method[28] achieves this by ignoring weak third- and higher-order multiple-scattering paths. The method applies to overlayers or reconstructed layers with a superlattice or an incommensurate lattice or a disordered lattice on a perfect substrate. It recognizes the fact that only a very limited set of plane waves contribute significantly to the detected intensities: many sets of beams (plane waves) can be neglected in the calculation. As a result, the dependence of the computation cost on the unit-cell area A falls from the conventional second or third power to no dependence in the extreme limit, and to a simple proportionality in more practical approaches. In other words, the computation cost can be made almost independent of the unit-cell area or even the presence of disorder. BSN can in addition be very effectively combined with the KSLA method described below.

In BSN two important sets of beams (i.e. plane waves) are identified and included in a multiple-scattering calculation, as shown in Fig. 1. The first beam set is simply the collection of all "integer-order" beams, due to the incident beam and its diffraction by the substrate lattice. The second beam set is a copy of the first beam set, but shifted in reciprocal space parallel to the surface. The shift is such that one of the shifted beams travels in the direction of the detector. Another way to define this second beam set is to choose an emission direction and take only those beams which the substrate can diffract into that direction. For instance, if a c(2×2) overlayer lattice is present and a half-order beam intensity is to be calculated, the first beam set consists of all integer-order beams, while the second beam set consists of all half-order beams. A typical calculation will be repeated for each set of beams, the number of such sets being the ratio of the unit cell areas of the superlattice and the substrate lattice.

With BSN (coupled with KSLA, see below) a number of complex molecular adsorbate structures has been determined. These are the coadsorption of benzene with CO on Pd[29], Rh[30,31] and Pt(111)[32] (cf. Fig. 2), or of ethylidyne with CO or NO on Rh(111)[33]. For example, with Rh(111)-(3×3)-C_6H_6+2CO each (3×3) unit cell contains one benzene molecule and 2 CO molecules. Ignoring the weakly-scattering H atoms still leaves 10 overlayer atoms in each cell, with 30 unknown coordinates. This unit cell gives $3 \times 3 = 9$ times more plane waves than does the clean Rh(111)-(1×1) surface, leading to 9 times larger layer diffraction matrices in the plane-wave representation. Renormalized forward scattering would therefore take approximately $9^2 = 81$ times more computer time than for the clean surface, and layer doubling would take approximately $9^3 = 729$ times more than

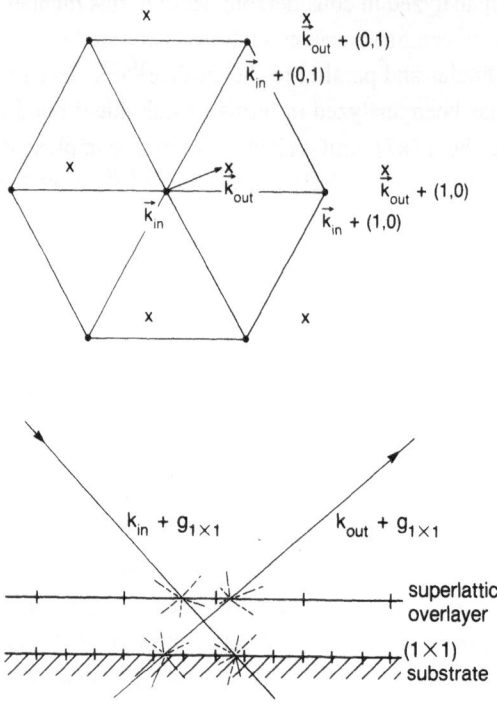

Fig. 1. The two beam sets used in the beam-set-neglect method are shown. In the top diagram, the dots represent beams (plane waves) obtained from the incident beam by diffraction from the substrate lattice. The crosses represent a similar set of beams obtained from the outgoing detected beam. These two sets are schematically shown in the bottom diagram, emphasizing scattering at the overlayer and separately at the substrate.

Substrate	(Gas Phase)	Pd(111)	Rh(111)		Pt(111)
Surface Structure		(3x3)-C_6H_6 + 2CO	(3x3)-C_6H_6 + 2CO	c($2\sqrt{3}\times4$)rect-C_6H_6 + CO	($2\sqrt{3}\times4$)rect-$2C_6H_6$ +4CO
The Structure of Benzene	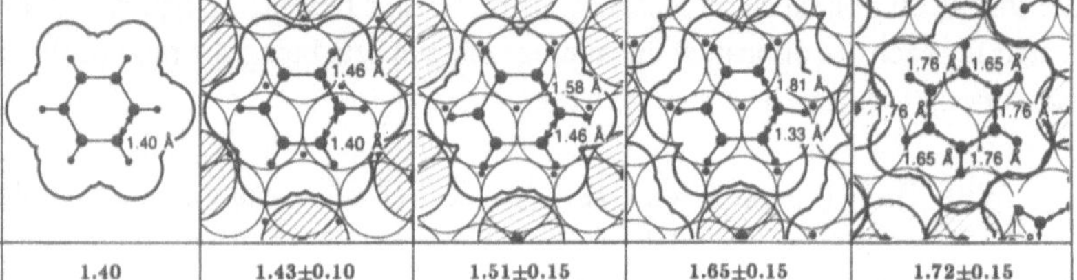 1.40 Å	1.46 Å 1.40 Å	1.58 Å 1.46 Å	1.81 Å 1.33 Å	1.76 Å 1.65 Å 1.76 Å 1.76 Å 1.65 Å 1.76 Å
C_6 Ring Radius (Å)	1.40	1.43±0.10	1.51±0.15	1.65±0.15	1.72±0.15
d_{M-C}(Å)	-	2.39±0.05	2.30±0.05	2.35±0.05	2.25±0.05
γ_{CH}(cm^{-1})'	670	720-770	780-810		830-850

Fig. 2. Coadsorption structures of benzene and CO on three metal substrates, compared with the gas phase benzene molecule, in plan view. CO molecules are shown dashed. Hydrogen positions are guessed. Benzene C-C distances are labeled, while the average benzene ring radius and the metal-C distances are indicated below, together with the umbrella-mode frequency from HREELS.

for the clean surface. With beam set neglect, one works only with double-sized matrices, resulting in only 4 or 8 times the clean-surface computation time, respectively, in the limit where one calculates only two sets of diffracted beams; this cost increases only proportionally to the number of additional beam sets calculated for comparison with experiment (in the current example there is a total of $3 \times 3 = 9$ such beam sets).

4.3 Kinematic Sub-Layer Addition (KSLA)

When two atoms in a layer are relatively distant from each other, LEED electrons have a smaller probability of scattering from both in succession, i.e. no significant multiple scattering path will connect those two atoms. (We can ignore indirect scattering paths via other layers, as such paths are taken into account by the plane-wave layer-stacking schemes). More generally, if clusters of atoms are well separated in a layer, as often is the case in a molecular layer, no significant multiple scattering will connect these clusters either. This implies that the scattering from such clusters can to a good approximation be added together by simple kinematic combination of amplitudes obtained separately from the different clusters. As a result, computer cost decreases dramatically. This we call "kinematic sublayer addition" (KSLA)[28].

The KSLA approximation is particularly effective in the case of the coadsorption of different species, such as that of benzene and CO discussed above. The diffraction properties of benzene and CO can be calculated independently, including all multiple scattering within each molecule, and then simply summed with a proper kinematic phase factor that includes their relative position in space. An additional advantage is that this simple sum can be done repeatedly for different relative positions of the molecules. The resulting layer can then be added to the substrate relatively efficiently by means of renormalized forward scattering or layer doubling. And this can also be combined with beam set neglect for greatest efficiency.

5. INCOMMENSURATE OVERLAYERS

An incommensurate overlayer has a two-dimensional periodic lattice which is independent of that of the substrate: it may have a different lattice constant and/or a different crystallographic orientation, such that the combined system has no common periodicity. This situation occurs frequently with overlayers that are strongly cohesive and can ignore the periodicity of the substrate on which they lie: for instance, graphite and oxides or other strong compounds form overlayers with their own bulk lattice, which in general does not match that of the substrate.

In terms of diffraction, incommensurate systems imply in principle that scattering can occur in all directions, not just in directions that correspond to the reciprocal lattices of the substrate or the overlayer. In the kinematic (single-scattering) limit, diffraction is in fact limited to those two reciprocal lattices: one then only obtains the two sets of diffraction spots. Multiple scattering, however, allows all combinations of reciprocal lattice vectors from the two incommensurate lattices. This creates an infinitely dense continuum of diffraction directions. Nevertheless, thanks to the rapid decay of contributions from higher-order multiple scattering, the kinematic pattern dominates, and some double-diffraction spots are also seen, while all higher-order spots are practically undetectable.

Beam set neglect has provided a convenient solution to the problem of calculating diffracted intensities from incommensurate overlayers[34]. By using exactly the same arguments as above, one can easily show that acceptably accurate calculations can be performed with just the two sets of beams defined earlier for beam set neglect. Again, the effect is to ignore weak third- and higher-order multiple-scattering paths. This approach has been successfully applied to the structure determination of a graphite layer grown from hydrocarbon decomposition on a Pt(111) substrate[34], as illustrated in Fig. 3.

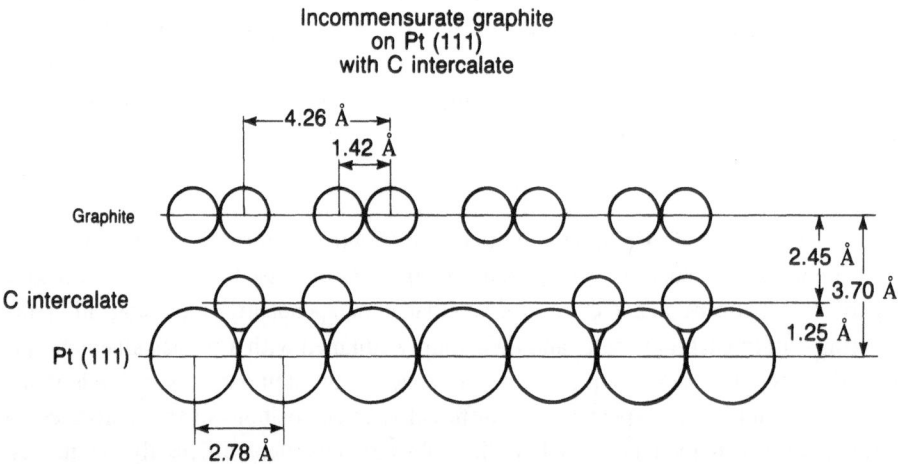

Fig. 3. Structural diagram of a graphite layer adsorbed on Pt(111), with an intercalated chemisorbed carbon layer. The graphite layer is incommensurate with the substrate, while the individual carbon atoms are bonded in hollow sites of the substrate.

6. DISORDERED OVERLAYERS: DIFFUSE LEED

6.1 Lattice-Gas Disorder

While many surfaces can be ordered with suitable sample preparation, many others are disordered. They may have disordered phases under certain conditions, or they may never order under any conditions. The LEED theory has been developed to solve the structure of some of these very interesting surfaces, principally those that have lattice-gas disorder[12,14]: an important example is the case of an atomic or molecular overlayer on a perfectly periodic substrate, in which all adsorbates occupy identical sites on the substrate. For instance, we may consider a disordered layer of oxygen on Ni(100), in which all oxygen atoms are bonded identically in four-fold hollow sites. Thus, we have identical short-range order around all adatoms, but no long-range order between adatoms.

In the absence of long-range order, diffraction is possible into all directions, yielding diffuse diffraction patterns: hence the term "diffuse LEED". If an ordered substrate is present, it can produce a set of sharp diffraction spots that is superimposed on the diffuse pattern.

To better understand how order and disorder affect the diffraction of electrons by a surface, we make a fundamental distinction between "lattice" and "basis"[35]. For an ordered surface, the lattice describes its periodicity alone, i.e. the shape, size and orientation of the unit cell, but not its contents. Thus, the lattice describes only the long-range structure. It is responsible for the presence and directions of sharp LEED beams and spots in the case of an ordered surface. The spot positions are not affected by multiple scattering. We may call this long-range lattice-induced contribution the "structure factor".

The "basis" is the set of atoms that is contained within any unit cell, together with their individual positions and scattering properties. The basis therefore describes the short-range order. The basis primarily affects the intensities of the LEED beams, and these also depend strongly on any multiple scattering. The intensities therefore depend in particular on the relative positions of the basis atoms through the multiple scattering: this is an important point, as we shall see shortly. Therefore, intensities are often said to be "dynamical". We may call this short-range basis-induced contribution the "form factor". Thus, the diffraction pattern is primarily determined by long-range order, while the diffraction intensity is primarily determined by short-range order.

One may qualitatively view the observed LEED intensities as the product of the basis-induced form factor and the lattice-induced structure factor. This product relationship becomes exact for the dilute lattice gas. In fact, most practical disordered overlayers are close enough to the limit of the dilute lattice gas for this product relationship to hold well: the criterion is that multiple scattering between adsorbates be weak, which has been verified for cases like a half monolayer of oxygen atoms on Ni(100). Thus, for a lattice gas, the final pattern can be viewed as the product of a kinematic structure factor reflecting the long-range disorder and a dynamical form factor reflecting the unique local short-range order[14,35].

For the two-dimensional lattice gas (we thus exclude defects like steps that induce a three-dimensional disorder), the structure factor can be shown to be independent of energy for constant parallel momentum transfer. As a result, there is an advantage in measuring diffuse intensities as a function of energy, but at constant parallel momentum transfer (as if one were tracking an imaginary diffraction spot as a function of energy): then the structure factor is constant and it can be ignored. Consequently, this mode of measurement is starting to be adopted in diffuse LEED, yielding "diffuse I-V curves".

If diffuse intensities are measured at constant energy as a function of exit angle (as has been the custom so far), then the unknown structure factor must be removed. This can be done for purely two-dimensional disorder by taking the logarithmic derivative of the intensity with respect to energy, keeping a constant parallel momentum transfer, since then the structure factor will divide out: this leads to the so-called Y-function[12] that replaces the intensity. Then a LEED calculation that allows all electron exit directions can simulate the experimental data and give access to all the desired short-range structural information. No long-range information about the overlayer is needed in this simulation. A disadvantage of measuring the logarithmic derivative is that the signal-to-noise ratio is degraded through the subtraction of nearly-equal intensity distributions. Another disadvantage is that the normally unknown inner potential (strictly speaking: muffin-tin constant) is less easily fit to experiment than with diffuse I-V curves.

6.2 Three-Step Method

Two schemes for diffuse LEED calculations are in use. The first, more exact scheme takes a cluster approach and is quite similar to the NEXAFS (XANES) problem[14]. This three-step approach first allows an incident plane wave to scatter in all possible ways through the substrate until it first reaches a disordered adsorbate: only plane waves are needed in this step to describe the scattering between atomic layers. In the second step, the wave that reaches the adsorbate is first expressed in spherical waves that converge on the adsorbate. These are allowed to scatter from the adsorbate and then in all possible ways through the surface until they return to the same disordered adsorbate. Finally, in the third step, these spherical waves are propagated again as plane waves in all possible ways through the substrate to the detector. This three-step approach is most suitable for simulating scanned-angle intensities at fixed energy. It can also be generalized to other forms of disorder: random vacancies, interstitial or substitutional impurities, etc.[36]. A related approach has also been developed for disordered steps, i.e. stepped surfaces with a random terrace width[37].

6.3 Beam Set Neglect

The second method, which is less accurate but speedier, uses the beam-set-neglect approach[14,35]. In the second step above, the spherical waves are replaced by the two sets of plane waves defined in the beam-set-neglect method (namely the set based on the incident beam and the set based on the detected direction, cf. section 4.2). The three steps of the first method can then be combined into a single efficient plane-wave step identical to conventional LEED methods. This approach is better adapted to the case when energy-dependent diffuse intensities (diffuse I-V curves) are measured at fixed parallel-momentum transfer.

6.4 Applications

Diffuse LEED studies have so far analyzed the structures of disordered overlayers of oxygen on W(100)[12,13] and on Ni(100)[38], CO_2 on Ni(110)[39], CO on Pt(111)[40] and C_6H_6 on Pt(111)[41].

The case of benzene on Pt(111) is illustrated in Fig. 4. Benzene could not be ordered on Pt(111) even when cooled to 170K and despite saturation coverage (coadsorption of CO does induce good ordering, however). Many adsorption geometries

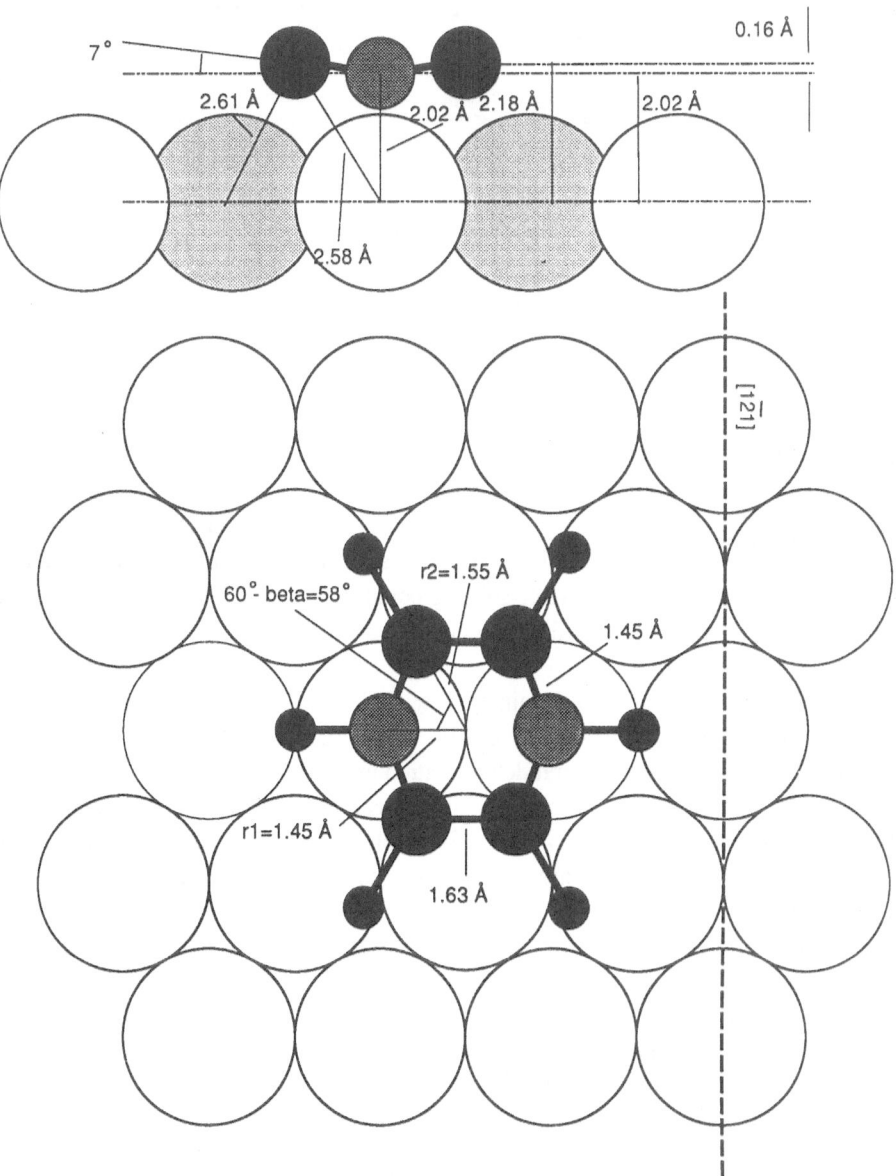

Fig. 4. Adsorption structure of disordered benzene on Pt(111), shown in side view (top panel) and top view (bottom panel). Hydrogen positions are guessed. C-C, metal-C and ring center-C distances are labeled, as well as selected interlayer spacings.

and molecular distortions were tested with the beam-set-neglect method (ignoring hydrogen and assuming a rigid substrate), including mixtures of different adsorption sites, whose diffuse intensities were assumed to simply add up. The "bridge site" shown in fig. 4 emerged as the best choice; it can be oriented three ways on the Pt(111) substrate, because of its 3-fold rotational symmetry: again, the intensities due to the 3 orientations were added.

7. COMPLEX RELAXATIONS

7.1 The Problem

For many structures, it is not difficult to determine or even guess at the approximate solution, but it is much harder to fix the details. For instance, most adatoms on close-packed metal surfaces are well known to adsorb in high-coordination sites with metal-adatom bond lengths that can easily be predicted to within 0.1 Å; but the adatoms most likely induce some relaxations within the substrate, which are complex in the sense that many atomic coordinates must be determined. Such relaxations generally are smaller than about 0.1 Å. Even if knowledge of the relaxations is not desired, to obtain the adatom-metal bond length to a given accuracy with LEED requires that the substrate relaxations also be determined to at least the same accuracy. That is because atomic coordinates are correlated in the diffraction process. Thus, accuracy of results naturally leads to complexity of results. Also, solving complex structures requires accuracy, in order to distinguish between many different possible models.

Conventional structure determination with LEED suffers in such a situation because all plausible combinations of atomic relaxations must be tested individually. Trial-and-error methods for structural determination by LEED will work up to only 4 or 5 adjustable parameters. They become prohibitively cumbersome when more than about 5 structural parameters must be determined. Even relatively simple surface structures like Ni(100)-c(2×2)-O require the fitting of at least 6 parameters if symmetrical adsorbate-induced relaxations are taken into account; asymmetries would raise that number to at least 15, if one only relaxes atoms in the overlayer and the first two substrate layers.

One approach to determining many structural parameters is to use a least-squares minimization or steepest-descent of an R-factor (the R-factor measures the disagreement between theory and experiment[6]). Starting from a guessed structure, the fit to experiment can be iteratively refined by adjusting the atomic coordinates. This can be done by numerically calculating partial derivatives with respect to each structural variable, or by using suitable approximations to these derivatives.

7.2 Tensor LEED

One approach for solving complex structures is to approximate LEED diffraction amplitudes as being linear expansions from those for a nearby surface geometry that was treated exactly. The tensor LEED technique[13,42,43] can provide the linear approximation. Tensor LEED yields excellent computation times, especially if many deviating structures around the reference structure are explored, because the linear expansion itself is a very simple operation. It appears to give excellent results for structural distortions less than about 0.4 Å, such that complex relaxations can be rapidly explored.

Tensor LEED is a perturbative approach to the calculation of LEED intensities. One starts by defining a reference structure - a particular surface structure which one guesses to be as close as possible to the actual surface structure. One then distorts this surface by moving some of the atoms to new positions. In this way one generates a trial structure which is related to the reference structure by a set of atomic displacements. Examples of such a pair of reference and trial surfaces might be an unrelaxed adsorbate structure and a relaxed version of the same structure, or an unreconstructed surface and a particular displacive reconstruction, respectively.

To first order, the difference between the amplitudes of a LEED beam scattered from the reference and trial surface, δA can be written as an expression which is linear in the atomic displacements which generate the trial structure. Thus, if one moves N atoms through $\delta r_{ij}(i = 1,...,N, j = 1,2,3)$, then:

$$(1) \qquad \delta A = \sum_{i=1}^{N}\sum_{j=1}^{3} T_{ij}\delta r_{ij}.$$

The quantity T is a tensor which depends only upon the scattering properties of the reference structure and can be calculated by performing what is essentially a full dynamical calculation for this structure. Once T is known then the diffracted intensities for many trial structures can be evaluated extremely efficiently by summing eq. (1) after substituting the appropriate set of atomic displacements.

This so-called linear version of tensor LEED is limited to atomic displacements of less than 0.1 Å, beyond which eq. (1) becomes a poor approximation. In this case one can appeal to a more sophisticated version of the theory, one which allows displacements of up to 0.4 Å, by reformulating eq. (1) as:

$$(2) \qquad \delta A = \sum_{i=1}^{N}\sum_{L,L'} T_{LL'} \cdot R_{LL'}(\delta r_{ij}).$$

$$(3) \qquad R_{LL'}(\delta r_{ij}) = j_l(\kappa\delta r_{ij})Y_{lm}(\delta r_{ij})j_{l'}(\kappa\delta r_{ij})Y_{l'm'}(\delta r_{ij}),$$

In eq. (2) the sum over the three cartesian coordinates has been replaced with a sum over angular momenta $L = (l, m)$ and $L' = (l', m')$, while the actual displacements of eq. (1) have been replaced by a function R of those displacements, consisting of the product of spherical Bessel functions and spherical harmonics. For small argument the decrease of the magnitude of the Bessel functions with order effectively cuts off the expansion. This, and the fact that R is a symmetric matrix, limits the number of terms on the LHS of eq. (2) to around 37 for the magnitude of atomic displacements for which this equation remains valid. Consequently eq. (2) is almost as straightforward as eq. (1) to evaluate and is the preferred formula for most situations.

The relative simplicity of the mathematical operations required to evaluate eqs. (1) or (2) and thus intensities from many trial surfaces has important computational implications. Firstly, the calculation is extremely fast compared to conventional full dynamical methods. For instance, by using tensor LEED theory, the computational time per trial structure can be reduced by a factor of 50 for a simple surface such as Cu(100) and by a factor of 10,000 for a p(2×2) overlayer system. Secondly, the time taken to evaluate intensities by tensor LEED is independent of the presence or lack of symmetry within any given trial structure. Therefore we can consider highly asymmetric systems with no loss of efficiency. These structures are largely inaccessible to conventional methods due to the large volume of parameter space associated with such systems and the inability to exploit time-saving symmetries. This is especially important if we are to use an automated structure search since we cannot predict in advance that the path to be taken through parameter space by the optimization procedure will pass through only symmetrical trial structures.

Tensor LEED was first applied to the structural determination of disordered oxygen on W(100) (in combination with diffuse LEED)[13], to determine not only the oxygen position, but also the structural distortions induced in the substrate. But it has become specially powerful in combination with an automated search technique, described in the next section.

7.3 Automated Search

Using tensor LEED, one can easily develop an automated structure search procedure. The use of tensor LEED in an automated structure strategy has been implemented by Cowell et al[44], as well as by Rous et al[45]. One makes an initial guess at the structure (this is the reference structure), which requires a fully-dynamical calculation. Then, as many structural parameters as one wishes can be relaxed and adjusted to experiment, by means of the tensor LEED expansion. This can be combined with a steepest-descent algorithm which automatically finds a nearby minimum in an R-factor. If no nearby minimum is found within about 0.4 Å, the method at least points in the direction where a minimum can be found, and the procedure can be restarted with a better reference structure.

With this method, the structures of Mo(100)-c(2×2)-C and Mo(100)-c(2×2)-S have been analyzed[46]. The 5 inequivalent atoms in the 3 topmost layers were allowed to relax, giving 15 adjustable parameters, in addition to the inner potential. The method was also applied to the reconstructed β-SiC(100)-c(2×2)[47] (see Fig. 5) and β-SiC(100)-p(2×1)[48] surfaces, relaxing 18 coordinates, as well as to molecule-induced restructuring in Rh(111)-(2×2)-C_2H_3, in which 30 coordinates were fitted to experiment[49] (see Fig. 6).

Another automated search strategy has been proposed by Moritz and Adams[50]. This approach uses conventional LEED theory, and applies it only to a small set of intensities at discrete energies (such as 5 to 10 intensity points at energies 15 eV apart in each beam). These intensities are compared with the corresponding experimental data through a new discrete-energy R-factor, after beam-by-beam scaling to remove the unreliable absolute intensity scales. The automated search uses a least-squares fit procedure with simultaneous refinement of all structural parameters. The intensity derivatives with respect to the varied parameters are needed and can be calculated numerically. Tensor LEED could also be used to obtain these derivatives.

This approach has been used to analyze the adsorbate-induced reconstructions of Ni(110)-(1×2)-H and Ni(110)-(2×1)-O, allowing 8 structural parameters to be determined with good accuracy[50,51].

7.4 Other Perturbations

In addition to determining atomic positions one can also generalize the above optimization procedures to search for non-structural parameters. This is especially important in the case of the inner potential (muffin-tin zero), variations of which are strongly correlated with the values of all structural parameters.

It is also possible to use tensor LEED to extend the search to surface properties such as the Debye temperature of each atomic species and anisotropic vibrations in the near-surface region.

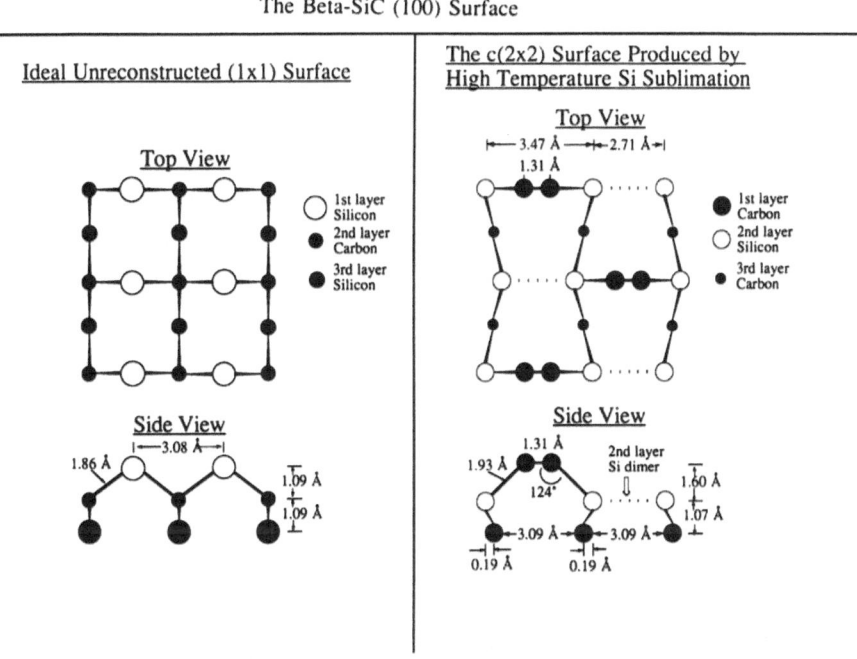

Fig. 5. Structure of the c(2x2) reconstruction of β-SiC(100) at right, in top and side views, compared with an unreconstructed, unrelaxed bulk termination at left. Visible are C-C pairs parallel to the surface and a second-layer Si-Si dimerization.

Thereby, search schemes are available which require a computation time proportional to the number of unknown parameters, rather than exponential in that number - an immense improvement over conventional search strategies. However, the R-factor minimum found by automated search cannot be guaranteed to be the global minimum, i.e. may not always give the correct answer; this is a limitation of any procedure involving diffraction. It is often necessary to start the automated tensor LEED from several very different reference structures, e.g. different adsorption sites, and see which one gives the best final fit.

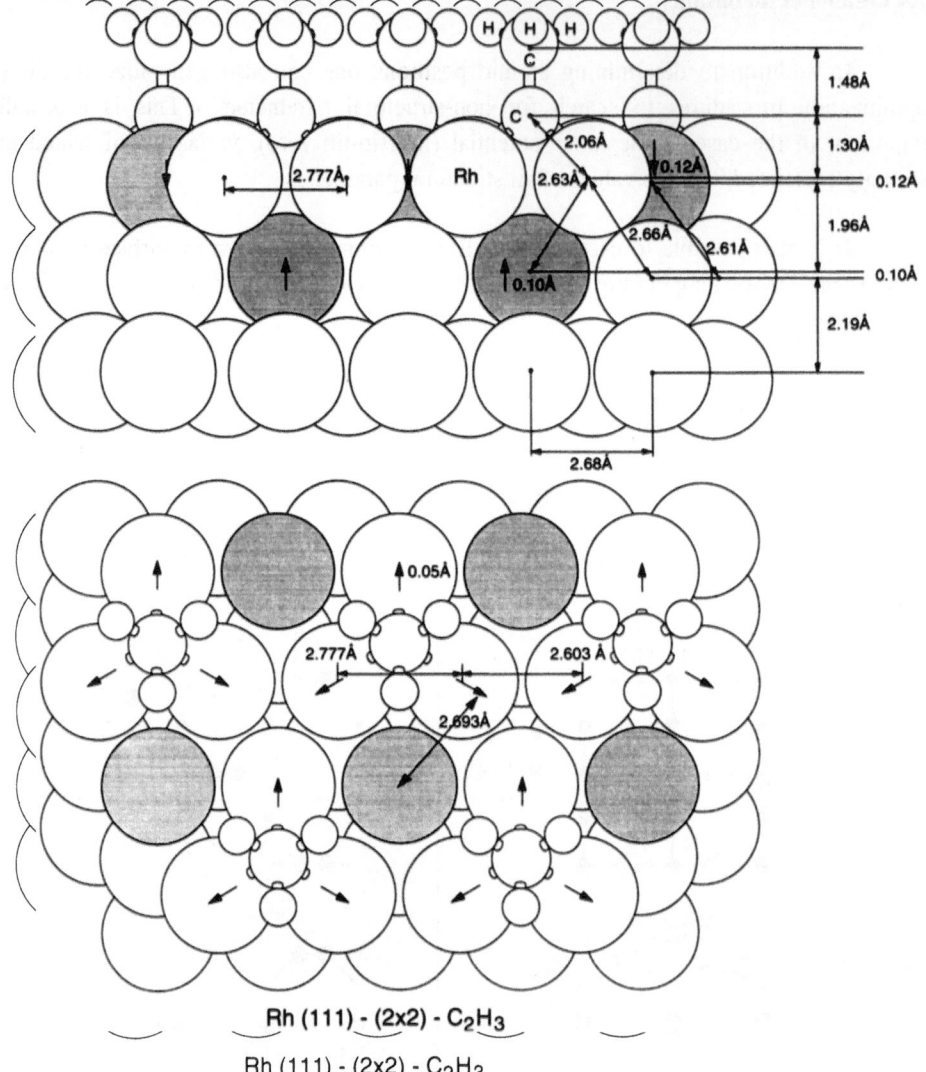

Rh (111) - (2x2) - C$_2$H$_3$

Rh (111) - (2x2) - C$_2$H$_3$

Fig. 6. Structure of ethylidyne on Rh(111), including adsorbate-induced relaxations in the metal. The side view (at top) shows metal relaxations perpendicular to the surface (gray atoms have been displaced as shown by arrows). The top view (below) shows metal relaxations parallel to the surface, in the direction of the arrows.

Since tensor LEED can treat a weakly-scattering atom as a perturbation (relative to vacuum), it should be possible to efficiently search for atoms like hydrogen: this is especially valuable in hydrogen-containing admolecules, because the often numerous hydrogen atoms add many unknown structural parameters (3 per atom), thereby vastly complicating the structural search. One can also envisage searching for compound structures (as in alloys), where one component is viewed as a perturbation relative to the other component, i.e. the reference structure is elemental rather than a compound. This can be generalized to many kinds of point defects, together with any relaxations of atomic positions around the point defects.

7.5 Direct Methods

Tensor LEED also leads to direct methods to solve surface structures[52-54]. This approach tries to entirely circumvent the need for a structural search. Squaring and taking the absolute value of either eq. (1) or eq. (2), then truncating to first order in atomic displacements, one obtains corresponding linear expansions for the diffracted intensity. In either case, the left-hand side (the reflected intensity of the actual structure) can be obtained from experiment. On the right-hand side, the tensor is known from a nearby guessed reference structure by performing a full dynamical calculation for that guessed structure. Thus one can find the displacement linking the guessed structure with the actual structure by solving a straightforward set of linear equations. If the linear expansion from guessed structure to actual structure is accurate, the structure is solved in one step. Otherwise, the procedure can be iterated to the required accuracy.

There are several implementations of this idea, allowing different information to be extracted from experimental data. The simplest one starts from eq. (1) and leads to a simple matrix equation to be inverted[52,54]. But it requires a very good guess for the structure, since eq. (1) is not accurate beyond about 0.1 Å from the guessed structure. A better approximation uses eq. (2), which extends the range of validity to about 0.4 Å[53,54], but this version has not yet proven as reliable.

However, one can also expand and retain terms to all orders, rather than only the linear terms. Then a series is obtained in the powers of the displacement components[54]. There are usually enough data points to determine not only the displacements, but their powers as well, thus delivering the moments of the probability distribution of the atomic locations. These provide additional information to describe thermal and other site disorder, including anharmonic vibrational effects, diffusion and multi-site adsorption. In fact, the probability distribution of the atomic locations can be related to the potential energy function within which the surface atoms move, thus determining it.

A further implementation treats added atoms, e.g. light atoms, as a perturbation[54]. If the reference (guessed) structure is periodic parallel to the surface, the change in diffracted intensity will have the same periodicity and can be expanded in a Fourier series, whose terms can be obtained in a way similar to the one described above for the power-series expansion. This approach is particularly useful for "delocalized" atoms like hydrogen, which can move about with large amplitudes across the surface. Their position distribution can be well described by Fourier components.

The direct method has been applied to several systems so far, using earlier determined or easily guessed structures as starting points for the reference calculations. In its simplest implementation, it was used to analyze the structures of clean Rh(110) and clean W(100), as well as the adsorbate system Rh(110)-(1×1)-2H[52]. More sophisticated implementations were used to investigate the adsorbate structures Ni(100)-c(2×2)-O (including substrate relaxation)[53] and Rh(100)-c(2×2)-O[54].

8. STEPPED SURFACES

8.1 The Problem

Relatively little structural information is available so far for stepped surfaces, especially when the terraces between steps are wide. Some structural information is available from LEED up to the terrace width of the fcc(331) surface, which has terraces with a width of 2 atomic diameters[1,55]. This is in part due to the problem that conventional LEED theories diverge for stepped surfaces and so cannot be used.

The cause of the divergence in LEED lies in the plane-wave representation of the wavefield between atomic layers[6]. When atomic layers (parallel to the surface) are narrowly spaced, as they are in stepped surfaces, the plane-wave expansion requires the inclusion of a large number of "evanescent" plane waves (these are non-oscillatory damped waves that decay slowly enough to transmit electrons from one atomic layer to the next), in addition to the oscillatory "propagating" plane waves. The total number of plane waves n is approximately proportional to the square of the inverse of the interlayer spacing, which spacing in turn is proportional to the inverse of the absolute length $u = \sqrt{(h^2 + k^2 + l^2)}$ of the surface's Miller-index vector (hkl). Thus n scales approximately as $h^2 + k^2 + l^2$, leading rapidly to convergence difficulties for high-Miller-index surfaces. In practice, with most common materials, this limits the use of the plane-wave representation to spacings larger than about 1 Å. Even the most robust plane-wave algorithms, layer doubling and the Bloch-wave method[3,6,17], often diverge for interlayer spacings below about 1 Å; for reference, Cu(331) has a bulk interlayer spacing of 0.83 Å.

A composite-layer approach is possible when several layers are closely-spaced, but separated from other layers by a sufficiently large spacing[4,6]. This situation often occurs with reconstructions and overlayers on low-Miller-index surfaces. Then the small spacings can be limited to the reconstructed layers or adsorbate layers, which can be combined into a new composite layer, while plane waves are used between the composite layer and the other layers.

However, with stepped surfaces no wide layer spacings are present to justify the composite-layer approach. One needs to abandon plane waves in favor of spherical waves.

Spherical-wave schemes are always more computer-intensive than are plane-wave schemes. Therefore, hybrid schemes have been proposed, which push down the minimum spacing that can be handled. For instance, Jepsen proposed a scheme in which waves approaching an atomic layer are expressed in plane waves, while those leaving an atomic layer are expressed in spherical waves. In this way, the minimum spacing is halved[56]. A related approach combines two slabs of layers through a doubling procedure in which the slabs are separated by another layer[57]. The scattering between the two slabs (ignoring the sandwiched intermediate layer) is treated with plane waves, while the scattering involving the sandwiched layer is expressed in spherical waves. Here also, the minimum spacing has been approximately halved.

8.2 The Real-Space Multiple Scattering Method

For smaller interlayer spacings, plane waves must be abandoned altogether. This approach is taken in a recent method[58] based on a real-space multiple scattering theory (RS-MST) which was first developed for and applied to electronic band-structure problems[59]. RS-MST uses the principle of the removal invariance which holds for semi-infinite periodic lattices: removing a layer from the free end of such a lattice does not change its reflectivity (except for a trivial phase factor), because the resulting surface is identical to the original one (except for being displaced with respect to it by the thickness of a single layer). This removal invariance provides a self-consistency condition for the reflectivity, which can be solved numerically. More generally, one can say that the full scattering t-matrix of the system is invariant with respect to the removal of any finite number of planes from the surface. The removal invariance property is then used to construct a self-consistent equation for the full t-matrix, Green function and consequently the reflectivity of the surface.

This method exploits the semi-infinite periodicity of the unrelaxed surface, corresponding to the ideal bulk termination of the crystal. In general, however, the surface will relax or be covered with adsorbates. Such deviations from the ideal semi-infinite bulk lattice can be incorporated in at least two ways. One may calculate the diffraction by joining the semi-infinite ideally-terminated bulk to a slab of a few atomic planes representative of the deviating surface, all in the spherical-wave representation. Alternatively, for relatively small deviations from the bulk structure, one may consider the actual surface as a structural distortion of the bulk termination amenable to treatment by a perturbative approach such as tensor LEED. Either way, the crucial ingredient is a description of the ideal bulk termination. Once this has been obtained, the diffraction from the actual surface can easily be solved.

Jepsen's method[56] was applied, for example, to the Fe(210) surface (whose terraces are two atomic diameters wide), with bulk interlayer spacings of 0.641 Å and surface interlayer spacings as small as 0.50 Å. The newest method has been applied to the Pt(210) surface[60], which is not only stepped, but heavily kinked (i.e. the step edges are not straight but ragged), cf. Fig. 7. The bulk interlayer spacing in this surface is 0.8765 Å, which is reduced to 0.675 Å at the surface; however, Pt is a more difficult case, since multiple scattering is unusually strong in this metal.

9. ELECTRON HOLOGRAPHY

Electron holography is the newest addition to the long list of surface structural tools. It bears close similarities with LEED at the fundamental level of electron-atom scattering. However, it provides a radically different approach to the extraction of surface structural information through direct imaging. Because of its novelty, electron holography deserves here a more explicit and elementary treatment than LEED.

9.1 Basic Idea

Holography makes a reference wave interfere with a scattered wave: the two-dimensional map of the interference pattern is the hologram. By shining a wave similar to the reference wave (e.g. time-reversed) through the hologram, one can reconstuct an image (and a twin image) of the original scattering object.

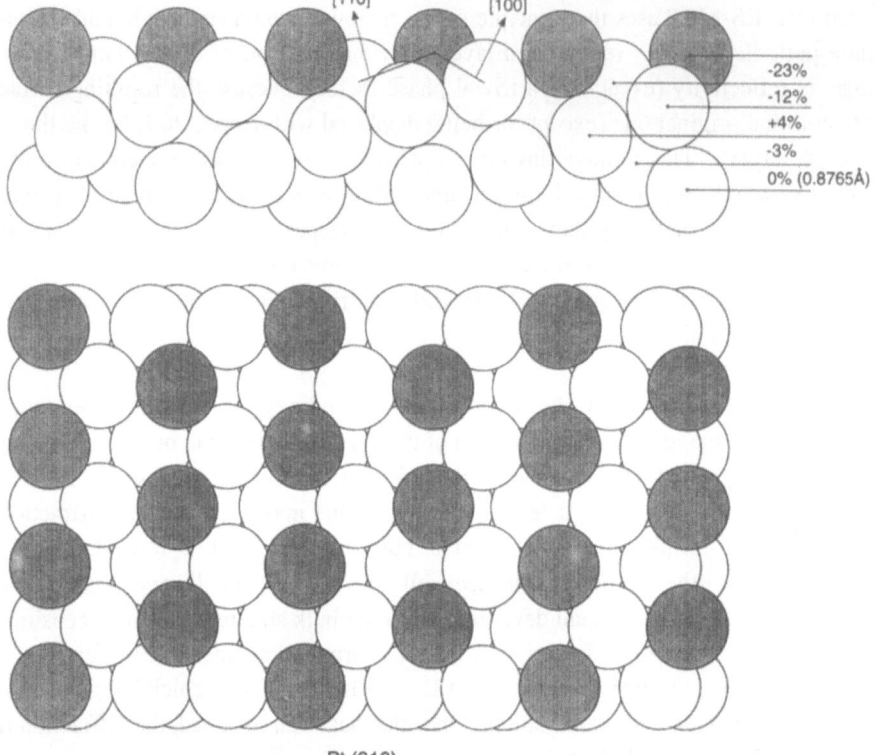

Fig. 7. Side view (at top) and top view (below) of the Pt(210) surface, indicating changes in the four topmost interlayer spacings (relative to the bulk spacing). Relaxations parallel to the surface are found to be negligible, within the error bars.

This procedure is familiar in optical holography with laser light. But the holographic idea, originated by D. Gabor in 1948, was in fact aimed at using electrons rather than light, and the intention was to image the atomic structure of crystals[61]. The image reconstruction process would require electron lenses that yield an accuracy better than 1 Å, the size of crystalline details of interest: however, such accuracy is still unattainable today (except with very-small-angle deviations, as used in electron microscopy).

It was more recently realized by Szöke[62] and Barton[63] that an electron emitted from a surface atom (e.g. a photoelectron) could serve as the reference wave. That wave would scatter from nearby atoms and interfere with the reference wave at a detector to produce the two-dimensional electron hologram, as a function of the polar and azimuthal exit angles. But instead of using a real wave for imaging, the reconstruction process was proposed to be carried out numerically in a computer. The imaging then basically amounts to a simple two-dimensional Fourier transformation of the measured hologram.

Let us illustrate this mathematically for a simple case: one atom at the origin $r = 0$ emitting an s-wave, and one nearby scattering atom at location r_1. The emitter creates a spherical reference wave

$$\frac{e^{ikr}}{r},$$

where k is the wavenumber and r is the radial distance from the emitting atom. This wave will scatter from the second atom, with a scattering amplitude f, generating a second spherical wave

$$\frac{e^{ikr_1}}{r_1}f\frac{e^{ik|\mathbf{r}-\mathbf{r}_1|}}{|\mathbf{r}-\mathbf{r}_1|}$$

centered on the scatterer. At the detector, at a large distance $r = R \gg r_1$, the sum of reference and scattered waves can be well approximated by

$$\psi(\mathbf{R}) = \frac{e^{ikR}}{R} + f\frac{e^{ikr_1}e^{ikR}e^{-i\mathbf{k}\cdot\mathbf{r}_1}}{r_1 R}$$

The hologram is then the corresponding probability distribution (or "intensity")

$$(4) \qquad I(\mathbf{k}) = \psi * \psi(\mathbf{R}) = \frac{1}{R^2}\left[1 + f*\frac{e^{-ikr_1}e^{i\mathbf{k}\cdot\mathbf{r}_1}}{r_1} + c.c. + \frac{|f|^2}{r_1^2}\right],$$

where the "$c.c.$" term represents the complex conjugate of the preceding term, and wavevector $\mathbf{k}$ has the same direction as $\mathbf{R}$ and additionally specifies the energy. A calculated plot of $I(\mathbf{k})$ is shown in Fig. 8 for a carbon emitter and an oxygen scatterer,

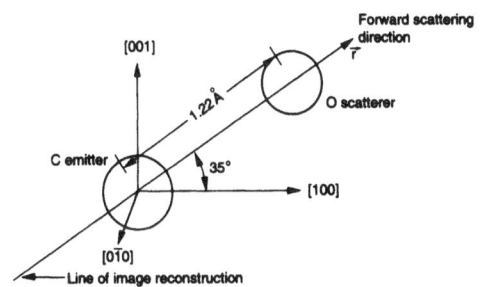

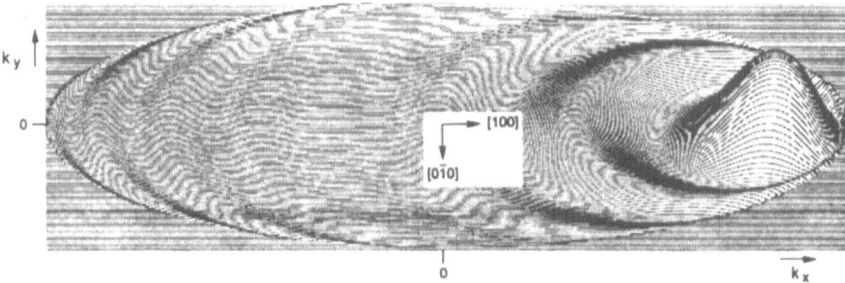

Fig. 8. Calculated hologram (lower panel), representing the flat projection of a hemisphere of exit directions, due to s-wave emission from carbon in a CO molecule (sketched in the upper panel). The CO axis is oriented at 55° from a fictitious surface normal, which is centered in this plot against exit momentum parallel to the surface (the polar angle range extends to 85° from normal emission). The emitted electron energy is 1206 eV and no thermal effect is included. The peak at right is the forward focusing peak oriented along the C-O axis.

253

representing a fixed CO molecule: it consists of a forward peak (directed from C to O) centered on a polar angle of 55° from the fictitious surface normal, surrounded by fringes. The forward peak is mainly due to an angle dependence of f and is not useful for imaging atoms (it does, however, conveniently indicate the interatomic direction). The fringes, on the other hand, contain all the relevant structural information, including interatomic distances, since the fringes represent the beat between reference wave and scattered wave.

Let us perform the image reconstruction by Fourier transformation over the two-dimensional space of detected directions $\mathbf{k}$, leaving the magnitude k fixed (as determined by the fixed energy):

$$(5) \qquad A(\mathbf{r}) = \iint d^2k \, I(\mathbf{k}) \, e^{-i\mathbf{k}\cdot\mathbf{r}}.$$

The image can be plotted as $|A(\mathbf{r})|$ at all three-dimensional positions $\mathbf{r}$; Fig. 9a shows a one-dimensional cut through this three-dimensional image for the case of the CO molecule. The Fourier transformation is equivalent to shining a suitable new wave related to the reference wave through the hologram[63].

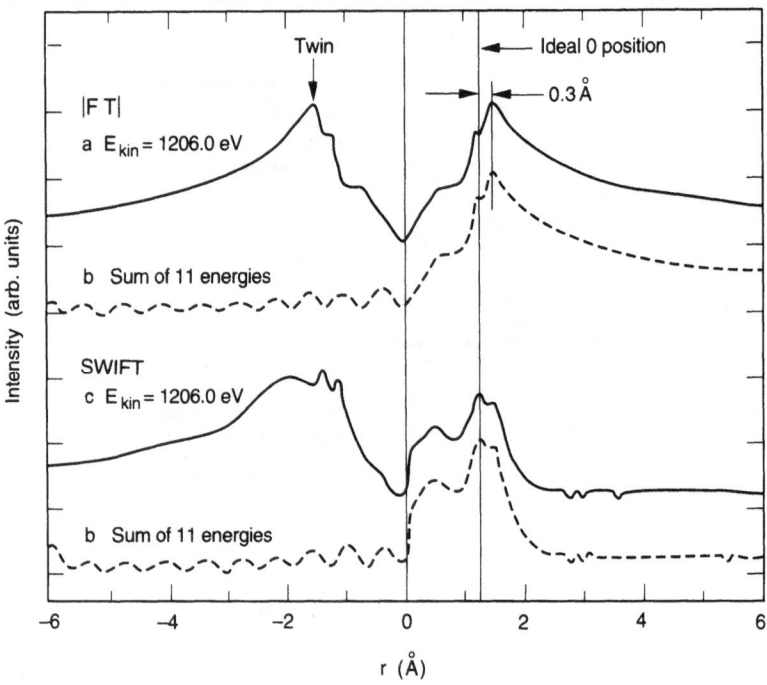

Fig. 9. Real-space images reconstructed from the hologram shown in Fig. 8 for a CO molecule: only the 1-dimensional image along the C-O axis is shown, with the C position at $r = 0$ and the expected O position at positive r. (a) Simple Fourier transform, following eq. (5), with $|FT| = |A|$. (b) Same as (a), but with phased summation over 11 energies centered on 1206 eV. (c) As (a), but using SWIFT, following eq. (8). (d) Same as (a), but combining the summation over 11 energies and SWIFT, following eq. (9).

We now ask what the meaning of this "image" is. To that end, we consider the meaning of the Fourier transformation of eq. (5). Suppose we have, in general, a distribution $S(\mathbf{r})$ of point sources, each emitting a spherical s-wave; this distribution shall be weighted and phased, so as to simulate the effect of multiple scattering, which in effect creates additional coherent electron sources. We thus allow no scattering, only direct emission to the detector. Also we assume that we can measure the *amplitude* and Fourier transform it. The measured amplitude becomes at large R:

$$\psi(\mathbf{k}, \mathbf{R}) = \frac{e^{ikR}}{R} \iiint d^3r \, S(\mathbf{r}) \, e^{-i\mathbf{k} \cdot \mathbf{r}}.$$

Backtransforming recovers $S(\mathbf{r})$, approximately:

$$S(\mathbf{r}) = \iint d^2k \, \psi(\mathbf{k}, \mathbf{R}) \, e^{-i\mathbf{k} \cdot \mathbf{R}},$$

(one may extend the double integral to a triple integral, by also integrating over energy, as will be discussed in section 9.5). We have recovered the initial source distribution, which is the ideal goal of holography. In practice, this ideal goal can of course not be achieved, due to a number of factors. One of them is the use of the absolute value of the amplitude squared; others are atomic phase shifts, finite integration range, etc. Thus, the actual process based on experiment yields a fictitious image $S'(\mathbf{r})$ that compensates for all those effects. In other words, the actual image $S'(\mathbf{r})$ is an *effective* point source distribution such that its emitted complex *amplitude* $\psi'(\mathbf{k}, \mathbf{R})$ would equal the measured real *intensity* $I(\mathbf{k})$: one sees an image that represents a set of effective emitters, placed in such a way that they produce a wavefield which is equal to the measured hologram intensity.

We can also discuss the actual image by analyzing how it comes about in our CO example, cf. Fig. 9a. Let us initially assume that the atomic scattering amplitude f is independent of scattering angle, so that it can be extracted from the integrals. We obtain four terms from eq. (5), corresponding to the four terms in eq. (4), respectively. The first term is structureless and uninteresting. The second we call real image:

$$(6) \qquad A_r(\mathbf{r}) = \frac{f^*}{R^2 r_1} e^{-ikr_1} \iint d^2k \, e^{i\mathbf{k} \cdot \mathbf{r}_1} \, e^{-i\mathbf{k} \cdot \mathbf{r}}.$$

This integral singles out the fringes in the hologram and converts them to an image of the scatterer. If the integration could be carried out over all $\mathbf{k}$-space in three dimensions, it would produce a delta-function at $\mathbf{r}_1$, giving an infinitely sharp image of the scattering atom, as one would ideally hope. In fact, one obtains a broad peak, as seen in Fig. 9a.

The third term we call twin image:

$$(7) \qquad A_t(\mathbf{r}) = \frac{f}{R^2 r_1} e^{ikr_1} \iint d^2k \, e^{-i\mathbf{k} \cdot \mathbf{r}_1} \, e^{-i\mathbf{k} \cdot \mathbf{r}}.$$

This integral would likewise produce a peak at position $-\mathbf{r}_1$: the twin image, which is symmetrically identical to the real image of eq. (6). It also singles out the fringes in the hologram, producing the symmetrical twin peak in Fig. 9a. Note that this twin image is due to the loss of the absolute phase of the wavefunction.

The fourth term is called the self-image: it is uninteresting in this 2-atom example (as long as f does not depend on the scattering angle).

Our discussion so far has shown that it is in principle easy to obtain a computed "real image" of the environment of an emitting atom in a surface. A first and important question is the achievable resolution, i.e. the sharpness of the features that represent individual atoms. We shall discuss this next. However, it is clear that there are practical complications which must be faced to turn electron holography into a useful tool for determining atomic positions. We shall address these before evaluating different experimental approaches and finally reviewing solutions to overcome some of the complications.

9.2 Image Resolution

Clearly, the image resolution should improve with increasing electron energy, as the electron wavelength decreases. But the length of the data range available from experiment also strongly affects the sharpness of Fourier transforms: a larger data range will improve resolution.

Suppose the data range is limited to a circular polar cap defined by all polar angles within h of the normal exit direction, and including all azimuthal angles. Then the resolution perpendicular to the surface is given by[64] $\Delta z = 2\pi/[k(1-\cos\theta)]$, while the resolution parallel to the surface is given by $\Delta x = \pi/(k\sin\theta)$. The denominators in these expressions represent the lengths of the ranges of accessible momentum values. This yields the ideal resolution, disregarding practical complications to be discussed shortly. (More generally, one could measure a circular polar cap about any exit direction: then the resolutions Δz and Δx apply to the directions along the cap axis and perpendicular to it, respectively.)

As an example, for a 1000 eV photoelectron (wavelength 0.4 Å) with $\theta = 40°$ we get $\Delta z = 1.66$ Å and $\Delta x = 0.30$ Å. A measured aperture of $\theta = 90°$ (i.e. the full hemisphere emerging from the surface, which is difficult to achieve in practice) would improve this to $\Delta z = 0.39$ Å and $\Delta x = 0.19$ Å. Such a parallel resolution Δx may be acceptable, but a perpendicular resolution Δz of over 1Å is not useful. Higher energies would help, but not much, because of the slow square-root relationship between wavenumber k and energy E. And, unfortunately, higher energies lead to weaker experimental signals that become difficult to detect. It appears that most experiments of this kind will probably have to remain below about 1000 eV. Thus, the resolutions quoted here must at present be considered the best achievable by imaging. As such, the technique of electron holography would mainly serve to directly, but only qualitatively, image relative atomic positions, without yielding bond lengths competitive with LEED and other accurate techniques.

9.3 Complications

A number of phenomena complicate the holographic reconstruction process described above and degrade the resolution. We shall address here: the twin image; the self-image that arises especially when imaging a periodic lattice; phase shifts and angle dependence of the atomic scattering amplitude f; multiple scattering of electrons from atom to atom; and multiple simultaneous emitters giving overlapping holograms.

We have already described the origin of the twin image, very familiar in optical holography. The twin perturbs the image interpretation when it overlaps with the real image of another atom. This would happen when two atoms are positioned approximately on opposite sides of the emitting atom: then the two real images nearly coincide with the two twins, making it difficult to tell where the real atoms actually are. This is not a problem if the emitting atom is an overlayer atom which only has neighbors on the substrate side: then all twins show up on the vacuum side of the emitter and can be ignored. A procedure for reducing twin images is desirable: this has now been achieved, as we shall see later.

Another complication is the self-image. In its simplest form the self-image is due to the fourth term in eq. (4). It is the Fourier transform of the atomic scattering factor f^*f, and is a broad function centered on the origin $r = 0$: this is harmless. However, when there are many nearby scatterers, this term includes a sum over all differences of atomic positions: it then produces additional artificial images as if each scatterer acted as an emitter[65]. These images reinforce each other when the atoms are ordered on a periodic lattice. Since the self-image scales with $|f|^2$, it tends to be strongest for heavy atoms. It is also strong when forward scattering is involved: f is usually strongly forward peaked, and frequently $|f|$ exceeds 1 near the forward direction, i.e. the scattered wave can be stronger than the reference wave.

The angular dependence of f is a serious complication that has received considerable attention[65-68]. Its main effect is apparent when the scatterer lies between emitter and detector: then strong forward focusing of the emitted electron wave produces a large peak in the hologram in the direction from emitter to scatterer (see fig. 8). This forward-focusing peak has a FWHM of typically 10 - 20°. (The Debye-Waller factor, valid for small vibration amplitudes, would further sharpen this, while on the other hand, large-amplitude rotational vibrations would broaden and weaken the forward peak). The factors f and f^* can therefore not be removed outside the integrals for A_r and A_t in eqs. (6) and (7). Worse is the fact that the phase of f depends strongly on scattering angle. It is common to find a difference of over 2π in the scattering phase between forward and backward scattering. The effect is to change the density of fringes and their locations in the hologram. This can displace the real and twin images of a scatterer by 1 Å or more, depending on the atomic element; the phases are such that the image is moved away from the emitter. Moreover, the displaced peak can be further broadened along the emitter-scatterer axis by over 1 Å. Such errors in imaging are visible in Fig. 9a and are clearly intolerable. Good progress has however been achieved in partially correcting these errors for simple systems, as we shall discuss later.

The displacement of the image due to the scattering phase shift can be visualized in terms of the well-established concept of forward focusing. In essence, the electron wave coming from the emitter converges beyond the scatterer into a fuzzy focus. This focus can be regarded as a source of emitted waves, which then gets imaged by the Fourier transformation: one obtains an image of the fuzzy focus rather than of the atom, since the electrons are perceived to originate in the focus.

Another effect of the angular dependence of f is the weakness of the scattered wave at large angles from the forward direction (especially after the Debye-Waller factor has been included). The contribution from each scatterer to the hologram is thus limited in angle: only a cone of perhaps 45° half-angle is strongly sensitive to a given scatterer. This is important in connection with the possible resolution of the reconstructed image. The aperture h used in section 9.2 to quantify the resolution cannot be larger than this cone half-angle. Another consequence is that the so-called "back-scattering" mode of electron holography, in which overlayer-emitted electrons are back-scattered by the substrate toward the detector, is faced with a very weak signal that is difficult to measure[69].

The next complicating factor in electron holography is multiple scattering of the electrons from one atom to another. The effect of multiple scattering is always to cause additional wavelets to leave the various scattering atoms for the detector[70]: these wavelets interfere with the reference wave, creating additional collections of interference fringes, and also coherently interfere with the "single-scattering" fringes discussed so far. The hologram thus becomes complex and the Fourier transform may be seriously perturbed. One manifestation is similar to the self-image: each scattering atom can act as an emitter and produce an additional hologram due to its particular neighborhood. Here again, a partial cure has been proposed and successfully tested for simple structures (see below).

At the energies of interest (300 - 1000 eV), multiple scattering occurs mostly in the forward direction. The special case of multiple forward scattering deserves particular mention because it is frequently encountered and because it surprisingly turns out to reduce the unwelcome influence of forward focusing discussed above[65,68]. Multiple forward scattering occurs when an emitter finds itself aligned with a chain of atoms. This applies to emitters in lattice positions several layers below a crystalline surface: they can emit electrons along chains of atoms toward the surface. In this instance it is found that "defocusing" takes place. The strong forward focusing is rapidly degraded as the electron wave scatters from one atom after another, thereby reducing the angle dependence of the scattering phase and magnitude. The forward peak is then replaced by a less peaked but more complex set of fringes that on average smooth out the effects of the forward peak.

Finally, we turn to multiple simultaneous emitters. One can chemically select a particular emitter element with photoemission and Auger emission, but all atoms of that element will be able to emit during the experiment. If all the emitters have the same atomic environment (i.e. the same neighborhood structure, including its orientation), only one common hologram results. However, if different emitters have different environments, each environment will produce individual holograms that add together incoherently. This happens, for instance, with metal atoms emitting from different depths in the surface, or with inequivalent carbon atoms in a molecule, etc. It can then become very hard to disentangle the contributions from the inequivalent atoms.

9.4 Electron Holography Experiments

Several sources of electrons have been proposed for performing holography at surfaces. First, photoemission from core levels can be used[63,68,71], which provides chemical selection and in which each atom emits independently of the others. The initial state is then often well known, e.g. an s-wave or a p-wave. This initial state is important, as it is the reference wave (we assumed an s-wave in section 9.1): it must be known to permit the image reconstruction. A drawback of photoemission is the need for synchrotron radiation to accumulate the large necessary database.

Auger electrons can also be used[67,71]. They have the same characteristics as photoelectrons, except that at low energies the initial state is unknown and unmeasurable; at high energies (> 500 eV) an s-wave is a good approximation. Auger electrons are convenient in that no synchrotron radiation is needed. This approach seems the most promising at present.

The inelastic scattering of electrons can also serve for electron holography[72,71], the scattered electron (also called Kikuchi electron) behaving similarly to a photoelectron or Auger electron. However, the initial state is largely unknown. And no chemical selectivity is possible.

More promising than elastic electron scattering is diffuse LEED[73,74], in which the wave leaving a disordered atom serves as the initial state. The randomness of the disorder assures the incoherence of the electrons leaving different disordered atoms. Chemical selectivity is not available, except when only one atomic element is disordered. Here again the initial state may be unknown, but it may be smooth enough to allow imaging, or it may be calculated in some cases. Experimental difficulties with this approach are the presence of sharp undesired diffraction spots due to the ordered substrate and the problem of separating out the elastic from the inelastic diffuse intensity.

To date, holograms have been measured with photoelectrons, Auger electrons, Kikuchi electrons and diffuse LEED. However, most of the research into the potential of electron holography has employed simulated holograms, calculated with existing electron scattering programs.

9.5 Improvements in Image Reconstruction

We shall now review two solutions that have been proposed to circumvent a number of the complications faced in electron holography. This field is in full development and it should be expected that new ideas will emerge to complement or replace the currently proposed methods.

The first solution addresses the problem of the forward-scattering peak of f, together with its phase[66-68]. The hope is to remove both the displacement and the broadening of the atomic images.

Let us rewrite eq. (6) for $A_r(\mathbf{r})$ with f^* included within the integral, since it is a strong function of the direction of $\mathbf{k}$:

$$A_r(\mathbf{r}) = \frac{1}{R^2 r_1} e^{-ikr_1} \iint d^2k \, f^* \, e^{ik.r_1} \, e^{-ik.r}.$$

If we had first divided $I(\mathbf{k})$ of eq. (4) by f^*, the factor f^* would not appear in this integral and one could get much closer to the desired ideal imaging of the technique as implied by the simple integral of eq. (6). The suggestion thus is to evaluate

$$(8) \qquad A(\mathbf{r}) = \iint d^2k \, \frac{I(\mathbf{k})}{f^*} \, e^{-i\mathbf{k}\cdot\mathbf{r}}.$$

rather than eq. (5). This method is called the scattered-wave included Fourier transform (SWIFT)[66,68].

The effectiveness of this forward-focusing correction has been borne out for simple systems, including our CO example, as shown in Fig. 9c: the real image is in fact shifted back to within 0.1 Å of its correct position, while its resolution is close to the ideal value. (One might object that f^* must be known for this approach to work: this is correct, but f^* varies relatively slowly and smoothly with chemical element, and one usually has a good feeling for which atoms are involved; f^* can thus be precalculated relatively easily; likewise the thermal vibration effects can be approximately guessed.)

A welcome by-product of SWIFT is that the twin image is displaced even further away from its ideal position, revealing it as a twin. This results from the mismatch of phases introduced by the correction in the transform for the twin: the corresponding integrand contains the factor f/f^*, which throws the phases off. Unfortunately, the twin image is also enhanced in this process.

Note that in eq. (8) for SWIFT, the division by f^* is introduced to correct for the forward focusing by atoms to be imaged. Thus, f^* should in principle be fixed by the (a priori unknown) direction of the emitter-scatterer axis $\mathbf{r}_1$. Eq. (8) relaxes this requirement by allowing the argument of f^* to be the integration variable $\mathbf{k}$. Thus, SWIFT does not require prior knowledge of atomic directions.

A similar approach which does use knowledge of the direction of the emitter-scatterer axis $\mathbf{r}_1$ is the phase-shift correction method[67]. This method essentially divides $I(\mathbf{k})$ by f^* fixed to the emitter-scatterer axis. The axis directions are deduced (rather accurately for simple systems) from the directions for the forward-focusing peaks. This approach appears to work equally well as SWIFT for simple cases.

The second solution proposed to improve holographic imaging uses additional energies. So far we have only discussed the use of a single hologram at one electron energy. A phased sum or integral over hologram transforms generated at different electron energies can improve resolution, can help suppress twin images and can reduce perturbations due to multiple scattering[64,69,70]. Let us again rewrite the expression for $A(\mathbf{r})$, but now we add integration over the magnitude k of $\mathbf{k}$, and we include a phase factor e^{ikr} (we also include the SWIFT correction for consistency, although refs. 64, 69 and 70 make different choices):

$$(9) \qquad A(\mathbf{r}) = \iiint d^3k \, \frac{I(\mathbf{k})}{f^*} \, e^{ikr} e^{-i\mathbf{k}\cdot\mathbf{r}}.$$

The reason for this form can be seen by writing the new expression for $A_r(\mathbf{r})$:

$$(10) \qquad A_r(\mathbf{r}) = \frac{1}{R^2 r_1} \iiint d^3k \, e^{-ikr_1} \, e^{ikr} \, e^{i\mathbf{k}\cdot\mathbf{r}_1} \, e^{-i\mathbf{k}\cdot\mathbf{r}}.$$

The phase factor e^{ikr} beats against the factor e^{-ikr_1} to single out the condition $r = r_1$, reinforcing the real image at $\mathbf{r} = \mathbf{r}_1$ (in other words: this factor phase-locks onto the real image). In the analogous expression for the twin image the new factor e^{ikr} again throws phases off, severely degrading the twin-image quality, as illustrated in Fig. 9d (Fig. 9b shows the effect of using just the phased energy summation). It can also be shown that the multiple-scattering contributions are simultaneously reduced[70]: the reason is that their phases vary differently than the phases of the single-scattering terms that produce the real image and so are also not phase-locked with the phase factor e^{ikr}.

Another benefit of additional energies is that the resolution perpendicular to the surface can be enhanced when the aperture half-angle θ is less than 90°. This is because the range of accessible k_z values (perpendicular to the surface) can be extended with additional energies, giving a larger "effective perpendicular aperture" and thereby increased resolution[64].

With these improvements it is likely that good holographic imaging can be obtained from experiment, at least for relatively simple surface structures. For these the resolution may reach the ideal limit presented in section 9.2.

10. CONCLUSIONS

We have contrasted the oldest and newest techniques for surface structure determination: LEED and electron holography. LEED is a very mature technique that has been extended to determine complex structures. Electron holography is in its infancy: it shows promising possibilities, especially for certain classes of structures that provide relatively simple holograms, but it may never reach the degree of accuracy and completeness of LEED.

The improvements in LEED methods described here have already enabled impressive increases in complexity and accuracy for surface structural determination. Complexity is found with large-unit-cell structures, disordered and incommensurate overlayers, for example. Accuracy is often the result of allowing for greater complexity, as when adsorbate-induced relaxations are investigated. In either case, larger databases are required to extract the additional information. The larger databases can be produced rapidly with video LEED. Diffuse LEED measurements can be performed with either video LEED or digital LEED.

The structures that have benefited most from these advances have been reconstructions of clean surfaces, relaxations induced by adsorbates and molecular adsorption. Important structural determinations that are becoming routine include adsorption on semiconductor surfaces, which before were generally too complex to solve. Other applications that are becoming more accessible are surfaces of compounds (including insulators with the help of digital LEED) and stepped surfaces (including adsorbates thereon).

The new methods developed for LEED are for the most part also applicable to other surface structure techniques, such as photoelectron diffraction and near-edge x-ray absorption fine structure (NEXAFS). This is because the same electron multiple scattering principles apply. These techniques will thus also benefit from generalizations to more complex structures and the ability to deliver more accurate results.

Acknowledgements

This work was supported in part by the Director, Office of Energy Research, Office of Basic Energy Sciences, Materials Sciences Division of the U.S. Department of Energy under Contract No. DE-AC03-76SF00098. Supercomputer time was also made available by the Office of Energy Research of the U.S. Department of Energy. A part of the theoretical development was funded by the Army Research Office. NATO support for participation in the ASI on Surfaces and Interfaces is also gratefully acknowledged.

References

1. J.M. MacLaren, J.B. Pendry, P.J. Rous, D.K. Saldin, G.A. Somorjai, M.A. Van Hove and D.D. Vvedensky, "Surface Crystallographic Information Service: A Handbook of Surface Structures", D. Reidel, Dordrecht (1987).
2. P.R. Watson, J. Phys. Chem. Ref. Data **16**:953 (1987), and ibid **19**:85 (1989).
3. J.B. Pendry, "Low-Energy Electron Diffraction", Academic Press, London (1974).
4. M.A. Van Hove and S.Y. Tong, "Surface Crystallography by LEED", Springer-Verlag, Berlin, Heidelberg, New York (1979).
5. M.A. Van Hove and G.A. Somorjai, "Adsorbed Monolayers on Solid Surfaces", Structure and Bonding, Vol. 38, Springer-Verlag, Berlin, Heidelberg, New York, (1979).
6. M.A. Van Hove, W.H. Weinberg and C.-M. Chan, "LEED: Experiment, Theory and Structural Determination", Springer-Verlag, Berlin, Heidelberg, New York (1986).
7. G.A. Somorjai, "Chemistry in Two Dimensions", Cornell University Press, Ithaca, New York (1981).
8. P. Heilmann, E. Lang, K.Heinz and K. Müller, Appl. Phys. 19, 247 (1976); E. Lang, P. Heilmann, G. Hanke, K. Heinz and K. Müller, Appl. Phys. **19**:287 (1979).
9. D.F. Ogletree, G.A. Somorjai and J.E. Katz, Rev. Sci. Instr. **57**:3012 (1986).
10. P.C. Stair, Rev. Sci. Instrum. **51**:132 (1980).
11. D.F. Ogletree, R.Q. Hwang, U. Starke, G.A. Somorjai and J.E. Katz, to be published.
12. K. Heinz, D.K. Saldin and J.B. Pendry, Phys. Rev. Lett. **55**:2312 (1985).
13. P.J. Rous, J.B. Pendry, D.K. Saldin, K. Heinz, K. Müller and N. Bickel, Phys. Rev. Lett. **57**:2951 (1986).
14. D.K. Saldin, J.B. Pendry, M.A. Van Hove and G.A. Somorjai, Phys. Rev. **B31**:1216 (1985).
15. K. Heinz, K. Müller, W. Popp and H. Lindner, Surf. Sci. **173**:366 (1986).
16. H. Ibach and S. Lehwald, Surf. Sci. **176**:629 (1986).
17. L.J. Clarke, "Surface Crystallography: An Introduction to LEED", J. Wiley, London (1985).
18. J. Rundgren and A. Salwén, Comput. Phys. Commun. **9**:312 (1975).

19. W. Moritz and D. Wolf, Surf. Sci. **163**:L655 (1985).

20. S.Y. Tong, H. Huang, C.M. Wei, W.F. Packard, F.K. Men, G. Glander and M.B. Webb, J. Vac. Sci. Technol. **A6**:615 (1988).

21. P. Fery, W. Moritz and D. Wolf, Phys. Rev. **B38**:7275 (1988).

22. H. Huang, S.Y. Tong, J. Quinn and F. Jona, Phys. Rev. **B41**:3276 (1990).

23. C.M. Wei, H. Huang, S.Y. Tong, G.S. Glander and M.B. Webb, Phys. Rev. **B42**:11284 (1990).

24. H. Huang, S.Y. Tong, W.S. Yang, H.D. Shih and F. Jona, Phys. Rev. **B42**:7483 (1990).

25. H. Huang, C.M. Wei, H. Li, B.P. Tonner and S.Y. Tong, Phys. Rev. Lett. **62**:559 (1989).

26. H. Huang, C.M. Wei, H. Li, B.P. Tonner and S.Y. Tong, Phys. Rev. Lett. **64**:1183 (1989).

27. S.Y. Tong, H. Huang, C.M. Wei, in "Chemistry and Physics of Solid Surfaces", Eds. R. Vanselow and R.F. Howe, Springer, Heidelberg, Berlin, New York (1990), p. 395.

28. M.A. Van Hove, R.F. Lin and G.A. Somorjai, Phys. Rev. Lett. **51**:778 (1983).

29. H. Ohtani, M.A. Van Hove and G.A. Somorjai, J. Phys. Chem. **92**:3974 (1988).

30. M.A. Van Hove, R.F. Lin and G.A. Somorjai, J. Am. Chem. Soc. **108**:2532 (1986).

31. R.F. Lin, G.S. Blackman, M.A. Van Hove and G.A. Somorjai, Acta Crys. **B43**:368 (1987).

32. D.F. Ogletree, M.A. Van Hove and G.A. Somorjai, Surf. Sci. **183**:1 (1987).

33. G.S. Blackman, C.T. Kao, B.E. Bent, C.M. Mate, M.A. Van Hove and G.A. Somorjai, Surf. Sci. **207**:66 (1988).

34. Z.P. Hu, D.F. Ogletree, M.A. Van Hove and G.A. Somorjai, Surf. Sci. **180**:433 (1987).

35. M.A. Van Hove, in "Chemistry and Physics of Solid Surfaces VII", R.F. Howe and R. Vanselow, Eds., Springer-Verlag, Berlin, Heidelberg, New York (1988), p. 513.

36. P.J. Rous and J.B. Pendry, Surf. Sci. **155**:241 (1985).

37. P.J. Rous and J.B. Pendry, Surf. Sci. **173**:1 (1986).

38. U. Starke, P.L. de Andres, D.K. Saldin, K. Heinz and J.B. Pendry, Phys. Rev. **B38**:12277 (1988).

39. G. Illing, D. Heskett, E.W. Plummer, H.-J. Freund, J. Somers, Th. Lindner, A.M. Bradshaw, U. Buskotte, M. Neumann, U. Starke, K. Heinz, P.L. de Andres, D.K. Saldin and J.B. Pendry, Surf. Sci. **206**:1 (1988).

40. G.S. Blackman, M.-L. Xu, D.F. Ogletree, M.A. Van Hove and G.A. Somorjai, Phys. Rev. Lett. **61**:2352 (1988).

41. A. Wander, G. Held, R.Q. Hwang, G.S. Blackman, M.-L. Xu, P. de Andres, M.A. Van Hove and G.A. Somorjai, Surf. Sci. **249**:21 (1991).

42. P.J. Rous and J.B. Pendry, Surf. Sci. 219, 355(1989); Surf. Sci. **219**:373 (1989).

43. P.J. Rous and J.B. Pendry, Comput. Phys. Commun. **54**:137 (1989); Comput. Phys. Commun. **54**:157 (1989).

44. P.G. Cowell, M. Prutton and S.P. Tear, Surf. Sci. **177**:L915 (1986).

45. P.J. Rous, M.A. Van Hove and G.A. Somorjai, Surf. Sci. **226**:15 (1990).

46. P.J. Rous, D. Jentz, D.G. Kelly, R.Q. Hwang, M.A. Van Hove and G.A. Somorjai, in "The Structure of Surfaces III", Eds. S.Y. Tong, M.A. Van Hove, K. Takayanagi and X.D. Xie, Springer, Berlin, Heidelberg, New York (1991), p. 432.

47. J.M. Powers, A. Wander, P.J. Rous, M.A. Van Hove and G.A. Somorjai, Phys. Rev., in press

48. J.M. Powers, A. Wander, M.A. Van Hove and G.A. Somorjai, to be published.

49. A. Wander, M.A. Van Hove and G.A. Somorjai, Phys. Rev. Lett. **67**:626(1991).

50. G. Kleinle, W. Moritz, D.L. Adams and G. Ertl, Surf. Sci. **219**:L637 (1989).

51. G. Kleinle, W. Moritz and G. Ertl, Surf. Sci. **238**:119 (1990).

52. J.B. Pendry, K. Heinz and W. Oed, Phys. Rev. Lett. **61**:2953 (1988).

53. K. Heinz, W. Oed and J.B. Pendry, Phys. Rev. **B41**:10179 (1990).

54. J.B. Pendry and K. Heinz, Surf. Sci. **2303**:137 (1990).

55. F. Jona and P.M. Marcus, in "The Structure of Surfaces II", J.F. van der Veen and M.A. Van Hove, Eds., Springer-Verlag, Berlin, Heidelberg, New York (1988), p. 90.

56. D.W. Jepsen, Phys. Rev. **B22**:5701 (1980).

57. P. Pinkava and S. Crampin, Surf. Sci. **233**:27 (1990).

58. X.-G. Zhang, P.J. Rous, J.M. MacLaren, A. Gonis, M.A. Van Hove and G.A. Somorjai, in "The Structure of Surfaces III", Eds. S.Y. Tong, M.A. Van Hove, K. Takayanagi and X.D. Xie, Springer, Berlin, Heidelberg, New York (1991), p. 144.

59. X.-G. Zhang and A. Gonis, Phys. Rev. Lett. **62**:1161 (1989).

60. X.-G. Zhang, M.A. Van Hove, G.A. Somorjai, P.J. Rous, D. Tobin, A. Gonis, J.M. MacLaren, K. Heinz, M. Michl, H. Lindner, K. Müller, M. Ehsasi and J.H. Block, Phys. Rev. Lett. **67**:1298 (1991).

61. D. Gabor, Nature **161**:777 (1948).

62. A. Szöke, in "Short Wavelength Coherent Radiation: Generation and Applications", Eds. D.J. Attwood and J. Baker, AIP Conf. Proc. No. 147, Am. Inst. Phys., New York (1986).

63. J.J. Barton, Phys. Rev. Lett. **61**:1356 (1988).

64. S.Y. Tong, H. Huang and H. Li, in "Advances in Surface and Thin Film Diffraction", Eds. P.I. Cohen, D.J. Eaglesham and T.C. Huang, Mat, Res. Soc. Symp. Ser. **208** (1991), p. 13.

65. S. Thevuthasan, G.S. Herman, A.P. Kaduwela, R.S. Saiki, Y.J. Kim and C.S. Fadley, Phys. Rev. Lett. **67**:469 (1991).

66. D.K. Saldin, G.R. Harp, B.L. Chen and B.P. Tonner, to be published.

67. S.Y. Tong, C.M. Wei, T.C. Zhao, H.Huang and H. Li, Phys. Rev. Lett. **66**:60 (1991).

68. S. Hardcastle, Z.-L. Han, G.R. Harp, J. Zhang, B.L. Chen, D.K. Saldin and B.P. Tonner, Surf. Sci. **245**:L190 (1991).

69. J.J. Barton and L.J. Terminello, in "The Structure of Surfaces III", Eds. S.Y. Tong, M.A. Van Hove, K. Takayanagi and X.D. Xie, Springer, Berlin, Heidelberg, New York (1991), p. 107.

70. J.J. Barton, Phys. Rev. Lett., to be published.

71. Z.-L. Han, S. Hardcastle, G.R. Harp, H. Li, X.-D. Wang, J. Zhang, and B.P. Tonner, subm. to Surf. Sci.

72. G.R. Harp, D.K. Saldin, and B.P. Tonner, Phys. Rev. Lett. **65**:1012 (1990).

73. D.K. Saldin and P.L. de Andres, Phys. Rev. Lett. **64**:1270 (1990).

74. M.A. Mendez, C. Glück and K. Heinz, Phys. Rev. Lett., to be published.

MAGNETIC PHASE TRANSITIONS AT AND BETWEEN INTERFACES

B. L. Gyorffy and C. Walden

H. H. Wills Physics Laboratory
University of Bristol
Tyndall Avenue
Bristol B56 6LQ

INTRODUCTION

Magnetism at defects, like surfaces, interfaces, grain boundaries and dislocations have been studied for a long time. Largely, this is the case because such inhomogeneous circumstances are unavoidable features of all interesting magnetic systems. However, there is another point of view. Magnetic structures associated with inhomogenieties may be studied in their own rights as examples of problems in the Statistical Mechanics of inhomogenous systems [Pekalski and Sznajd, 1984]. This field is relatively new and is still in the process of conceptual development [Binder, 1983, Taub et al, 1991]. Consequently, key examples are of particular general interest. For an elementary introduction to the rich variety of magnetic phenomena which occurs only at surfaces and interfaces the reader is refereed to the very readable, recent, book of Kaneyoshi [1991]. Here we shall consider only two of these from the second point of view mentioned above. The first phenomenon we shall study is magnetic wetting [Walden and Gyorffy, 1988]. We shall examine the circumstances where and how it may occur within the framework of a Landau theory which is appropriate to the case of vector magnetization displaying cubic anisotropy. We shall stress the intrinsic interest of critical wetting in this model [Walden et al, 1990]Secondly, we shall focus on the phenomena which is usually referred to as Exchange, or unidirectional, Anisotropy [Meiklejohn, 1962]. Our principle purpose will be to show that it is the magnetic analog of the well known phenomenon of Capillary Condensation [Rowlinson and Widom, 1982, Evans, 1991]. We shall argue that a recognition of this connection throws some useful light on the subject. Evidently, the current dramatic advances in the fabrication and characterization of metallic, magnetic thin films and multilayers (Janker et al. 1989, Fert et al., 1991) lend particular impetuous to the above considerations. The prospects of accurate experiments on these new high quality structures are encouraging and hence progress on important conceptual questions can be expected. Also,

Equilibrium Structure and Properties of Surfaces and Interfaces
Edited by A. Gonis and G.M. Stocks, Plenum Press, New York, 1992

fortunately, magnetism in these systems is of considerable technological interest in connection with magnetic information storage [de Wit, 1992]. This explains the grand scale of current research in this field.

I. MAGNETIC WETTING

I.1 The Scalar (Ising) Case

Wetting is a very general phenomena at inhomogenieties, like surfaces, which may be preferentially covered by one of two, or more, thermodynamic phases [Croxton, 1986]. The wetting transition, which is attracting considerable current interest in statistical mechanics, takes place near where such phases coexist in the bulk [Dietrich, 1988]. It manifests itself by a sudden change in the nature and extent of the coverage as some thermodynamic variable, like temperature, or chemical potential, approaches its critical value. In the most studied case the phases are liquid and gas [Croxton, 1986]. Here, we wish to call attention to a magnetic analogs of these remarkable phenomena.

To simplify the exposition, during the following introductory remarks we shall confine our attention to a model with only one (Ising like) component of magnetization, $M_z(x)$ which we shall denote by $m(x)$. Later we shall switch to a Laundau theory with cubic anisotropy for describing the magnetization vector $\vec{M}(\vec{r})$.

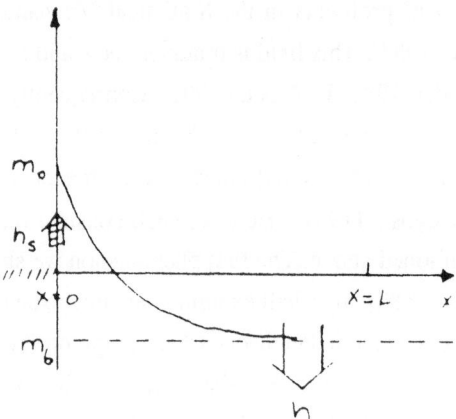

Fig. 1. Spatial decay of the surface magnetization m_0, h_S, and h are the surface and bulk external magnetic fields respectively.

Consider a semi-infinite ferromagnet with surface conditions such that the energetically favoured surface magnetization $m_0 = m(x = 0)$ is up as shown in Fig. 1. Let us now, as usual, probe this system with a constant magnetic field h, pointing down, and define the spontaneous (bulk) magnetization, below the Curie temperature T_C, by

$$m_b(T,h) = \lim_{h \to 0} \lim_{V \to \infty} \frac{1}{V} M(T,h,V) \tag{1}$$

266

where V is the volume of the semi-infinite sample and $M(T,h,V)$ is its magnetization at the temperature $T < T_C$ and field $h < 0$. Also, let the inhomogeneous magnetic state be defined by the magnetization density $m(\vec{r})$, to be determined later and assume that it depends only on the x-coordinate. Then, for a finite sample of surface area A and width, in the x-direction, L we may write

$$M(T,h,V) = Alm_b + A\int_0^L dx\left(m(x) - m_b\right) \qquad (2)$$

Namely,

$$\lim_{V \to \infty} \frac{1}{V} M(T,h,V) = m_b(T,h) + \lim_{L \to \infty} \frac{m_s(T,h,L)}{L} \qquad (3)$$

where the excess magnetization $m_S(T,h)$, the coverage, is defined as

$$m_s(T,h) = \int_0^\infty dx\left(m(x) - m_b\right) \qquad (4)$$

The first point to note about the above relations is that $m_S(T,h)$ is finite and therefore $\lim_{L \to \infty} \frac{1}{L} m_s(h,T,L) = 0$. Thus, the presence of the surface does not change the bulk magnetization defined in Eq 1.

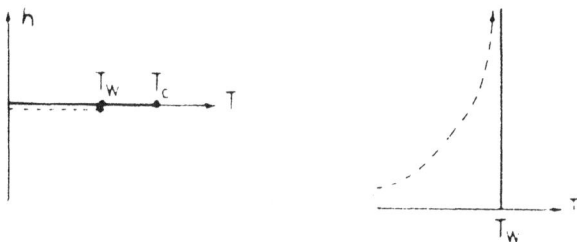

Fig. 2. Critical wetting at the wetting temperature $T_W < T_C$ as T increases along the h = o line in the h, T plane.

The second more central point is that, the above statement notwithstanding, m_S is a new, surface, thermodynamic variable. Evidently, it is associated with the existence of an infinite surface. Moreover, it turns out to behave in a remarkably singular fashion at selected points in the (T,h) plane. In short, corresponding to the two different thermodynamic limits: $V \to \infty, A \to \infty$ a piece of matter will be in a surface phase as well as in a bulk phase and there will be phase transitions with respect to each kind of phases. The following section will be a brief account of such phase transition for a model which we shall be studying later.

I.2 Critical Wetting

Surprisingly, for $T<T_W$ the spontaneous surface magnetization $m_s(T) = \lim_{h \to 0} m_s(T,h)$ is finite but for $T_W < T < T_C$ it is infinite as shown in Fig.. 2. We say that the surface has been wetted, as T approaches T_W from below, by the up spin phase. As will be shown presently this remarkable event is due to the fact that the surface lowers the free energy of the 'up spin phase' compared with that of the 'down spin phase' through the action of the surface field h_S. The temperature T_W is called the wetting temperature and, as is clear from Fig.. 2, it is a singular point of $m_S(T)$. Interestingly, although

$$\lim_{T \to T_w} m_s(T) = \infty$$

$$\lim_{h \to 0} \lim_{L \to \infty} \frac{1}{L} m_s(T,h) = 0 \tag{5}$$

and hence the bulk magnetization and its singularities, at T_C for instance, are unaffected by above surface phase transition, conventionally referred to as critical wetting [Dietrich, 1988].

I.3. Complete Wetting

This is another route to the same wet state as above. It is depicted schematically in Fig. 3.

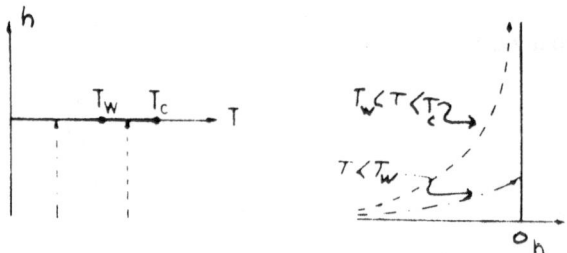

Fig. 3. Complete wetting as h $\to$ o at T > T_W.

I.4 First Order Wetting

The divergence of $m_S(T)$ at T_W is reminiscent of the behaviour of the inverse order parameter in a continuous (2nd order) phase transition. Thus it should not come as a surprise that critical wetting has a first order wetting version. This is illustrated in Fig. 4.

I.5 Surface Thermodynamics

To lend credence to describing the singularities of $m_S(T,h)$ as phase transitions we now introduce the basic elements of surface thermodynamics. To begin with, in analogy with the surface magnetization $m_S(T,h)$, we define the surface (Gibbs) free energy per surface area, σ_S, as follows

$$\lim_{V \to \infty} \frac{F}{V} = f_b + \frac{1}{L} \sigma_s(T,h,h_s) + 0\left(\frac{1}{L^2}\right) \tag{6}$$

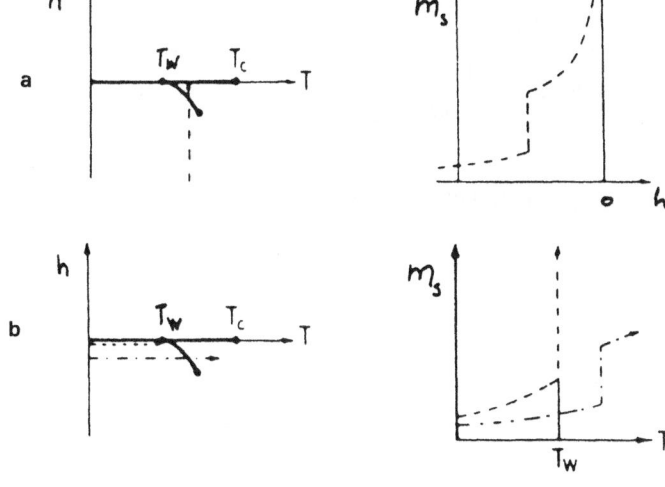

Fig. 4. First order wetting as on crossing the prewetting line along T = constant, a), and h = constant, b), trajectories.

where f_b is the constant, bulk, free energy density. The surface free energy so defined is often referred to as the surface tension.

A straight forward generalization of the Landau theory for the bulk suggests the following basic equations for describing the surface.

$$m_s(T,h,h_s) = \int_0^\infty dx(m(x) - m_b) \tag{7}$$

$$\sigma_s(T,h,h_s) = \int_0^\infty dx\left(f(T,h,h_s;[m(x)]) - f(T,h,h_s,[m_b])\right)$$

$$+\frac{1}{2}Cm^2(0) - h_s m(0) \tag{8}$$

where $f(T,h,h_S;[m(x)])$ is a generalized free energy density functional of the magnetization $m(x)$. The equilibrium state of the system is described by the function $m(x)$ which minimizes σ_S. The thermodynamic surface tension is the value of σ_S at this minimum. Thus a theoretical model consists of specifying the functional f. This we shall do in the next section. Here we merely wish to record some general features of the theory.

The second law of thermodynamics for a surface works out to be

$$d\sigma_s = sdT - m_s dh - m_o dh_s \tag{9}$$

This implies

$$m_s = -\left(\frac{\partial \sigma_s}{\partial h}\right)_{T,h_s} \text{ and } m_0 = -\left(\frac{\partial \sigma_s}{\partial h_s}\right)_{T,h} \tag{10}$$

Moreover, we can define the susceptibilities

$$\chi_s = \left(\frac{\partial^2 \sigma_s}{\partial h^2}\right)_{T,h_s} \quad \text{and} \quad \chi_{00} = -\left(\frac{\partial^2 \sigma_s}{\partial h_s^2}\right)_{T,h} \tag{11}$$

Finally the analogs of the bulk critical exponents β,γ,δ, identified in the asymptotic forms:

$$m_b(t) \sim t^\beta; \chi_b(t) \sim t^{-\gamma}, m_b(h) \sim h^{\frac{1}{\delta}} \tag{12}$$

where $t = \dfrac{T-T_c}{T_c}$ is the reduced temperature, may be defined as follows

$$m_s(t) \sim t^{\beta_s}; \quad \chi_s(t) \sim t^{-\gamma_s}, \quad m_s(h) = h^{\frac{1}{\delta_s}} \tag{13}$$

and

$$m_0(t) \sim t^{\beta_0}; \chi_{00}(t) \sim t^{-\gamma_0}, m_0(h_s) \sim h_s^{+\frac{1}{\delta_0}} \tag{14}$$

where t is an appropriate surface reduced temperature. The above systematic correspondences amply justifies the practice of talking of surface phase transitions, and surface critical phenomena.

I.6 The Mean-Field Theory of the Wetting Transition

An efficient approach to the problem at hand is to assume a specific phenomenological form for the Guizburg-Landau free energy functional $f(T,h;[m(x)])$ in Eq 7. Following the classic precedent, set by van der Waals [Rowlinson and Widom, 1982] we take that to be

$$f(T,h;[m(x)]) = \frac{1}{2}\left(\frac{dm}{dx}\right)^2 + \frac{1}{2}\alpha(T)m^2 + \frac{1}{4}bm^4 - mh \tag{15}$$

The corresponding Euler-Lagrange equation

$$\frac{\delta\sigma_s}{\delta m(x)} = 0 \tag{16}$$

yields

$$-m''(x) + am(x) + bm^3(x) - h = 0 \tag{17}$$

and

$$m'(0) = Cm(o) - h_s \tag{18}$$

where $m''(x)$ and $m'(x)$ are the first and second derivatives, respectively, of $m(x)$ with respect to x.

Multiplying Eq. 16 by $m'(x)$ it is easy to show that $-\frac{1}{2}(m')^2 + \Psi(m)$,

where

$$\psi(m) = \frac{1}{2}am^2 + \frac{1}{2}bm^4 - hm \tag{19}$$

is a 'constant of the motion'. Since in the bulk, $x \to \infty$, $m(x) \to m_b$ and $m'(x) \to 0$ we find

$$-\frac{1}{2}m'^2 + \psi(m) = \psi(m_b) \tag{20}$$

This equation, with the boundary condition given in Eq. 18, can be solved by graphical construction, e.g., plotting

$$m'(x) = sgn(m_b - m(x))\sqrt{2[\psi(m) - \psi(m_b)]} \equiv Q(m) \tag{21}$$

and

$$m'(0) = Cm(0) - h_s \equiv Y(m) \tag{22}$$

where the choice of $sng(m_b - m(x))$ in Eq. 21 is dictated by the requirement that for $C > O$ $m(x)$ decays rather than increases as x moves into the bulk.

A selection of interesting cases of $Q(m)$ and $Y(m)$ are shown in Fig.. 5 evidently the equilibrium profile $m(x)$ starts with its surface value $m(0)$ at the point where $Q(m)$ and $Y(m)$ cross and travels along the $Q(m)$ curve as $x \to \infty$ until the bulk value m_b is reached at $Q(m_b) = 0$. The dilemma, in Fig. 5c, where $Q(m)$ and $Y(m)$ cross at several points is resolved by the equal area construction. For further details see [Dietrich, 1981].

To select out, for comment, but one of the many interesting predictions of the above theory we note that the thin-thick transition across the prewetting line in Fig. 4 corresponds to the change from the magnetization profile in Fig. 5a to that in Fig. 5c. As indicated the distance of the 'up-down' interface, 1, goes to infinity as $h \to 0$ in a characteristic logarithmic fashion. This transition is called complete wetting and, in general, it is described by the asymptotic relation

$$\ell \cong \ell nh \tag{23}$$

where we have used Eq 12 and the fact that for large l $m_s \sim |m_b| l..$ The above logarithmic results implies that the mean field exponent $\beta_S = 0$.

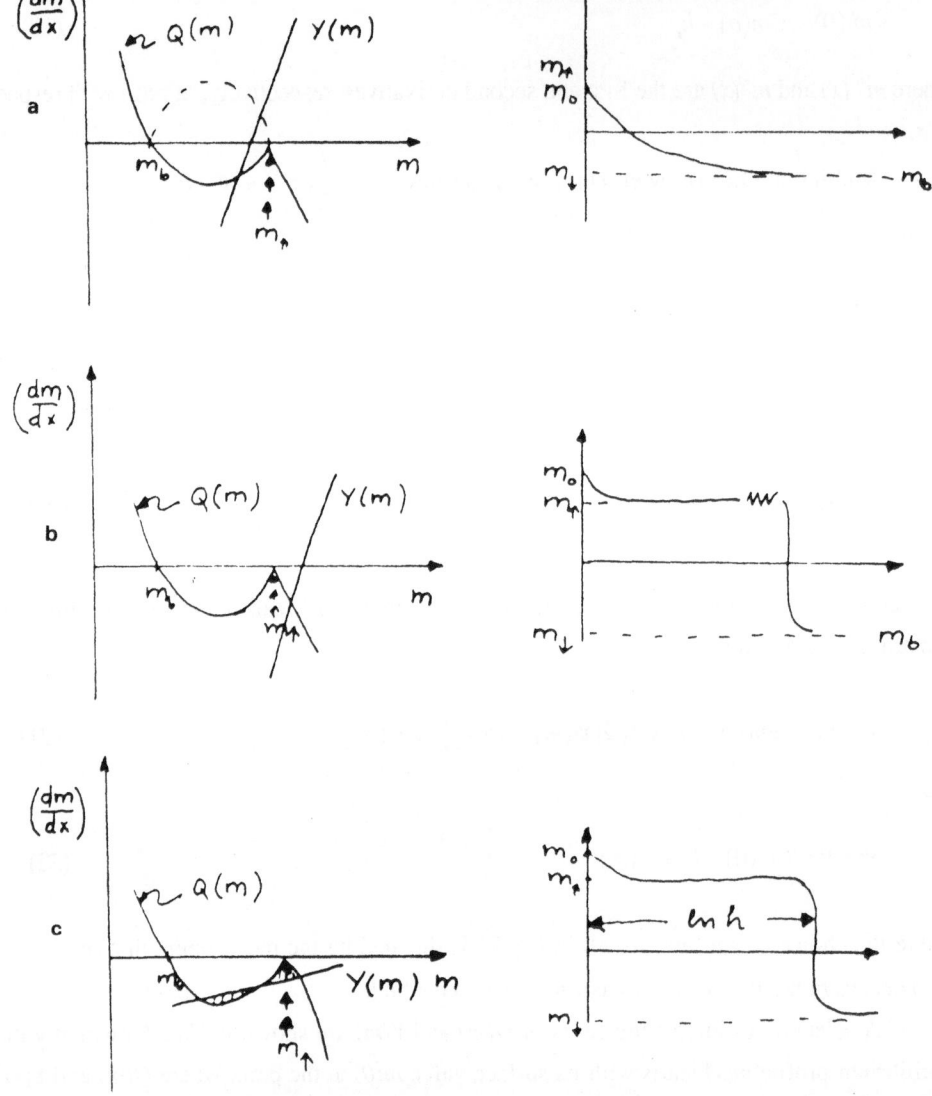

Fig. 5. Phase portraits of the magnetization profile $m(x)$ at the surface $x = 0$ for non wetting a) critical wetting b) and complete wetting points in parameter space.

Similar arguments can be used to show that for $C \geq \xi_b^{-1}(T_w)$, where $\xi_b(T)$ is the bulk correlation length, critical wetting, discussed in Sec. I.1, occur. In this case l again is a useful variable replacing m_S as the order parameter of the phase transition. Using the above mean field theory one finds

$$l = -\xi_b \ln\left(\frac{T_w - T}{T_w}\right) \tag{24}$$

This implies that the surface critical exponent β_S, defined in Eq 13, is $\beta_S = 0$.

As is well known, from the point of view of critical point phenomena [Stanley, 1972]. Ising magnets and liquids fall into the same universality class. Indeed, the whole discussion of this section, the magnetic language notwithstanding, has been discovered (Cahn, '77 Ebner and Saam, '77) and developed in reference to liquids. Moreover, all observations, so far, of wetting transitions have been made in that context. Of course this should not occasion a surprise since there are no genuinely Ising Ferromagnets. However, the Ising case discussed above should prompt the idea that wetting like phenomena can also occur in vector models which describe genuine Ferromagnets. In the next section we shall demonstrate that this indeed is the case.

I.7 The Vector Model

In a model with vector magnetization, $\vec{M}(\vec{r})$, the wetting phenomena acquires a rich variety of new features. Generalizing the argument of the previous section we base our discussion on a Landau theory [Landau and Lifshitz, 1960] with the following surface free energy per unit area:

$$\sigma\left[\vec{M}(\vec{r})\right] = \mu\left\{\int_{-\infty}^{0} dz\left[\frac{1}{2}\xi_0^2\left|\frac{d\vec{M}}{dz}\right|^2 + f\left(\vec{M}(z)\right) - f\left(\vec{M}_B\right)\right] + \xi_0 f_s\left(\vec{M}(0)\right)\right\} \tag{25}$$

where

$$f(\vec{M}) = \frac{1}{2}t\left(M_x^2 + M_y^2\right) + \frac{1}{4}\mu\left(M_x^4 + M_y^4\right) + \frac{\lambda}{2}M_x^2 M_y^2 - \vec{H}\cdot\vec{M}$$

$$f_s(\vec{M}) = \frac{\xi_0 c}{2}\left(M_x^2 + M_y^2\right) - \vec{H}_s\cdot\vec{M} \tag{26}$$

ξ_0 is the zero temperature coherence length, t is the reduce temperature $(T - T_c)/T_c$, λ is the measure of cubic anisotropy, c is the surface enhancement and $\vec{H}_s$ is the surface magnetization parameters. Without serious loss of generality we take $M_z = 0$. The technical problem is to minimize $\sigma_s[M]$ Eq. 24 subject to the boundary conditions implied by Eq. 25.

The anisotropy described by λ is an important feature of the model. Without it we are using a Heisenberg model for which the Bloch wall surface energy (tension) vanishes and hence the interface separating an up and down phases have infinite width. For $\lambda \neq 1$ (the isotropic case) this width is finite and hence wetting is a possibility. Unfortunately the problem is now very difficult. To make progress we pick a particular value of λ which simplifies matters. For $\lambda = 3$ the Euler-Lagrange equations separate into two independent nonlinear equations for the variables

$$P = M_x + M_y \text{ and } Q = M_x - M_y \tag{27}$$

and these can be solved by the same graphical construction as was used in Sec. II. Following Walden and Gyorffy [1989] we note that in this case there are two prewetting lines in the

273

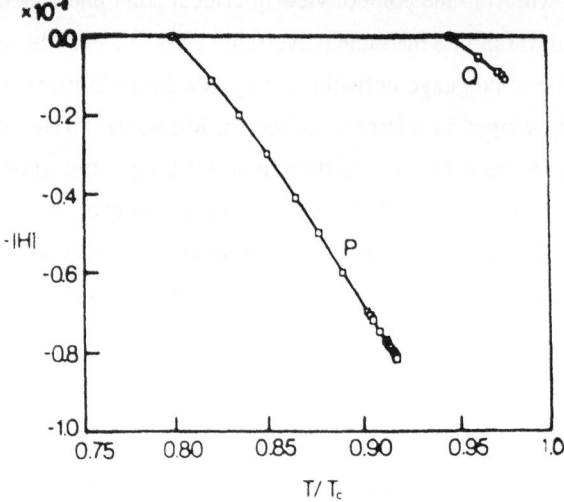

Fig. 6. The two prewetting lines in the vector model with cubic anisotropy, $\lambda = 3$.

phase diagram. For a particular case of surface conditions, $\vec{H}_s$ and c, these are displayed in Fig. 6. The competing profiles, $M_x(z)$ and $M_y(z)$ can be represented by trajectories in the M_x vs M_y plane. Examples near both lines are shown in Fig. 7. The corresponding profiles, $M_x(z)$ and $M_y(z)$, are also shown in Fig. 8. The physics of these solutions is that on crossing the prewetting line the magnetization profile suddenly changes from one to the other which previously was merely a metastable solution. Interestingly, the wall is wetted by *spin sideways* phase (domain) across the P-line and by a *'spin up phase'* (domain) across the Q-line.

For $\lambda \neq 0$ the Euler-Lagrange equations for $M_x(z)$ and $M_y(z)$ cannot be decoupled. However, by a combination of numerical integrations and general arguments we have been able to establish the main features the very complex phase diagram. For instance if the surface angle between $\vec{M}$ and y axis: $\theta_s \geq \dfrac{3\pi}{4}$ partial wetting is the stable phase for all $T < T_C$, if, however, $\dfrac{\pi}{4} < \theta_s \leq \dfrac{3\pi}{4}$ there is wetting by the spin sideways phase. In the range $\dfrac{\pi}{4} < \theta_s \leq \theta_s^c(\lambda)$ there are two transitions: firstly by a spin-sideways and then by a spin-up phase on further increasing the temperature at constant bulk external fields. The range in which there are two transitions shrinks to zero as $\lambda \to \infty$.

I.8 Critical Wetting in a Vector Model

Along the $H = O$ line in the phase diagram of Fig. 3 continuous wetting transition (critical wetting) can occur as T approaches the wetting temperature T_W if the surface conditions are appropriate. A remarkable feature of the cubic anisotropy model is that the critical exponents which govern this phase transition appear to be non-universal [Walden et al., 1990].

274

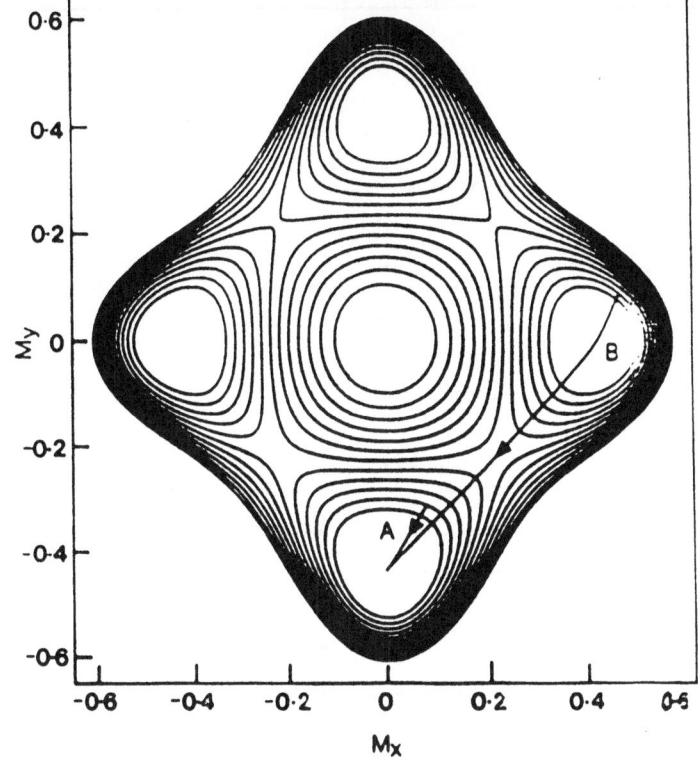

Fig. 7. Typical trajectories of the magnetization profiles in the My, Mx plane. Note the presence of the four maxima in the constant energy contours. These define the easy axis in the bulk.

At critical wetting the thickness of the wetting layer $l_o \sim \left(T_w - T\right)^{-\beta s}$. The surface energy $\Delta\sigma_s\left(l_o\right)$ and the coherence length parallel to the surface $\xi_{\parallel}(T)$ are also singular. Namely, $\Delta\sigma_s(T) \sim \left(T_w - T\right)^{2-\alpha_s}$ and $\xi_{\parallel}(T) \cong \left(T_w - T\right)^{\nu_{\parallel}}$. Walden et al. [1989] have found that, in the above mean field theory, the critical exponents β_s, α_s and $\nu_{\parallel}$ depend on λ. Namely, they are non universal. Specifically, for $3<\lambda<9$

$\beta_s = 0$ (logarithmic growth)

$$\alpha_s = 2 - \left[1 - \sqrt{\frac{\lambda}{\lambda-1}}\right]^{-1}$$

$$\nu_{\parallel} = \frac{1}{2}\left[1 - \sqrt{\frac{2}{\lambda-1}}\right]^{-1} \tag{28}$$

They argued that this curious behaviour was the result of two independent scaling lengths, ξ_x and ξ_y, entering the problem. This situation was first encountered and investigated by Hauge and Olaussen [1989] in a model describing a liquid-gas system at a wall. However, the liquid-gas model they studied is too artificial to have an experimentally realizable system associated

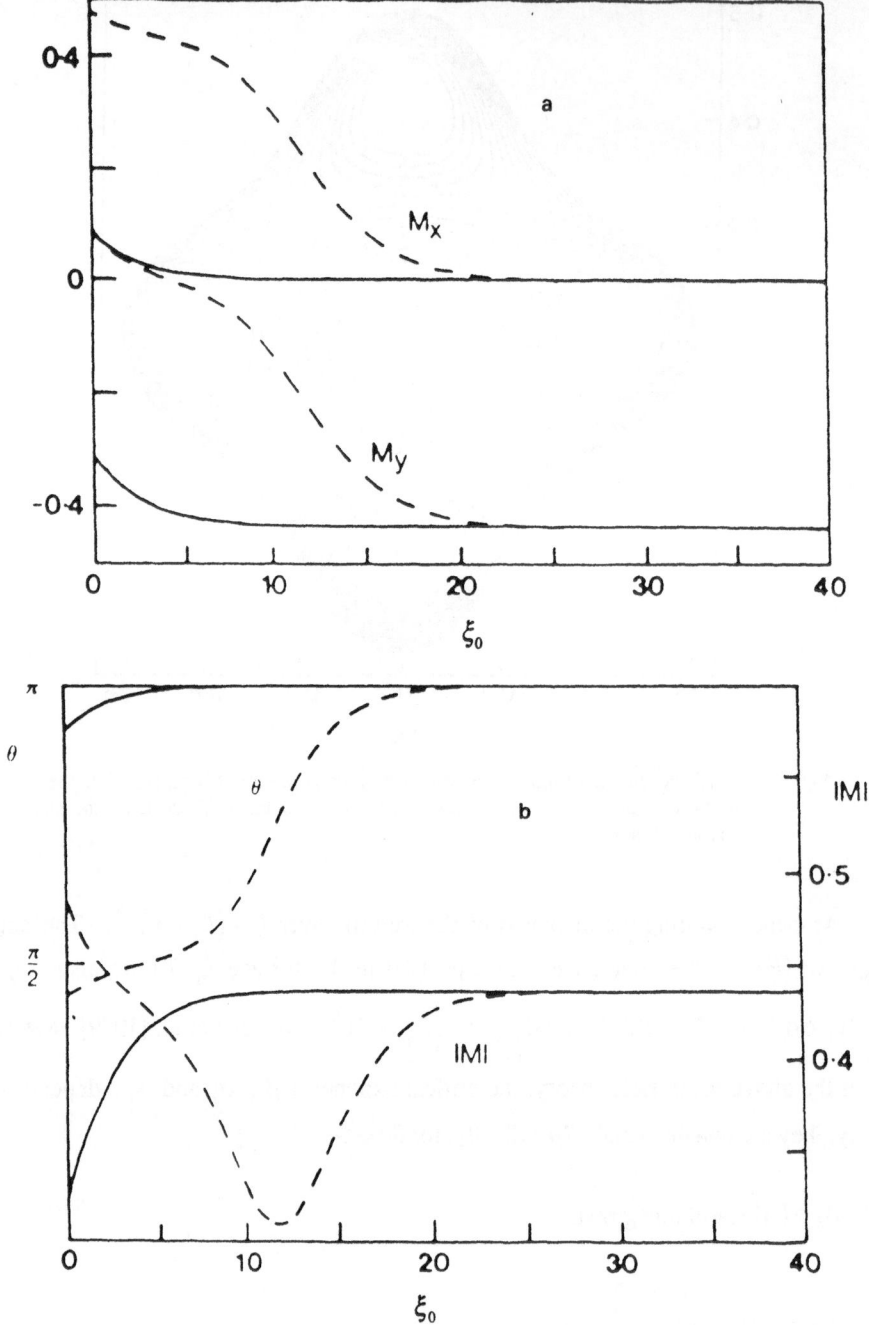

Fig. 8. Variation of the competing magnetization distributions near a surface, with a surface field, for a vector model. Note the existence of a wetting layer in one (dashed line) while the other (full line) describe simple healing.

with it. It is hoped that the present cubic anisotropy model, which gives a realistic description of many transition metal magnets, will generate experimental interest within the context of magnetic films or bylayers. Evidently, an observation of non-universal critical wetting exponents would be an important contribution to the statistical mechanics of inhomogeneous systems.

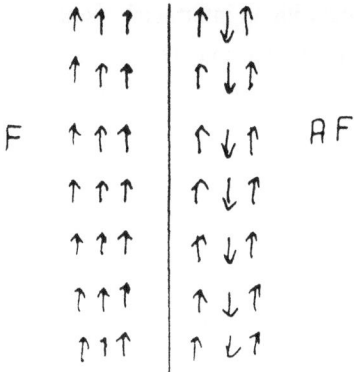

Fig. 9. Schematic picture of a Ferromagnetic (F) and Anti Ferromagnetic (AF) bylayer on which the experimental results in Fig. 10 have been measured.

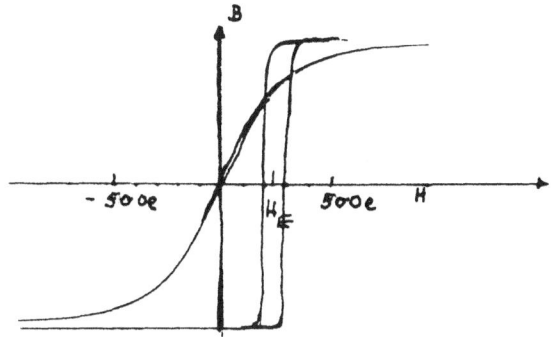

Fig. 10. The B - H loop of an Exchange-Biased film shown schematically in Fig. 9. The Ferromagnet is a 100Å $Fe_{20}ni_{80}$ (permalloy) layer while the Antiferromagnet is Fe.5Mn.50 and 250Å $H_E = 23.5$ 01.

II. EXCHANGE ANISOTROPY AND CAPILLARY CONDENSATION

II.1 The Phenomena of Exchange Anisotropy

Exchange anisotropy is a name given to a group of phenomena which occur in bylayers where one side is a ferromagnet (F) and the other is an antiferromagnet (AF) as shown in Fig. 9. The discovery and early work in the field is well reviewed by Yelon [1971]. For a more recent account of the theoretical difficulties in interpreting the experimental facts see Malozemoff [88]. For clarity of presentation, in this paper we shall confine our attention to the much studied puzzle of the shifted $\bar{M}(\bar{H})$, hysteresis loop. An experimental example of this is displayed in Fig. 10.

The issue we wish to take up concerns the starting point for constructing theories of the above effect. An attractive way to proceed was put forward by Malozemoff [1988]. Assuming that the temperature T is well below the Neel temperature T_N of a, magnetically hard, antiferromagnet he studied the two configurations of the ferromagnetic layer shown in Fig. 9.

277

He supposed that he was dealing with an interfacial phenomena and chose to represent the energy difference, ΔE, between the two magnetic configurations in terms of the surface tensions $\sigma_{w\uparrow}$ and $\sigma_{w\uparrow}$ associated with antiferromagnetic wall and the spin $\uparrow$ and spin $\downarrow$ ferromagnets respectively:

$$\Delta E = \left(\sigma_{w\uparrow} - \sigma_{w\downarrow}\right) A_s \tag{29}$$

where A_S is the area of the corresponding interface. Note that the other, magnetically inert vacuum, side of the ferr-omagnetic film is assumed to make no significant contribution to ΔE. He then defined an exchange field H_E by the relation

$$\Delta E = M_F H_E L A_s \tag{30}$$

where M_F is the magnitude of the average magnetization of the ferromagnet within the volume LA_S.

From Eqs. 29 and 30 it follows that

$$H_E = \frac{\sigma_{w\uparrow} - \sigma_{w\downarrow}}{M_F L} \tag{31}$$

Moreover, it is physically plausible that H_E is the 'off-set' field in Fig. 10.

Malozemoff ['87,'88] also pointed out difficulties in assuming an atomically perfect F-AF interface with uniform interfacial exchange coupling J_i. The field cooled AF layer will then adopt either a "compensated" or "uncompensated" structure. In the former the spins are ordered such that the net magnetization in the first atomic plane of the antiferromagnet is zero, and one deduces that the exchange bias field H_E vanishes. In the latter case all spins in the first AF plane are, to a first approximation, aligned. Indeed, calculations of the electronic and magnetic structure by Lambin and Herman ['84] indicate that such a situation is expected to obtain for atomically flat Ni3Fe/FeMn (100) surfaces. From a simple bond breaking argument one then deduces H_E of the order $J_i/a^2 M_b L$ with 'a' the lattice parameter. Unfortunately, this estimate is about two orders of magnitude larger than the experimentally observed values, since there is no reason to expect J_i to be dramatically smaller than a typical exchange coupling.

As has been noted by Malozemoff ['87,'88], a reduction factor of approximately $\alpha / \sqrt{A/K}$, (A and K are the usual exchange and anisotropy constants) where neglecting factors of order unit, $\sqrt{A/K}$ is a typical domain wall width would bring H_E in line with experiment. Moreover, Mauri et al. ['87] have proposed a mechanism which, is based on the formation of a domain wall in the AF layer, whereby $\sigma_{AF,\downarrow} - \sigma_{AF\uparrow}$ is reduced to the surface tension of a domain wall $\sigma_{\uparrow\downarrow} \approx \sqrt{AK}$. Evidently, this argument supplies the requisite factor. Unfortunately, this mechanism cannot explain the observation of a bias field H_E of comparable magnitude for AF layers much thinner than a typical domain wall width. The possibility of domain wall formation in the F layer, of course, also exists though once again, L is typically too small to accommodate one [Tsang et al, '82]. The model which compares most favourably with experiment is that due to Malozemoff ['87, '88]. He imagines atomic roughness at the F-AF interface, which leads to randomness in the interfacial exchange J_i. On field cooling the sample this acts like a random field on the antiferromagnet and causes it to break up laterally into domains whose size is of the order of a domain wall width. Detailed analysis shows that such a mechanism also reduces $\sigma_{AF,\downarrow} - \sigma_{AF\uparrow}$ to approximately $\sqrt{AK}$. Such a model also appears to account for the 'magnetic training" effect, whereby the coercivity and loop shift display a gradual relaxation to some "stable" value on successive cycling of the magnetic field.

It is not our aim to address such issues here. Rather, we wish to point out that Eq. (31) is the magnetic analog of the Kelvin equation which governs capillary condensation of gases in narrow pores. We then use this observation to show that, at least in the limit of a thick F film, the surface tension of a domain wall is a thermodynamic upper limit for $\sigma_{AF,\downarrow} - \sigma_{AF\uparrow}$.

II.2 A Derivation of the Magnetic Kelvin Equation

For the purposes of illustration we restrict attention to the case where the ferromagnet is of the uniaxial type, and consider its behaviour under the application of a uniform field, H, parallel to the easy axis. Suppose the system is described by a coarse-grained Helmholtz free energy functional F[M(r),T]. In the bulk one then fiends uniform phases which correspond to local minima of the Gibbs free energy functional

$$G[M(r),T,H] = F[M(r),T] - \int dr M(r) H(\vec{r}) \tag{32}$$

and two such phases may coexist when their respective minima are equal. For the uniaxial system under consideration this occurs for H = 0. Varying H will lead to one of the two phases being favoured. However, provided the disfavoured phase still represents a local minimum of G, one may still have (metastable) domains of that phase, and this typically leads to hysteresis phenomena. Of course, the detailed mechanism by which magnetization reversal takes place will play a crucial role in determining the size and shape of the hysteresis loop, but this does not concern us.

Whilst in the bulk the magnetization within each phase is uniform, in the thin film geometry envisioned in Fig. 9 the equilibrium magnetization within a given phase will in general vary spatially and depend on the thickness L of the F layer. However, except for reports by Tsang et al.[1982] for F films with L~1μm there is no evidence for such behaviour. In addition, it will depend on the spin configuration $\{S_i^{AF}\}$ in the AF layer, which for the moment we imagine to be frozen in.

For a given L and $\{S_i^{AF}\}$ the F layer will adopt the configuration that minimizes the Gibbs free energy $G(H,T,L,\{S_i^{AF}\})$. It is convenient to divide G into bulk and surface contributions,

$$G = F[M_b] - VM_bH + 2\sigma\left(L,\left\{S_i^{AF}\right\}\right)A_s \qquad (33)$$

where Mb is the uniform bulk magnetization which minimizes (32), A_S is area of the F-AF interface, $V = LA_S$ is the volume of the F layer, and $\sigma\left(L,\left\{S_i^{AF}\right\}\right)$ is the sum of the free energies per unit area of the F-AF and F-vacuum interfaces.

For sufficiently large L we may envisage spin-up $(\uparrow)$ and spin-down $(\downarrow)$ configurations (Fig. 9) which are essentially superpositions of seminfinite $AF - (\uparrow or \downarrow)$ and vacuum $(\uparrow or \downarrow)$ configurations. Expanding the bulk terms for small H < O about bulk coexistence, one then has, to leading order

$$
\begin{aligned}
G^{(\downarrow)}(T,H,L) &= F[M_b] - VM_bH + \left(\sigma_{AF,\downarrow} + \sigma_{v,\downarrow}\right)A_s \\
G^{(\uparrow)}(T,H,L) &= F[M_b] + VM_bH + \left(\sigma_{AF,\downarrow} + \sigma_{v,\downarrow}\right)A_s
\end{aligned}
\qquad (34)
$$

so that coexistence now occurs for a field H_E given by

$$H_E = \frac{\sigma_{AF,\uparrow} - \sigma_{AF,\downarrow} + \sigma_{v,\uparrow} - \sigma_{v,\downarrow}}{2M_bL} \qquad (35)$$

Since there is no obvious reason why the boundary conditions at the free surface of the F film should break time-reversal symmetry, we may suppose that $\sigma_{v,\uparrow} = \sigma_{v,\downarrow}$. In this case (35) is identical to Malozemoff's relation (1), and we shall show presently it is the analog of the Kelvin equation.

II.3 Capillary Condensation

Although the macroscopic Kelvin's equation is usually derived in the context of liquids confined in a cylindrical vessel and it features the curvature of the miniscus [Evans et al., 1987], from the point of view of our present concern, it is more convenient to study the planar geometry depicted in Fig. 11. For this situation Kelvin's equation is given by:

$$\kappa_B T \ln\frac{P_{sat}}{P} = \frac{2\gamma_{1,g}\cos\theta_w}{L\left(\rho_1 - \rho_g\right)} \qquad (36)$$

where Psat is the saturation pressure, $\gamma_{1,g}$ is the liquid gas surface tension, θ_w is the wetting angle, ρl and ργ are the liquid and gas densities at coexistence respectively. The meaning of Eq. 36 is that it gives the pressure P < Psat at which the gas confined between the plates condenses. Using Young's equation

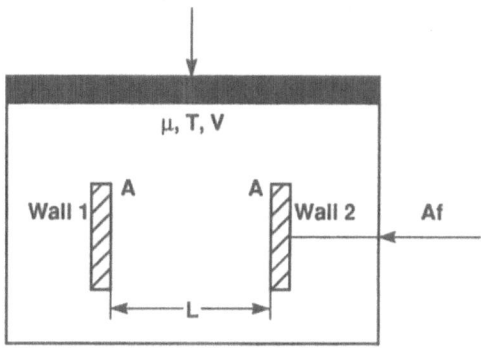

Fig. 11. System of two adsorbing plates (each with surface area A repelling with a force Af. The plates are separated by a distance H and are enclosed in a reservoir at fixed M, T, V.

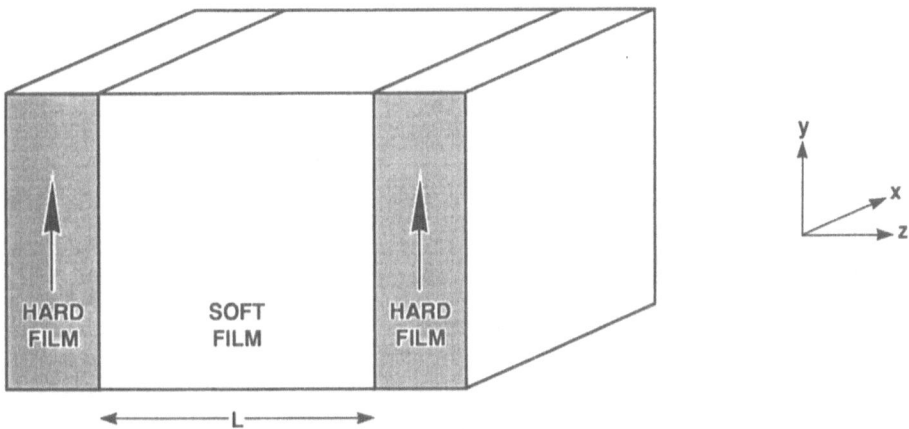

Fig. 12. The geometry considered. We assume the films are of infinite extent in both x- and y- dimensions.

$$\gamma_{s\gamma} = \gamma_{sl} + \gamma_{1\gamma}\cos\theta_w \tag{37}$$

which relates the solid-gas and solid-liquid surface tensions, γ_{sg} and γ_{sl} respectively, and assuming that the gas is nearly ideal we may rewrite [Evans et al., 1986] Eq. 36 as

$$\Delta\mu = 2\frac{\gamma_{sq} - \gamma_{sl}}{L(\rho_l - \rho_g)} \tag{38}$$

where $\Delta\mu$ is the deviation of the chemical potential from coexistence. Equation 38 is now easily recognized as the magnetic analog of Eq. 35.

II.4 Consequences of the Magnetic Kelvin's Equation

The first point to be made about the discussion in the last two sections is that it establishes Eq. 35 as an exact thermodynamic relation in the $L \to \infty$ limit and hence a sound, and very general, starting point for the study of thin film magnetism. To illustrate its consequences we studied a simple model with axial (quadratic) anisotropy. It is described by the surface tension functional

$$2\tilde{\sigma} = +\int_0^L dz\left\{\frac{\alpha'\,\xi_0^2}{2}\left|\frac{dM}{dz}\right| + f(M) - f(M_b)\right\}$$

$$+\xi_0 f_s(M(0)) + \xi_0 f_s(M(L)), \tag{39}$$

where

$$f(M) = \frac{1}{2}a'\,t|M|^2 + \frac{1}{4!}u|M|^4 + \frac{1}{4}a'\,\kappa M_x^2 - H\cdot M$$

$$f_s(M) = \frac{1}{2}a'\,\xi_o g|M|^2 - H_1\cdot M \tag{40}$$

Moreover, L = ferromagnetic soft film thickness; κ = uniaxial anisotropy parameter (easy y-axis for $\kappa > 0$); H = (O,Hy), applied field; $t = (T - T_c^{\kappa=0})/T_c^{\kappa=0}$, reduced temperature; M_b = bulk magnetization (for given H and t); ξ_0 = microscopic coherence length of the order of a lattice spacing, g = surface enhancement parameter (allows for different exchange coupling between spins near surface of soft film).

The most convenient geometry is shown in Fig. 12. We shall only deal with this and consider what happens as L becomes sufficiently large for the system to accommodate domain walls parallel to the xy-plane. In the analogous case for liquids the Kelvin equation must be modified to take account of the wetting films forming on the capillary wall. For the magnetic problem at hand this also happens and, as shown in Fig. 13, such films form. The

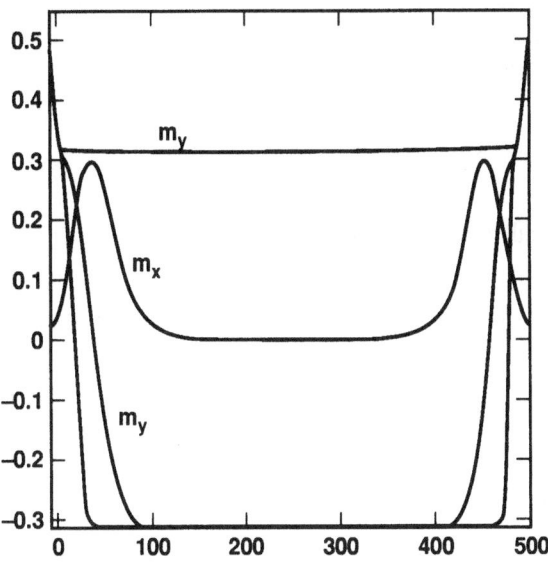

Fig. 13. Extremal magnetization profiles $m(z') \equiv M(z / \xi_0)\sqrt{u / 6a'}$ for $t = -0.1$, $\kappa = 0.01$, $\xi_0 g = 0.001$, $h_{1_y} \equiv \sqrt{u / 6a'} H_{1_y} / a' = 0.2$, $L' \equiv L / \xi_0 = 500$ and $h_y \equiv \sqrt{u / 6a'} H_y / a' = -0.01x10^2$. For the stable "Bloch" phase both m_x and m_y vary. For the (metastable) spin-up and (unstable) "Ising" phases, m_x is identically zero.

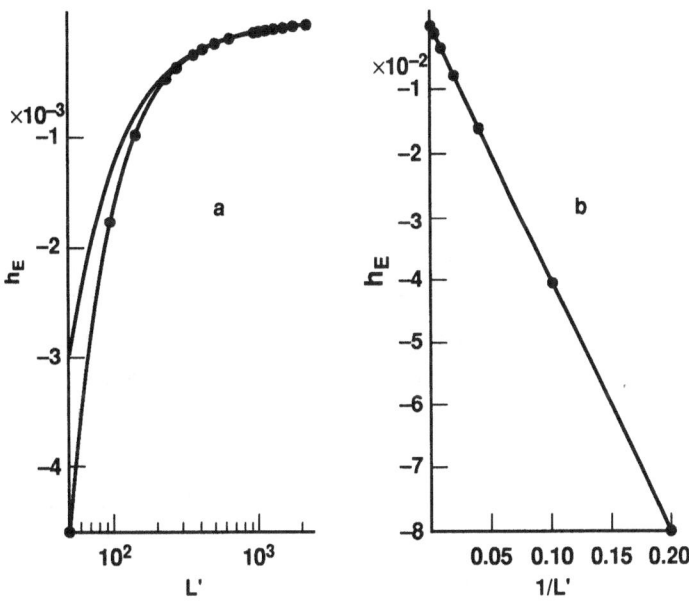

Fig. 14. Loop shifts $h_E \equiv \sqrt{u / 6a'} H_E / a'$ as a function of L' for (a) $\kappa = 0.01$, $t = -0.2$ (where wetting films obtain), and (b) $\kappa = 0.1$, $t = -0.4$ (partial wetting). The curve without dots represents the solutions of Eq. (35) and (41) as appropriate. The dots are the solutions of the Landau equation.

corresponding modification of the magnetic Kelvin Equation reads as follows

$$H_E \approx -\frac{\sigma_{\uparrow\downarrow}}{|M_{by}|(L - 2\ell)}$$ (41)

where ℓ is the wetting film thickness and

$$\sigma_{\uparrow\downarrow} = \text{domain wall surface tension} \sim \sqrt{A/K}$$ (42)

To test this formula we evaluated H_E as the value of H for which spin up $\left(\uparrow\right)$ and spin down $\left(\downarrow\right)$ phases (which are in general non uniform) have the same value of $\tilde{\sigma}$ by solving, numerically, the Euler Lagrange equation corresponding to minimizing $\tilde{\sigma}$ in Eq. 39. The results of such calculations compared with the predictions of Eq 41 and Eq 35 for two different anisotropies $\kappa = .01$ and $\kappa = .1$ are shown in Fig. 14. Indeed, Fig. 14a demonstrates that Eq. (41) provides a reasonable approximation only when L is large. On the other hand, when there are no wetting films present (partial wetting) Eq. (35) holds well even for L much less than the thickness of a domain wall $\xi_o / \sqrt{2\kappa}$ (Fig. 14b).

II.5 Ising- and Bloch-Like Phases

What is novel about uniaxial ferromagnetic systems is that, as one approaches T_C the character of the domain wall changes from Bloch-like (i.e. rotational) to Ising-like (i.e., with $M_x = o$ [Lowrie and Low, 1981]. For the present model we have shown that the bounded domain wall associated with a wetting film also exhibits such a transition.

In Fig. 15 the upper curve is the locus of points on which

$$\Gamma_x \equiv \int_0^L M_x dz$$ (43)

vanishes, and represents a chiral-symmetry-breaking transition. Linear stability analysis confirms that the Ising-like phase becomes unstable as one crosses this line. The lower curve corresponds to $H_E(t)$. As L is reduced one reaches a point where one can no longer speak in terms of individual wetting films and the symmetry-broken profiles acquire a nonzero M_x for all z between O and L.

For much larger anisotropy ($\kappa=0.1$) the topology of the phase diagrams for the same L as Fig. 15 is somewhat different, and exhibits a 'prewetting line" [Walden and Gyorffy, 1988; Walden et al., 1990] on which phases with thin and thick wetting films coexist. This is sketched in Fig. 16. One of us (CJW) would like to acknowledge financial support in the form of an EC BRITE fellowship.

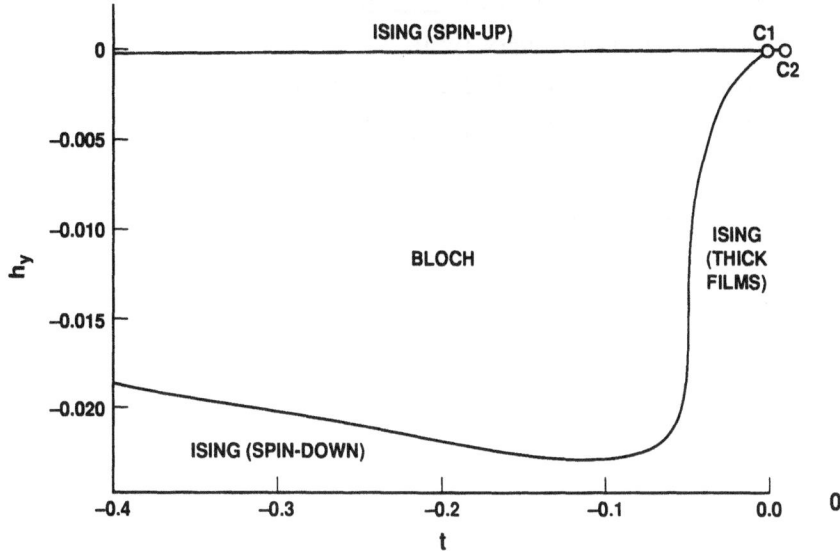

Fig. 15. Phase diagram in the $h_y t$-plane for g, k, h_{1y} and L' as in Fig. 13. C_1 is a critical endpoint, whilst C_2 is the (shifted (7)) critical point of the system.

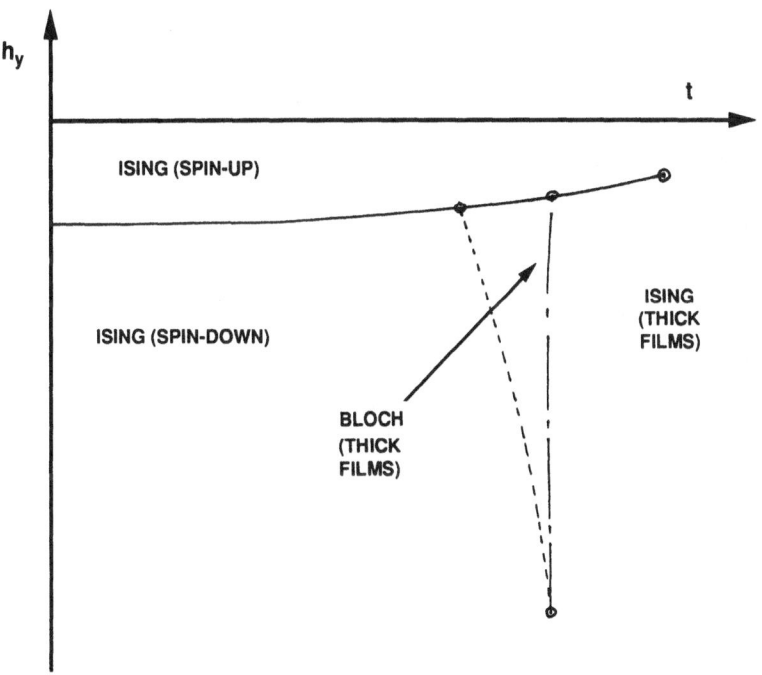

Fig. 16. Sketch of phase diagram for $\kappa = 0.1$ (other parameters as in Fig. 13). The dotted line is the "prewetting line." The dashed line is a line of second order Ising-Bloch transitions.

REFERENCES

Binder, K. in "Phase Transitions and Critical Phenomena", Vol 8, Eds., Dombe, C. and Lebowitz, J. L. (Academic Press, 1983).

Brézin, E., Halperin, B. I., Liebler, S., J. Phys. (Paris) 44, 775, (1983).

Croxton, C. A. (Ed) "Fluid Interfacial Phenomena" (Wiley, 1986).

Dietrich, S., in "Phase Transitions and Critical Phenomena" Vol 10 ,Eds. Dombe, C. and Lebowitz, J. L. (Academic Press, London, 1988).

Evans, R., Marini Bettolo Marconi, U. and Tarazona, P., J. Chem. Phys. 84, 2376 (1986).

Evans, R. in "Capillarity Today" Eds., Pete, G. and Sanfeld, A. (Springer-Verlag) pp 62 (1991).

Fert, A., Gunterodt, G., Heinrich, B., Marinero, E. E., Mauer, M., Eds., "Magnetic Thin Films, Multilayers and Superlattices" (Elsevier Science Publishers, 1991).

Fisher, D. S. and Huse, D. A., Phys. Rev. B32, 247 (1985).

Hauge, E. H., Phys. Rev. B33, 3322 (1986).

Hauge, E. H. and Olaussen, K., Phys. Rev. B32, 4766 (1985).

Jonker, B. T., Hermas, J. P., and Marinero (Eds) "Growth, Characterization and Properties of Ultrathin Magnetic Films and Multilayers" (Mater. Res. Soc. Symp. Proc. 151 Pittsburg, 1989).

Kaneyoshi, T., "Introduction to Surface Magnetism" (CRC Press, Boca Raton Florida, 1991).

Lambin, Ph. and Herman, F., Phys. Rev. B30, 6903 (1984).

Landau, L. and Lifshitz, E., "Electrodynamics of Continuous Media" (Pergamon Press, Oxford, 1960).

Lawrie, I. D. and Lowe, M. J., J. Phys. A:Math. Gen. 14, 981 (1981).

Meiklejohn, W. H. and Bean, C. P., Phys. Rev. 102, 1413 (1956); ibid 105, 904 (1957).

Meiklejohn, W.H., J. Appl. Phys. 33, 1328 (1962).

Malozemoff, A. P., Phys. Rev. B 35 3679; 37, 7673 (1988); J. Appl. Phys. 63, 3874 (1988). This differs by a factor of 2 from the result quoted by Malozemoff [2], since we have two surfaces which break timereversal symmetry.

Malozemoff, A. P., Phys. Rev. B35, 3679 (1987).

Malozemoff, A. P., J. Appl. Phys. 63, 3874 (1988).

Mauri, D., Kay, E., Scholl, D., and Howard, J. K., J. Appl. Phys. 62, 3047 (1987).

The domain wall phase transition in this model was first discussed by C. Montonen, Nucl. Phys. B112, 349 (1976), and independently by S. Sarker, S. E. Trullinger & A. R. Bishop, Phys. Rev. Lett. 59A, 255 (1976).

Nakanishi, H. and Fisher, M. E., J. Chem. Phys. 78, 3279 (1983).

A specific example is discussed by Ounadjela, K. and Suran, G., J. Appl. Phys. 63, 3244 (1988).

Pekalski, A. and Szmajd, J., (Eds) "Static Critical Phenomena in Inhomogeneous Systems" (Springer-Verlag, Berlin, 1984).

Rowlinson, J. S. and Widom, B., "Molecular Theory of Capillarity" (Calrendon Press, 1982).

Stanly, E., "Critical Point Phenomena."

Subbaswamy, K. R., and Trullinger, S. E., Phys. Rev. Al9, 1340 (1979) (1988).

Taub, H., Torzo, G., Lauter, H. J., and Fain, S-C. (Eds) "Phase Transitions in Surface Films." ASI Series B, Physics 267 (Plenum Press, 1991).

Tsang, C. and Lee, K., J. Appl. Phys. $\underline{53}$, p 2605 (1982).

We have previously discussed wetting in a magnetic context: Walden, C. J., Gyorffy, B. L., and Parry, A. O., Phys. Rev. B$\underline{42}$, 798 (1990).

Walden, C. J. and Gyorffy, B. L., J. Phys. Colloq. $\underline{49}$, C8-1635 (1988).

Walden, C. J., Gyorffy, B. L. and Parry, A. O., Phys. Rev. B $\underline{42}$ (1990).

de Wit, H. J., Rep. Progr. Phys. 55, pp 113, (1992).

Yeron, A. V. in "Physics of Thin Films", Vol 2, p 205, eds., Francombe, M. and Hoffmann, R. (Academic Press, New York, 1971).

SURFACE AND MAGNETIC EFFECTS IN CORE LEVEL PHOTOEMISSION

P. J. Durham

SERC Daresbury Laboratory, Warrington WA4 4AD, England

I. INTRODUCTION

Much of the interest in angle-resolved photoemission over the last ten years or more has revolved around the observation of valence or conduction band states[1], and the determination of their detailed dispersion curves. Nevertheless, angle-resolved photoemission from core levels also contains a great deal of useful information about the local electronic structure and the local atomic geometry in the bulk and at surfaces. In this note, two different aspects of photoemission from core levels are discussed. First, an *ab initio* method of calculating photoemission from shallow core levels (including the effects of surface potential shifts) is described and contrasted with the standard deconvolution treatment of experimental data. It is suggested that the finite bandwidth of semi-core levels may, under certain circumstances, influence the interpretation of the photoemission spectra. Second, an outline of spin-polarized photoemission from core levels in magnetic systems is given, and its relationship to the recently observed magnetic dichroism in core level photoemission is discussed. Finally, a connection between these two themes is suggested.

II. PHOTOEMISSION AND SURFACE CORE LEVEL SHIFTS

The effective potential felt by an electron in a surface atom differs from that in a bulk atom because the local atomic coordination and, in consequence, charge distribution is different. The size of this effect can be estimated for metals from an elementary calculation based on surface band-narrowing coupled with the requirement of local charge neutrality[2]. Since charge redistribution occurs mainly in the outer parts of the surface atom, the inner core levels will suffer a more or less uniform shift in binding energy, and this so-called surface core level shift (SCLS) can be observed in photoemission. In the simplest cases, the experimental spectra show not one but two core lines, corresponding to states on bulk and surface atoms. The standard way of analyzing such data is to deconvolute the spectrum, assuming some intrinsic line-shape (usually of the Doniach-Sunic form), into its component lines and thereby evaluate the SCLS. Often, such deconvolution procedures yield more than two component lines; then it is usually assumed that core levels in atomic layers adjacent to the surface layer itself ("underlayers") are also shifted relative to the bulk. It should be noted that SCLS values are not very large, and typically amount to tenths of an electron volt. This means that SCLS can only be resolved in photoemission from relatively shallow core levels whose lifetime widths are not too large. Examples are the 4f levels of the lanthanides and 5d transition metals, and the 4p levels of the early 4d transition metals, all of which tend to have binding energies in the 10–30 eV range.

Equilibrium Structure and Properties of Surfaces and Interfaces
Edited by A. Gonis and G.M. Stocks, Plenum Press, New York, 1992

289

Since SCLS measurements are, in fact, angle-resolved photoemission, it should be possible to calculate the core level spectra by the *ab initio* 1–electron methods developed to describe valence band spectra. Figure 1 shows some calculations of the 4p photoemission spectra of Y(0001) surfaces for two emission angles, and compares with experiment[3]. The calculations were made with the layer-KKR-based program NEWPOOL, using self-consistent potentials taken from a slab/supercell LMTO calculation to model the surface electronic structure. [Note that this was a non-relativistic calculation; the $4p_{1/2}$ and $4p_{3/2}$ contributions were assumed to be copies of the non-relativistic spectrum displaced by the experimental spin-orbit splitting and having a 1:2 intensity ratio.] Agreement with experiment is very satisfactory, and the increased surface contribution expected at the higher emission angle is particularly well reproduced.

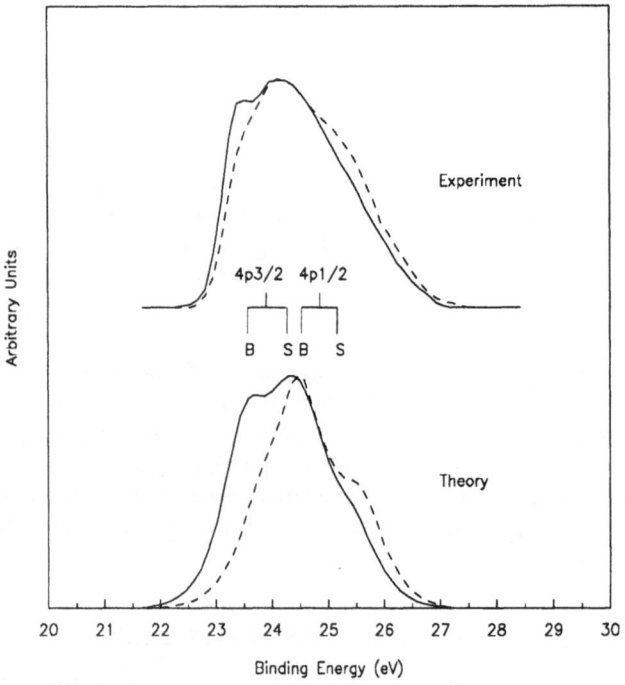

Figure 1. Y(0001) 4p photoemission emission spectra[3], normal emission (full curves) and 60° emission (dashed curves). Bulk and surface contributions from each of the spin-orbit split components are indicated schematically.

An interesting possibility in surface core level photoemission is to monitor the intensity of a surface core line as a function of photon energy. The energy of the photoelectron will then be changing and so will the amplitudes for it to be scattered by the surrounding atoms. One might then expect variations in the photoemission cross-section, related to the XAFS or XANES observed in x-ray absorption, which would carry information about the positions of neighbouring atoms — an aspect, clearly, of photoelectron diffraction. Again, this should be described by NEWPOOL-type *ab initio* 1–electron calculations. Figure 2 shows a set of such calculations for the 4p levels at a Nb(100) surface. [These calculations were performed with a fully relativistic version of NEWPOOL[4], and so the $4p_{1/2}$ and $4p_{3/2}$ contributions emerge naturally.] The "unbroadened" spectra actually used a very small broadening corresponding roughly to the long lifetime of the Nb 4p states. The broadened spectra included a more realistic estimate of the achievable experimental resolution.

Clearly the peak intensities change with photon energy. There is, however, a surprise. The calculations shown in figure 2 do not include any surface shift at all — they used a self-consistent potential from bulk LMTO calculations on all atomic layers including the surface — and yet they reveal several component peaks contributing to a complex line-shape. The origin of these peaks is in fact a band structure effect. Simple estimates of overlap integrals and full KKR calculations show that the 4p levels of Nb form a band about 1 eV wide, and when, as in figure 2, the photocurrent is calculated with very small broadening parameters one finds just the kind of direct transition features that are familiar from valence band spectra. In the simplest interpretation, these features change with photon energy because the direct transitions move in k-space. In this case, therefore, it would be dangerous to analyze spectra by attempting to deconvolute into a number of Doniach-Sunic lines — finite bandwidth effects could be mistaken for surface shifts. Indeed, for the standard deconvolution procedure to be valid, it is necessary that

(i) SCLS > intrinsic core level lifetime width (in order to resolve the shift at all), and
(ii) intrinsic core level lifetime width > core level bandwidth (so that the core hole can be assumed static, as in the Doniach-Sunic model).

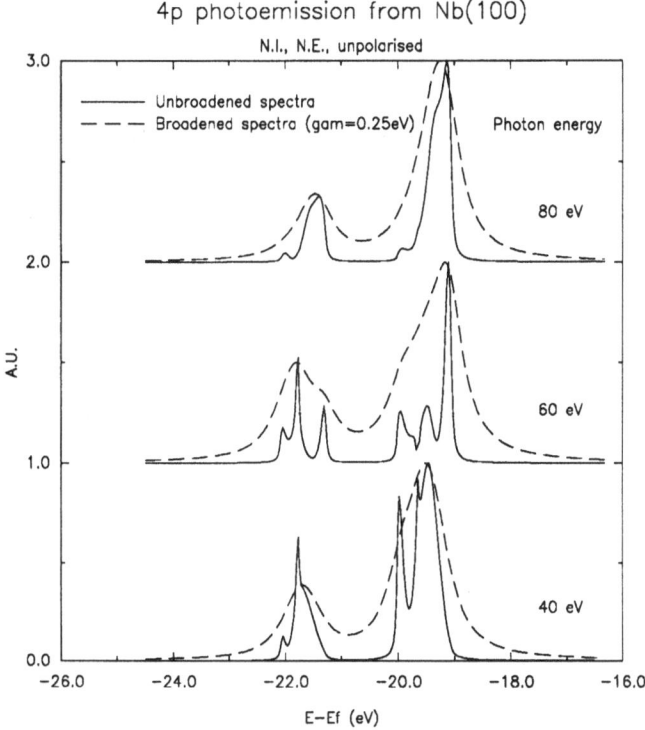

Figure 2. Calculated photoemission from Nb(100) 4p levels for 3 photon energies (normal incidence, normal emission).

As the core level becomes deeper, and more tightly-bound, the second condition becomes easier to satisfy but the first more difficult. The conditions are certainly not obviously fulfilled in the case of the Nb 4p levels, and I suspect the same is true for several other systems (including the 4f levels of the earlier 5d transition elements). However, as the example of the Y 4p photoemission shows, in such cases one can resort to full angle-resolved photocurrent calculations to analyze the data. Finally, it is interesting to note that band effects in semi-core levels are sometimes found to play a role in total energy studies of cohesion[5]; these semi-core levels are often the very states which should show SCLS in photoemission.

III. EXCHANGE AND SPIN-ORBIT COUPLING IN CORE LEVEL PHOTOEMISSION

We now turn from surface phenomena to the inner core levels of magnetic systems. The modern approach to itinerant magnetism is based on spin density functional theory, usually practised within the local (LSD) approximation. The many successes (and occasional failures) of this method are well documented; in particular, spin-resolved photoemission from the d-bands has been a key probe of the spin-polarised band structures of magnetic transition metals[6]. Recently it has become possible to make relativistic spin density functional calculations (ie. to solve the effective spin-polarised Dirac equation), mainly motivated by problems such as magneto-crystalline anisotropy. This development is also necessary to describe the core states of magnetic systems, since relativistic effects are always strong in the interior of atoms. Figure 3 shows schematically some calculations[7] of the core levels of ferromagnetic Fe with and without exchange splitting. For the deeper levels the exchange field removes m_J—degeneracy rather in the way an external magnetic field would operate in the Zeeman effect (and, indeed, perturbation theory works well for these states). There is, however, an important difference from the Zeeman effect, where the ordering of split levels goes with m_J. In figure 3 the ordering of exchange split levels goes with $<\sigma_z>$, where σ_z is the z-component of the electron spin and the z-axis points along the moment.

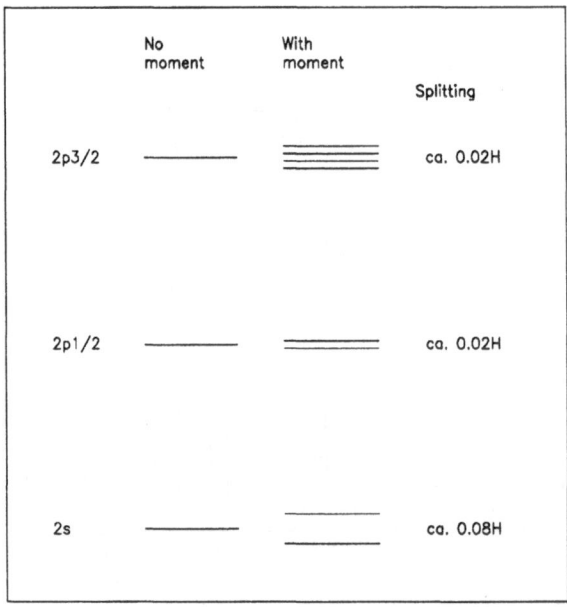

Figure 3. Schematic energy level diagram for core levels of Fe (after Ebert[7])

Figure 3 indicates that photoemission from, say the 2s or 3s states should give a two-peak spectrum, with the splitting of the peaks related to the local moment on the Fe atom. This possibility was recognized some time ago, in fact, and there have been numerous experimental studies of the split 3s photoemission from different Fe compounds and alloys[8]. The correlation of peak splitting with local Fe movement is, it has to be said, imperfect. Nevertheless, the underlying picture of a spin-up and a spin-down 3s peak has been beautifully confirmed in the spin-resolved photoemission experiment of Hillebrecht et al[9]. Spin-resolution is difficult and not routinely practised in photoemission studies, however, and an interesting alternative was demonstrated recently by Baumgarten et al[10] in their study of magnetic dichroism in photoemission from the 2p levels of Fe.

This effect can be understood by a simple argument beginning with a consideration of the 2p levels of a fictitious Fe atom with the local moment set to zero. We are thus dealing with the familiar 4–fold degenerate $2p_{3/2}$ states and the 2–fold degenerate $2p_{1/2}$ states, with wave functions whose spin-angular parts can be found in numerous text books. Let a photon of either left or right hand circular (LHC or RHC) polarization travelling in the z-direction be absorbed by the atom and emit an electron of either up or down spin in a direction $\hat{k}$ (all polarizations, spins and momenta are referred to a z-axis defined by the photon incidence direction). It is easy to calculate the rate of such photoemission processes from each of the 6 possible initial 2p states; the results are given in table 1, in which the units are $|M_{p\to d}|^2$ ($M_{p\to d}$ being the electron-photon matrix element between the radial 2p wave function and that of the final d state) and q is the ratio of p→s to p→d transition rates:

$$q = \frac{|M_{p\to d}|^2}{|M_{p\to s}|^2}$$

Table 1. Model calculations of spin-resolved photoemission from Fe 2p levels; the spherical harmonics $Y_{2,0}$ etc. have an implied argument of $\hat{k}$, the electron emission direction.

	m_J	$I^\uparrow$(RHC)	$I^\downarrow$(RHC)	$I^\uparrow$(LHC)	$I^\downarrow$(LHC)
$2p_{3/2}$	-3/2	0	$|q+Y_{2,0}|^2$	0	$|Y_{2,-2}|^2$
	-1/2	$1/3|q+Y_{2,0}|^2$	$2/3|Y_{2,1}|^2$	$1/3|Y_{2,-2}|^2$	$2/3|Y_{2,-1}|^2$
	+1/2	$2/3|Y_{2,1}|^2$	$1/3|Y_{2,2}|^2$	$2/3|Y_{2,-1}|^2$	$1/3|q+Y_{2,0}|^2$
	+3/2	$|Y_{2,2}|^2$	0	$|q+Y_{2,0}|^2$	0
$2p_{1/2}$	-1/2	$2/3|q+Y_{2,0}|^2$	$1/3|Y_{2,1}|^2$	$2/3|Y_{2,-2}|^2$	$1/3|Y_{2,-1}|^2$
	+1/2	$1/3|Y_{2,1}|^2$	$2/3|Y_{2,2}|^2$	$1/3|Y_{2,-1}|^2$	$2/3|q+Y_{2,0}|^2$

Assuming that q is negligible and taking the case in which the photoelectron is emitted along the photon incidence direction we find, summing over the contributions from $2p_{3/2}$ and $2p_{1/2}$ states that the total intensities I_{Total} and electron spin polarizations P, where

$$I_{Total} = I^\uparrow + I^\downarrow$$

and

$$P = \frac{I^\uparrow - I^\downarrow}{I_{Total}}$$

are given in table 2.

Table 2. Total intensity and spin-polarisation

	I_{Total}(RHC)	P(RHC)	I_{Total}(LHC)	P(LHC)
$2p_{3/2}$	20/3	-1/2	20/3	+1/2
$2p_{1/2}$	10/3	+1	10/3	-1

293

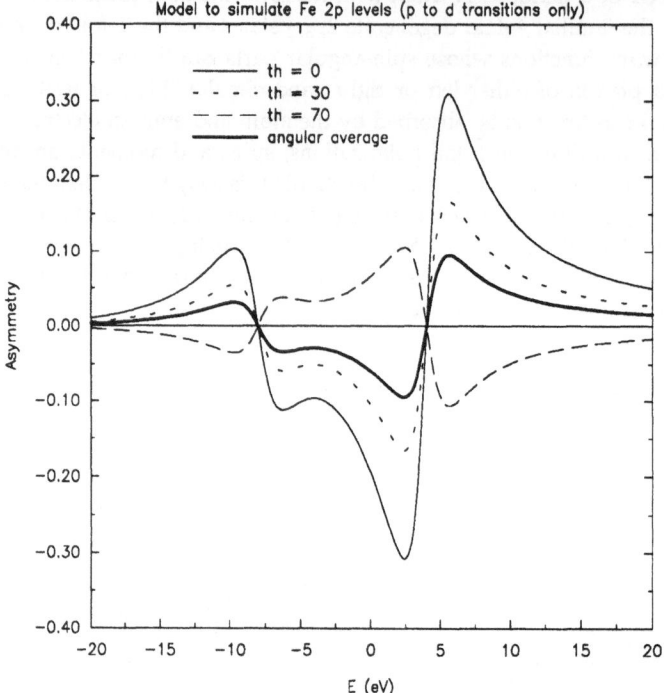

Figure 4. Model calculation of the dichroic asymmetry in Fe
2p photoemission for different electron emission angles (th).

Thus, in the absence of a local moment the experiment just analyzed shows a net spin polarization but no dichroism — the total intensities are the same for LHC and RHC light. Note, however, that for the different polarizations photoemission occurs from states with different m_J values; it is only because these states are degenerate that there is no dichroism. But, as figure 3 shows, the effect of exchange splitting is to lift this degeneracy, and this must therefore lead to different photoemission spectra for LHC and RHC polarizations. This is the origin of the magnetic dichroism observed by Baumgarten et al. Indeed, it is easy, on the basis of the above argument and the energy levels in figure 3, to simulate the dichroism when the photon is incident along the local moment direction; the resulting asymmetry curves (ie. I(RHC)-I(LHC)) are shown in figure 4. A full analysis, using the correct solutions of the spin-polarised Dirac equation and a relativistic treatment of the electron-photon interaction, has been given by Ebert et al[11]. The results agree well with experiment and are quite similar to figure 4. It is interesting to note that sign of the dichroism asymmetry for the $2p_{1/2}$ states is opposite to that of the $2p_{3/2}$ states. This is a consequence of the ordering of exchange-split levels with $<\sigma_z>$, as remarked above, and not with m_J.

Clearly, this dichroism experiment is in principle a useful method of observing exchange splitting without electron-spin-resolution. Indeed, a similar analysis to the above indicates that in some cases (eg. the $2p_{3/2}$ states) linearly polarized light may yield the same information if the dependence on electron emission angle is monitored. Some key physics issues remain to be resolved, however. The above argument was stated in terms of the solutions of the spin-polarised Dirac equation; the effect came from the description of the ground state core levels. But another school of thought[12], basing its arguments on atomic multiplet theory, attributes the spectral splitting to exchange effects in the final state. Further calculations and experiment are needed in order to gauge the accuracy of these alternative descriptions of core electrons in magnetic systems. This will be important not only in photoemission dichroism but in other related spectroscopies such as x-ray absorption and fluorescence[13].

IV. CONCLUDING REMARKS

Finally, we mention a speculation which may link the two apparently disparate topics discussed above. In discussing SCLS we suggested that the degeneracy of shallow core levels can be lifted by band structure effects. In discussing magnetic systems, we noted that core level degeneracies are lifted by exchange splitting, and that this could give rise to a dichroism in photoemission. This suggests that distinctive polarization effects, if not circular dichroism itself, may show up in photoemission from the shallow core levels of non-magnetic systems. Indeed, preliminary experiments by R. G. Jordan and coworkers (private communication) indicate that for a given emission angle photocurrent intensities vary with the polarization direction of incident linearly polarized light in a way which may signal such band structure effects. We hope to report more fully on this in a later publication.

ACKNOWLEDGMENTS

I thank R G Jordan, A M Begley, G Y Guo, K Purcell, H Ebert, B L Gyorffy, P Strange, E Seddon and H Padmore for useful conversations bearing on this paper.

References

1. eg. F. J. Himpsel, Adv. Phys. **32**,1 (1983).
2. eg. A. Zangwill, *Physics at Surfaces*, Cambridge University Press (1988).
3. R. G. Jordan, A. M. Begley, S. D. Barrett, P. J. Durham and W. M. Temmerman, Solid St. Commun. **76**, 579 (1990).
4. B. Ginatempo, P. J. Durham, B. L. Gyorffy and W. M. Temmerman, Phys. Rev. Lett. **54**, 1581 (1985).
5. eg. Z. Szotek, W. M. Temmerman and H. Winter, Physica **B165&166**, 275 (1990).
6. eg. E. Kisker, J. Magn. Magn. Mat. **45**, 23 (1984); P. J. Durham, J. B. Staunton and B. L. Gyorffy, J. Magn. Magn. Mat. **45**, 38 (1984).
7. H. Ebert, J. Phys.: Condens. Matter **1**, 9111 (1989).
8. J. F. van Acker, Z. M. Stadnik, J. C. Fuggle, H. J. W. M. Hoekstra, K. H. J. Buschow and G. Stroink, Phys. Rev. **B37**, 6827 (1988).
9. F. U. Hillebrecht, R. Jungblut and E. Kisker, Phys. Rev. Lett. **65**, 2450 (1990).
10. L. Baumgarten, C. M. Schneider, H. Petersen, F. Schäfers and J. Kirschner, Phys. Rev. Lett. **65**, 492 (1990).
11. H. Ebert, L. Baumgarten, C. M. Schneider and J. Kirschner, Phys. Rev. **B44**, 4406 (1991).
12. G. van der Laan, Phys. Rev. Lett. **66**, 2527 (1991); S. Imada and T. Jo, J. Phys. Soc. Japan **60** 2843 (1991); B.T.Thole and G. van der Laan, Phys. Rev. Lett. **67**, 3306 (1991).
13. G. Schütz, W. Wagner, W. Wilhelm, P. Kienle, R. Zeller, R. Frahm and G. Materlik, Phys. Rev. Lett. **58**, 737 (1987); P. Strange, P. J. Durham and B. L. Gyorffy, Phys. Rev. Lett. **67**, 3590 (1991).

GROUND STATE RESULTS OF FERROMAGNETIC MONOLAYERS

Stefan Blügel

Institut für Festkörperforschung des Forschungszentrums Jülich
D-5170 Jülich, Federal Republic of Germany

I. INTRODUCTION

The investigation of ultrathin magnetic films constitutes a quickly developing field in science. Diverse methods are used to prepare them and a wide, flexible, and novel set of experimental techniques is used for the measurements[1]. A variety of these artifical materials has been studied. Among them are magnetic ultrathin 3d transition metal films stabilized on various non–magnetic substrates[2]. They can be grown on single crystals epitaxially in a layer by layer mode. These overlayers represent realizations of well characterized two–dimensional (2D) itinerant magnets for which a direct comparison with theoretical predictions becomes possible. In these systems the electronic and magnetic structure, the magneto–crystalline anisotropy energy, the magnetic phase transition and micromagnetism are significantly different from bulk systems.

One of the most basic issues relating to the magnetism of ultrathin films is to understand which combination of overlayer and substrate material offers the possibility of a spontaneous magnetization in 2D. In fact, itinerant magnetism in three dimensions (3D) is observed for metals and compounds synthesized with elements of the 3d transition metal series; in particular: Cr, Mn, Fe, Co, Ni. In general, Co and Ni are ferromagnetic, Cr is antiferromagnetic, and Mn and Fe are ferromagnetic or antiferromagnetic depending on the crystal structure. With some exceptions such as e.g. the weakly ferromagnetic compounds $ZrZn_2$ [3], Sc_3In [4], magnetism for Sc, Ti, V, 4d, and 5d metals is not observed. The possibility of ferromagnetism of 4d metals was explored by Gunnerson[5] and Janak[6] on the basis of the Stoner model. They found that with intra–atomic exchange integrals I_{4d} of about 0.65 eV and local densities of states (LDOS) n_{loc} at the Fermi energy (E_F) varying from 0.32 states/spin/eV for Mo to 1.15 states/spin/eV for Pd, the Stoner criterion for a ferromagnetic instability, $n_{loc}(E_F)I > 1$ is never satisfied for any 4d metal. This was confirmed by ab–initio bandstructure calculations of Moruzzi and Marcus[7] where no magnetic solutions were found for any 4d bulk metal at equilibrium volumes. Following the general trend of decreasing localization of valence d wavefunctions when moving from the 3d to 4d and 5d series we find consequently an increase of the d band width and a reduction of LDOS $n_{3d} > n_{4d} > n_{5d}$ at the Fermi energy. Together with the fact that the exchange integral also decreases as $I_{3d} > I_{4d} > I_{5d}$ 5d magnetism becomes out of question.

Equilibrium Structure and Properties of Surfaces and Interfaces
Edited by A. Gonis and G.M. Stocks, Plenum Press, New York, 1992

297

Indeed almost all experimentally stabilized ultrathin films at one monolayer range on non–magnetic substrates (e.g. V/Ag, Cr/Ag, Mn/Ag, Fe/Ag, Fe/Pd, Fe/Au, Co/Cu, Co/Ag, Co/Pd, Co/Au[2]) were restricted to 3d–metals. Practically, no 4d or 5d system has been investigated.

Itinerant magnetism in 2D is not a priori restricted to those elements which exhibit magnetism in 3D. Because of the reduced coordination number of nearest neighbor atoms in a monolayer film the d band width in 2D is considerably smaller and correspondingly the LDOS at the Fermi energy is considerably larger than in 3D[8–11]. Thus the magnetic instability should occur for a much wider variety of transition metal elements. Following this line of argument it is clear that the strength of the d–d hybridization between monolayer and substrates is an additional parameter which controles the d–band width of the monolayer. For instance large bandgap material as substrate allows the formation of 2D monolayer bands within the bandgap of the substrate material. In this case the impact on the magnetization of the monolayer due to the substrate is expected to be small. The same is true for noble metal substrates, which have d–bands well below the Fermi energy. The width of the monolayer d–band is not significantly broadened by the monolayer–substrate d–d interaction, and magnetism is restricted to the monolayer. Increasing the d–d hybridization by choosing transition metals as substrates will lead to a considerable broadening of the monolayer bands until we have changed from the 2D limit to the semi–infinite regime. Superimposed on these simple dimensionality arguments of the d–d hybridization are van Hove singularities due to the interference of Bloch waves within the planar lattice of the 2D monolayer. They depend on the crystal structure, the symmetry and phase of the wave function and are due to the long range interactions within the monolayer. Therefore, their contributions to magnetism is referred to as 2D bandstructure effects. They are expected to play an important role in stabilizing a magnetic phase. In 3D they are for instance responsible for the magnetism of fcc Ni or the large susceptibility of Pd[12]. This view is supported by the fact that artifical bcc Ni does not show these bandstructure effects at the Fermi energy and remains non–magnetic.

The aim of the present paper is to present a coherent account of a systematic search for the existence of 3d, 4d and 5d monolayer ferromagnetism on Ag, Au, and Pd(001) substrates using ab–initio electronic structure theory. Thereby we are focussing on the trends within each d–metal series, between the d–metal series and on the importance of the substrate hybridization for monolayer magnetism. The impact on the monolayer magnetism due to hybridization with the substrate becomes visible by comparing the monolayer results on Ag with those on Au and Pd. Ag and Au have both d bands located well below the Fermi energy which keeps the broadening of the monolayer bands at a minimum. The d bands of Au are about 1eV closer to the Fermi energy than those of Ag, and in case of Pd its d bands are crossing the Fermi level. Therefore, Pd represents the limit of strong substrate hybridization, which is similar in strength to the one in the monolayer itself. In addition, since those substrates are often used in experimental realizations a check of the theoretical results can be expected.

II. THEORETICAL METHOD

The ab–initio results are obtained with the full–potential linearized augmented–plane–wave method (FLAPW)[13] for film geometry. Seven–layer (001) films consisting of five layers Ag or Au and one d–metal monolayer on each surface are considered. The validity of this structural model was confirmed by repeating some calculations with films consisting of seven layers of Ag and Au. The magnetic polarizability of Pd is much larger than that of Ag and Au. Therefore, d metal monolayers on Pd substrate are modelled by nine-layer (001) films consisting of seven layer Pd. A total number of about 2×50 symmetrized augmented plane waves per atom are used as variational basis set. Among the cut–off parameters inherent in the FLAPW–method the magnetic moment depends most strongly on the accuracy of the Brillouin–zone integration. The magnetic moments and total energy differences between ferro– and paramagnetic solutions were monitored by performing calculations with increasing number of $k_{||}$ points in the irreducible wedge of the 2D Brillouin–zone, until final convergence was obtained for 36 or 78 special $k_{||}$ [14], respectively. Core states are calculated full– and valence states are treated scalar–relativistically. The calculations apply density–functional theory using the local spin density approximation (LSDA) of von Barth and Hedin (vBH)[15], in the parameterization of Moruzzi, Janak and Williams (MJW)[16]. Some of the results have been crosschecked with the LSDA of Vosko, Wilk and Nusair (VWN)[17] as well as the one of vBH. The monolayer–substrate interlayer spacing was taken to be the average of their bulk lattice spacings $a_0^{d-sub} = \frac{1}{2}(a_0^d/2 + a_0^{sub}/2)$. All lattice parameters were chosen according to MJW[16].

III. RESULTS

The intensive calculations reveal that i) all 3d metal monolayers (Ti, V, Cr, Mn, Fe, Co, Ni) on Ag(001) substrate show ferromagnetic solutions[10]. This is very likely also true on a Au(001) substrate. ii) With the exception of Ti, monolayer magnetism was also predicted for the 3d monolayers on Pd[18]. iii) Among the 4d metals Tc, Ru, and Rh are ferromagnetic on Ag as well as Ru and Rh on Au[19], iv) among the 5d metals Os and Ir are magnetic on Ag, but only Ir is magnetic on Au[20]. v) No monolayer magnetism was found for 4d metals on Pd with the exception of Ru for which a rather small magnetic moment of $0.2\mu_B$ was calculated. Therefore, no 5d monolayer magnetism on Pd is expected. Thus with Ti, V; Tc, Ru, Rh; Os, and Ir on Ag or Au we found transition metals, which are magnetic in 2D and nonmagnetic in 3D.

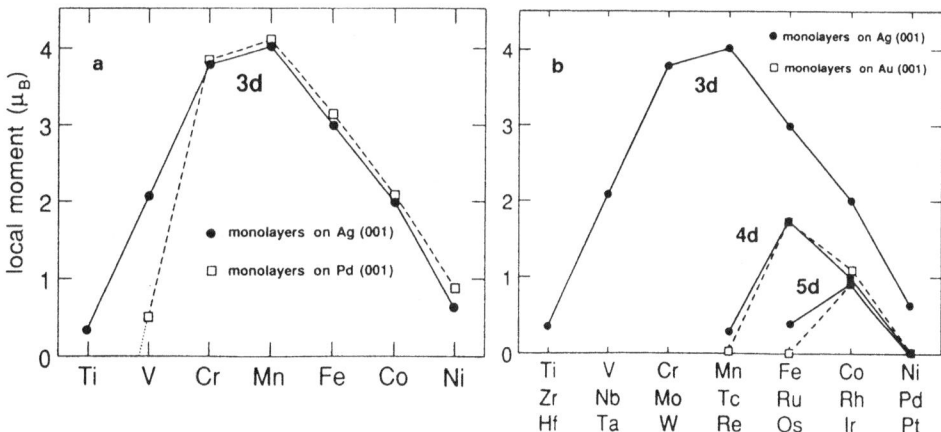

Fig. 1. Local magnetic moments as calculated for ferromagnetic a) 3d monolayers on Ag(001) (dark dots connected by full line), and 3d monolayers on Pd(001) (open squares connected by dashed line). b) 3d, 4d, and 5d monolayers on Ag(001) (dark dots connected by full line), and 4d, and 5d monolayers on Au(001) (open squares connected by dashed line).

The local magnetic moments are partly very large, not only for the 3d monolayers, but surprisingly also for the 4d and 5d ones (see Fig. 1). Among the 3d metals the largest local moment of about $4\mu_B$ was found for Mn independent from the substrate. It is interesting to notice that from Ni to Mn the magnetic moments increase in steps of $1\mu_B$, which is a consequence of the strong ferromagnetism[11]. The influence of the substrate is only substantial for Ti, V, and recognizable for Ni monolayers. When comparing the results of the local moments between 3d, 4d, and 5d monolayers on Ag(001) or Au(001) a remarkable trend is observed: The element with the largest magnetic moment among each transition metal series is shifted from Mn to Ru (isoelectronic to Fe) and at last to Ir (isoelectronic to Co), respectively. Following these trends we do not expect ferromagnetism for any other 4d or 5d metal on noble metal (001) substrates, and indeed Mo and Re remained nonmagnetic. The overall picture of monolayers on Ag and Au is the same, but as explained below the different substrate interactions cause Tc and Os on Au to be nonmagnetic and lead to a slightly larger moment for Rh. Pd and Pt are predicted to be nonmagnetic.

The magnetic moments of Ti ($0.34\mu_B$), Tc ($0.29\mu_B$), and Os ($0.34\mu_B$) on Ag are rather small and depend very likely on details such as e.g. the experimental growth conditions and monolayer relaxations which are neglected here. The change of the substrate thickness of Ag or Au, respectively, from 5 to 7 layers modified the magnetic moments in the monolayer by less than 2%. Os on Au was the only exception found so far. Magnetism of this system is extremly subtle. We found no magnetism for Os on 5 layers of Au but a moment of ($0.39\mu_B$) for Os on 7 layers of Au. For the later system, using 78 $k_\parallel$ points might not be sufficient to achieve converged results. In general, the LSDA of VWN and vBH give about 7% and 10% smaller magnetic moments, respectively, than the LSDA of MJW, but the final results do not change qualitatively.

IV. DISCUSSION

To understand these results we discuss the spin–split LDOS (Fig. 2) obtained for Mn, Fe, and Co and for their isoelectronic 4d and 5d elements Tc, Ru, Rh, and Re, Os, Ir on Ag(001), respectively, all resulting from ferromagnetic calculations. The major features in each panel arise from well defined monolayer d bands of these transition metal atoms. In this respect 3 points are noteworthy: (i) Comparing monolayers in the sequence 3d elements to 4d and to 5d one observes that the d band width broadens in steps of about 1 eV from about 1.5 eV to 3.5 eV. This is a consequence of the significantly increasing overlap of the d orbitals within the monolayer when moving from 3d to 5d transition metals. In accordance with the increasing d band width the exchange splitting is reduced substantially. One observes for instance an exchange splitting of about 3 eV, 1 eV, and ≈ 0 eV for Fe, Ru, and Os monolayers, respectively. (ii) In all cases the d band width is substantially narrower than that of the corresponding metallic bulk values. This is consistent with the argument that an increasing coordination number of d metal atoms causes band broadening and prevents the 4d and 5d bulk metals from being magnetic. (iii) Moving along each transition metal series the d bands of Co, Rh, and Ir monolayers are always smaller than the bands of their corresponding neighboring elements Fe, Ru, and Os or even Mn, Tc, and Re, respectively. This is due to the increasing localization of d wavefunctions for consecutive transition metal elements. d–band narrowing together with a reduction of the number of d holes for elements along each series makes ferromagnetism progressively more likely but with smaller magnetic moments. As a consequence Ir ($0.91\mu_B$) exhibits the largest magnetic

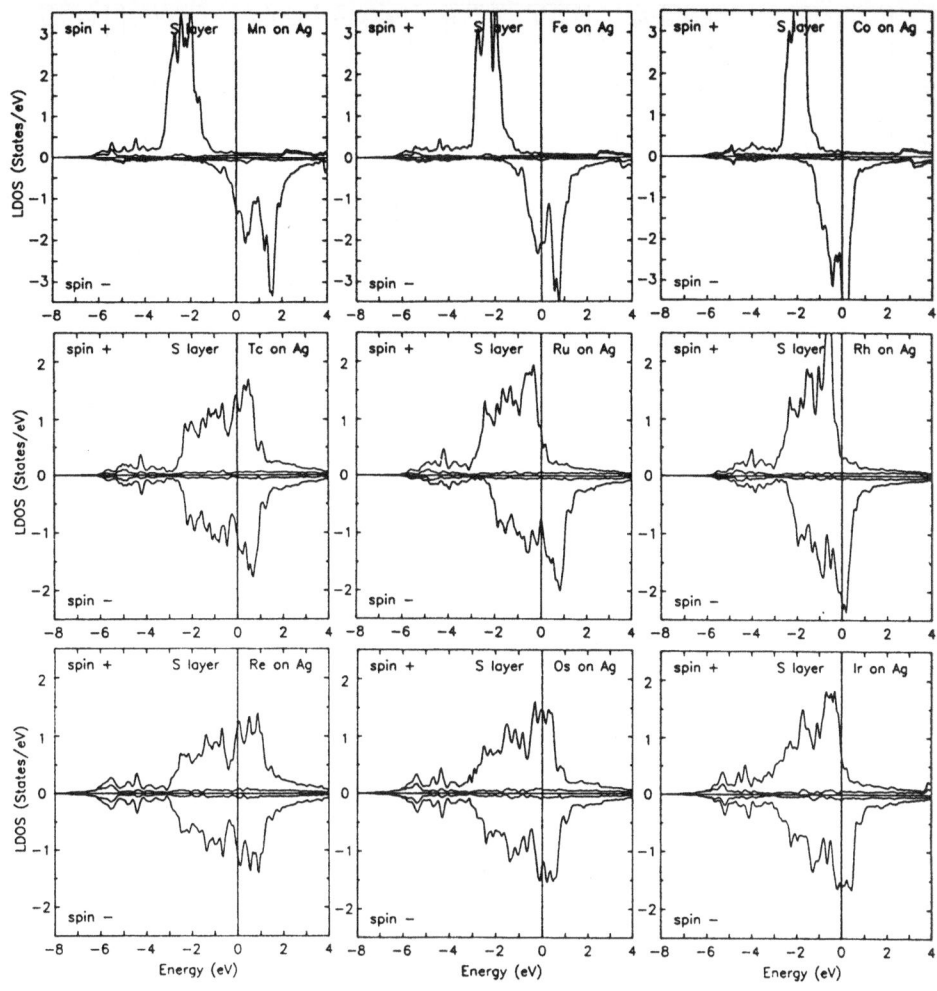

Fig. 2. Local densities of states (LDOS) together with the insignificant sum of $s + p + f$ contributions for ferromagnetic monolayers on Ag(001). Shown are the LDOS for the majority ($+$) and minority ($-$) spin direction. Notice that for Mn, Fe and Co the scale of LDOS is chosen differently.

moment among the 5d metals, for Os the bandwidth is already quite broad and only a small magnetic moment of $0.34\mu_B$ was found, while Re becomes nonmagnetic. The systems Ru ($1.73\mu_B$), Tc ($0.21\mu_B$), and Mo on Ag are analogous to Ir, Os, and Re, respectively, among the 4d metals, but shifted to the left in the periodic table.

Calculations of Zeller[11] and Oswald et al.[21] reveal local magnetic moments for nearly all single 3d impurities in Ag and Pd, respectively, with the largest moments for Cr and Mn. In similar calculations for 4d impurities in Ag[22] magnetism for Mo and Tc (isoelectronic to Cr and Mn, respectively) was found, but not for Ru or Rh. Furthermore, they found no magnetism for any 4d dimer cluster in Ag. This result can be interpreted that for 3d systems the local spin susceptibility is already divergent, while for the magnetism of 4d and 5d monolayers additional contributions of the nonlocal susceptibilities are absolutely crucial for the occurrence of magnetism. The dominance of the local susceptibility for 3d monolayer explains why their local moments resemble

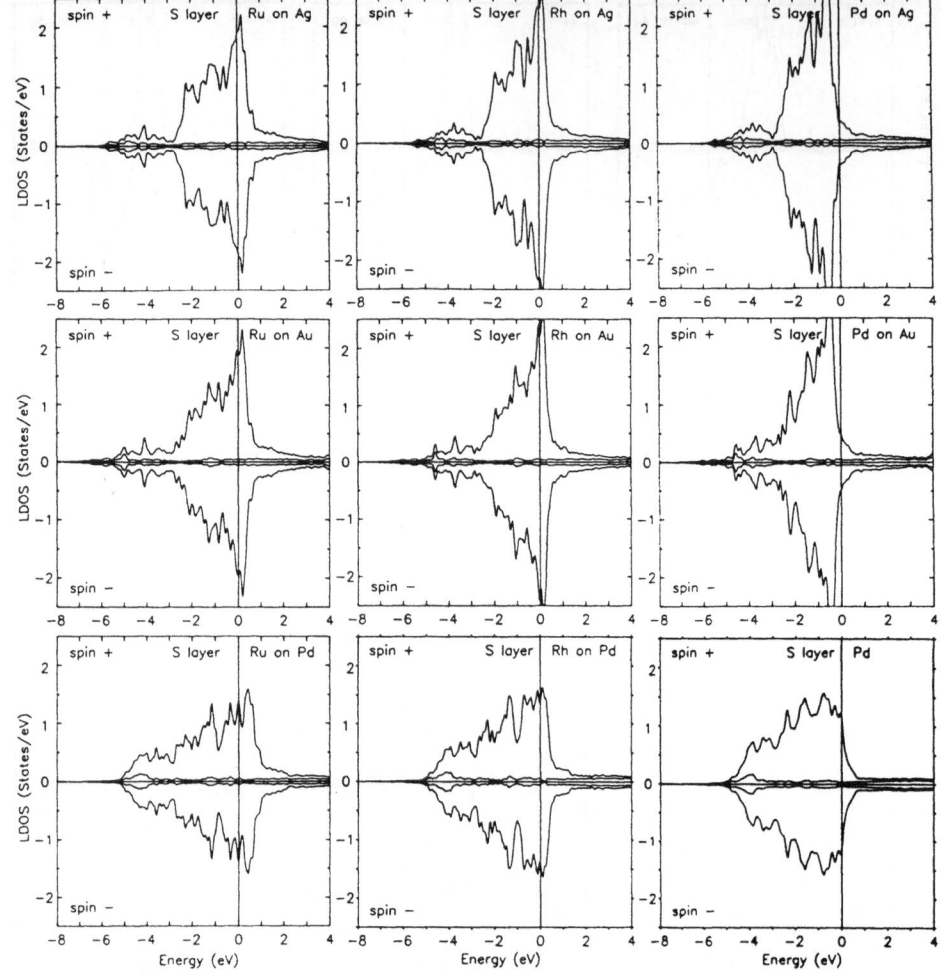

Fig. 3. Local densities of states (LDOS) together with the insignificant sum of $s + p + f$ contributions obtained from a paramagnetic calculation.

Hund's rule in Fig. 1. The nonlocal susceptibility reflects the nature of the 2D bandstructure. It contributes to sharp structures displayed in the paramagnetic LDOS of the monolayer atom, which are shown in Fig. 3 for Ru, Rh, and Pd on Ag, Au and Pd substrate, respectively. One finds that the LDOS for antibonding states (typically located in the upper third of the d band) is much higher than for bonding states. This 2D bandstructure effect stabilizes the 4d and 5d monolayer magnetism for late transition metals on noble metal substrates and explains together with the d-band broadening the absence of ferromagnetism of the early 4d and 5d monolayers. The overall picture of monolayers on Ag and Au is the same, but the hybridization with the Pd substrate is clearly visible. At the end of the 4d and 5d series Pd and Pt monolayers on Ag and Au are nonmagnetic again. The large hybridization between their d orbitals and the sp electrons of the noble metal substrates is the controlling parameter, which reduces already the moment of Ni as a monolayer on Ag. After all the large spin susceptibility of bulk Pd is a 3D bandstructure effect not reproduced for 2D layers on noble metals. Therefore, the LDOS at E_F for a Pd surface atom is much larger than for a Pd atom on noble metal substrates.

A comparison of the total energy difference between the ferromagnetic and paramagnetic solutions exhibits that the ferromagnetic state is always lower in energy than the paramagnetic one and should be the ground state for the late transition metals. The previously investigated competing c(2 × 2) antiferromagnetic configuration[10,23] is the ground state structure of the early transition metal monolayers within each series (e.g. V, Cr, Mn among the 3d series) leaving the ferromagnetic solution metastable. But, with the exceptions of probably Mo, Tc, and Re, the bands of the early 4d and 5d metals are already too broad for antiferromagnetism. For (111) monolayers different results may occur. Their coordination numbers of nearest neighbor atoms are increased by 50%, and their sixfold symmetries may change the 2D bandstructure effects.

The presented results are in good agreement with calculational results of Fu et al.[8] on the 3d monolayers and with the results of Zhu et al. for Pd on Ag(001)[24] as well as for Rh on Au(001)[25]. However, the results of Eriksson et al.[26] for Tc, Ru, and Rh and Ag(001) could not be confirmed. We are tempted to attribute the differences in the results to their use of a small number of $k_\parallel$ points or their ultrathin Ag films chosen in combination with the use of the fixed moment method. There is a lot of experimental confirmation of the monolayer magnetism of Fe, Co, Ni[1] on various substrates. As the results show, changing the substrate from Ag or Au to Pd causes drastic reductions of the 4d magnetic monolayer moments. Similar changes will occur for 5d monolayers. Thus deviations from the ideal growth mode, including layer by layer growth and perfect registry between monolayer and substrate, might prevent the experimental verification of some of the results. This might be one of the reasons why recent experiments failed to find magnetism for Rh on Ag and Au[27–29].

V. SUMMARY

We found ferromagnetism with large magnetic moments for 3d, 4d, and 5d transition metal monolayers. The most favorable condition for obtaining monolayer magnetism among the 3d, 4d, and 5d metals is shifted from Mn, to Ru (isoelectronic to Fe), and to Ir (isoelectronic to Co). This shift was explained as a consequence of the increasing d–d hybridization when moving from 3d to 5d metals and a decreasing d–d hybridization when moving within a transition metal series from the beginning to the end. Magnetism of 3d monolayers is essentially controlled by Hund's rule and to a certain extent independent from the substrate. Magnetism of 4d and 5d monolayers depends strongly on the substrate and was (with the exception of Ru) only found for monolayers on noble metals. The occurrence of magnetism for 4d and 5d monolayers is a truly 2D phenomenon. The intra–atomic, local susceptibility is not sufficient to drive a magnetic instability. Magnetism is supported by the nonlocal susceptibilities, which reflect the nature of the 2D bandstructure. This can lead to the fact that exchange splitting and magnetization disappear simultaneously at the Curie temperature. Because of the combination of large spin–orbit interaction and large local moments, the large overlap of the 4d and 5d orbitals and the fact that the magnetism disappears in 3D, interesting effects are foreseen for the magneto–crystalline anisotropies, critical temperatures and critical exponents of Ru, Rh, and in particular of Ir monolayers.

ACKNOWLEDGEMENT

The calculations were performed by the FLAPW program for thin films developed by the Northwestern group.

REFERENCES

1. L. M. Falicov, D. T. Pierce, S. D. Bader, R. Gronsky, K. B. Hathaway, H. J. Hopster, D. N. Lambeth, S. S. P. Parkin, G. Prinz, M. Salamon, I. K. Schuller, R. H. Victora, J. Mater. Res. **5**, 1299 (1990).

2. See the collection of papers, *Magnetism in Ultrathin Films*, edited by D. Pescia [Appl. Phys. A**49**, Nos. 5 and 6 (1989)], and the numerous references therein.

3. B. T. Matthias, R. M. Bozorth, Phys. Rev. **109**, 604 (1958).

4. B. T. Matthias et al., Phys. Rev. Lett. **7**, 7 (1961).

5. O. Gunnerson, J. Phys. F **6**, 587 (1976).

6. J. F. Janak, Phys. Rev. B **16**, 255 (1977).

7. V. L. Moruzzi and P. M. Marcus, Phys. Rev. B **42**, 10322 (1990).

8. C. L, Fu, A. J. Freeman, and T. Oguchi, Phys. Rev. Lett. **54**, 2700 (1985).

9. L. M. Falicov, R. H. Victora, J. Tersoff: In *The Structure of Surfaces*, ed. by M. A. van Hove, S. Y. Tong, Springer Ser. Surf. Sci. Vol **2** (Springer, Berlin, Heidelberg 1985).

10. S. Blügel, and P. H. Dederichs, Europhys. Lett. **9**, 597 (1989).

11. S. Blügel, B. Drittler, R. Zeller, and P. H. Dederichs, Appl. Phys. A **49**, 547 (1988).

12. K. Terakura, N. Hamada, T. Oguchi, and T. Asada, J. Phys. F **12**, 1661 (1982).

13. E. Wimmer, H. Krakauer, M. Weinert, and A.J. Freeman, Phys. Rev. B **24**, 864 (1981). M. Weinert, E. Wimmer, and A.J. Freeman, Phys. Rev. B **26**, 4571 (1982).

14. S. L. Cunningham, Phys. Rev. B **10**, 4988 (1974).

15. U. von Barth and L. Hedin, J. Phys. C **5**, 1629 (1972).

16. V.L. Moruzzi, J.F. Janak, A.R. Williams, Calculated Electronic Properties of Metals, Pergamon New York (1978).

17. S. H. Vosko, L. Wilk, and N. Nusair, Can. J. Phys. **58**, 1200 (1980).

18. S. Blügel, Europhys. Lett. **7**, 743 (1988).

19. S. Blügel, Europhys. Lett. **18**, 257 (1992).

20. S. Blügel, Phys. Rev. Lett. **68**, 851 (1992).

21. A. Oswald, R. Zeller, P. H. Dederichs, Phys. Rev. Lett. **56**, 1419 (1986).

22. K. Willenborg, R. Zeller, and P. H. Dederichs, Europhys. Lett. **18**, 263 (1992).

23. S. Blügel, M. Weinert and P.H. Dederichs, Phys. Rev. Lett. **60**, 1077 (1988).

24. M. J. Zhu, D. M. Bylander, and L. Kleinman, Phys. Rev. B **42**, 2874 (1990).

25. M. J. Zhu, D. M. Bylander, and L. Kleinman, Phys. Rev. B **43**, 4007 (1991).

26. O. Eriksson, R. C. Albers, and A. M. Boring, Phys. Rev. Lett. **66**, 1350 (1991).

27. H. Li, S. C. Wu, D. Tian, Y. S. Li, J. Quinn, and F. Jona, Phys. Rev. B **44**, 1438 (1991).

28. G. A. Mulhollan, R. L. Fink, and J. L. Erskine, Phys. Rev. B **44**, 2393 (1991).

29. C. Liu, S. D. Bader, Phys. Rev. B **44**, 12062 (1991).

QUASIELASTIC NEUTRON SCATTERING STUDY OF
ROTATIONAL DIFFUSION OF METHANE ADSORBED ON MGO(100)

J.M. Gay[a], D. Degenhardt[b], P. Stocker [a] and H.J. Lauter[c]

[a] CRMC2-CNRS[1] Département de Physique - Case 901 - 13288 Marseille Cedex 9 - France
[b] Hasylab/Desy - Notkestrasse 58 - 2000 Hamburg 52 - Germany
[c] Institut Laue Langevin, 156 X - 38042 Grenoble Cedex - France

The methane monolayer adsorbed on MgO(100) is a model system for studies of symmetry frustration. The square-symmetry of the substrate surface forces methane to pack with the same symmetry in a C 2x2 structure, whereas the densest natural packing is hexagonal. If the unit cell is presently known, the molecule orientation within the cell and its location with respect to the MgO surface are still unclear, and the previous experimental studies have been unable to draw firm conclusions on this question. Fundamental difficulties in model calculations come from the lack of precise theoretical or empirical basis for the parameters that characterize the molecule-surface potential. Various predictions result from these uncertainties of the potential. Most of the simulations deal with the orientation and position of a single-CH_4 molecule on top of the MgO(100) surface. Closely related to our experimental study is the recent molecular-dynamics simulation by Alavi[1]. Among the different configurations of CH_4 adsorbed on MgO that naturally arise, Alavi predicts that the stablest should be the tripod-down orientation with three H atoms sitting on the substrate surface and a C-H bound perpendicular to it. In addition, the simulation indicates a cog-wheeling motion defined as the equal and opposite rotation of nearest-neighbour molecules. This corresponds to the low-temperature orientationally ordered phase. A Kosterlitz-Thouless transition is predicted around 27 K resulting in a orientationally disordered phase with molecules freely rotating about the C-H threefold axis perpendicular to the surface.

Quasi-elastic neutron scattering (QENS) is a powerful tool for studies of rotational diffusion[2-4]. The neutron incoherent scattering law is indeed the Fourier Transform of the self-pair correlation function of the nuclei, and is therefore sensitive to the molecular diffusion. More precisely, one may check that the neutron energy (some meV) is in the same range as the energy loss or gain of the neutrons interacting with the molecules. For a solid phase, the QENS experiments are essentially rotational diffusivity measurements since the inter- or intra-molecular vibrations usually contribute to the incoherent scattering in the inelastic spectra further from the elastic peak, except a continuous Debye-Waller term. The QENS law splits in two terms, elastic and inelastic :

$$S(Q,\omega) = S^{el}(Q) \cdot \delta(\omega) + S^{inel}(Q,\omega) \qquad (1)$$

where Q is the scattering vector and $\hbar\omega$ the energy loss or gain.

Some classical models described in the literature[2] including isotropic brownian motion (the H atoms of the CH_4 molecule freely rotate on a sphere) and uniaxial motion (the

---

[1] Laboratoire associé aux Universités d'Aix-Marseille II et III.

Equilibrium Structure and Properties of Surfaces and Interfaces
Edited by A. Gonis and G.M. Stocks, Plenum Press, New York, 1992

305

H atoms rotate on a cylinder) have been tested against the data. It is worth reminding that QENS is particularly well adapted to studies with methane due to the large incoherent scattering cross-section of H. Straightforward from the experiments, one can measure the so-called Elastic Incoherent Structure Factor (EISF) which is

$$EISF(Q) = \frac{S^{el}(Q)}{\int_{-\infty}^{+\infty} S(Q,\omega)\, d\omega} \qquad (2)$$

The EISF deviating from 1 is the signature of rotational diffusion in solid phase and its Q-dependence is representative of the type of rotational motion.

The experiment reported here was performed at the Institute Laue-Langevin in Grenoble, France, using the time-of-flight spectrometer IN5. One CH_4 monolayer was adsorbed on a highly uniform MgO powder, and QENS spectra were recorded at several temperatures ranging from 20 K to 50 K.

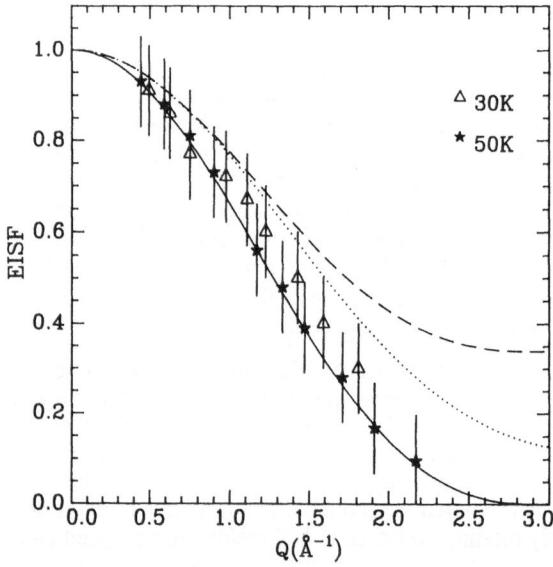

Fig. 1. Elastic Incoherent Structure Factor versus scattering vector Q. The experimental data at 30 K (Δ) and 50 K ($\star$) are compared to simulated EISFs from the isotropic rotation model (solid line) and the uniaxial continuous rotation tripod (dashed line) and dipod (dotted line) models.

The EISFs measured at 30K and 50K are reported in figure 1. The limited energy-range of the experimental study induces a cut-off in the intensity integration (see formula 2), that results in an overestimate of the EISF. This difficulty has been overcome by finding a model line which fits fairly well the data in the experimental range and which is used to extrapolate the data to infinity. Are also shown in figure 1 the simulated EISFs for isotropic rotation and uniaxial rotations in the tripod-down and dipod-down orientations. The tripod-down model corresponds to the Alavi molecular dynamics prediction. It clearly appears that this model does not fit the data. The first qualitative conclusion is therefore that the molecule do not execute rotation about a C-H bound. We may say that all the H-atoms are likely involved in the rotational diffusion resulting in a strong attenuation of the EISF at large Q.

An isotropic motion is possible, but we cannot rule out other possibilities like molecules in dipod-down orientation rotating around the twofold axis perpendicular to the substrate surface at low temperature (T≤30K). Such a dipod-down orientation was predicted as possible by other experimental measurements[5-7]. On the other hand, the EISFs at 40 K and 50 K are unambiguously the signature of free rotations. The QENS spectra measured at 20K do not feature the same gradual decrease of the wings as at 30K. This is likely due to the superimposition of some tunneling effects still present as observed at lower temperatures[7].

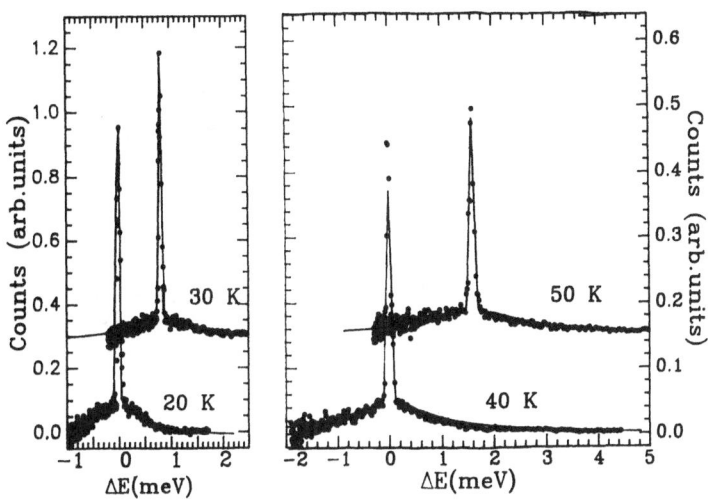

Fig. 2. QENS spectra at T = 20 K, 30 K (Q=1.81Å^{-1}), 40 K and 50 K (Q=1.91Å^{-1}). The spectra at 20K and 30K are measured with wavelength 6 Å and energy resolution 63 µeV. At 40K and 50K, they are 5 Å and 109 µeV, respectively. The wings are characteristic of rotational diffusion. Some tunneling effects[7] are still present at 20K in the ±0.5 meV energy range. The solid line is a simulation from the isotropic rotation model.

Besides the type of rotational diffusion deduced from the EISFs, the QENS measurements allow to determine the diffusion coefficient from the lineshape of the spectra. As shown in figure 1, the highest sensitivity to the type of rotational motion is obtained for the largest Q values, for which the inelastic component is relatively strong. Figure 2 shows the spectra at different temperatures, for Q=1.81 Å^{-1} (20K and 30K) and Q=1.91 Å^{-1} (40K and 50K). For all the studied temperatures, we tentatively fit the data for a full set of spectra with Q ranging from 0.1 to 2.1Å^{-1}, using the isotropic motion model described by formula 3.

$$S(Q,\omega) = j_0^2 (Qa) \cdot \delta(\omega) + \frac{1}{\pi} \sum_{\ell=1}^{\infty} (2\ell + 1) j_\ell^2 (Qa) \frac{D_r \ell (\ell+1)}{\left(D_r \ell (\ell+1)\right)^2 + \omega^2} \qquad (3)$$

where j_ℓ denotes the spherical Bessel functions. The rotational diffusion coefficients D_r used in the calculated lines of figure 2 are reported in Table I. They are similar to that of methane on graphite[3].

Table I . Rotational diffusion coefficient D_r (formula 3) and mean-squared displacements in the methane monolayer adsorbed on MgO (100).

T (K)	Dr (10^{11} s-1)	$<u^2>$ $Å^2$
20	3.5 ± 0.5	0.04 ± 0.02
30	4.8 ± 0.5	0.05 ± 0.02
40	6.5 ± 0.5	0.08 ± 0.02
50	8.0 ± 0.5	0.08 ± 0.02

Quasi-elastic neutron scattering, due to its particular sensitivity to proton motions, has been used for studying the rotational diffusion of methane/MgO between 20 K and 50 K. Recent molecular-dynamics simulations have suggested the tripod-down configuration of the molecule rotating around the C-H bound perpendicular to the surface. This experimental study unambiguously rules out the tripod-down configuration. Above 40 K, isotropic motion of the methane molecule is likely. At lower temperature, rotations are probably more complicated and could be intermediate between the free rotor case and uniaxial rotation around the two-fold axis perpendicular to the surface in the dipod-down configuration. These results are consistent with all the experimental low temperature studies of methane monolayer adsorbed on MgO and particularly the recent measurements of rotational tunneling at 4 K.

References

[1]A. Alavi, Phys. Rev. Lett. **64** , 2289 (1990); Mol. Phys. **71** , 1173 (1990).

[2] See, for instance : "Quasi Elastic Neutron Scattering", M. Bée Adam Hilger (1988).

[3] P. Thorel, J.P. Coulomb and M. Bienfait, Surf. Sci. **114**, L 43 (1982).

[4]J.P. Coulomb and M. Bienfait, J. Phys. **47**, 89 (1986).

[5]J.P. Coulomb, K. Madih, B. Croset and H.J. Lauter, Phys. Rev. Lett. **54**, 1536 (1985).

[6]D.R. Jung, J. Cui, D.R. Frankl, G. Ihm, H.Y. Kim and M.W. Cole, Phys. Rev. B **40**, 11893 (1989).

[7]J.Z. Larese, J.M. Hastings, L. Passell, D. Smith and D. Richter, J. Chem. Phys. to be published.

LOW ENERGY ION SCATTERING AT EXTREMELY LOW ION DOSES

R.G.van Welzenis, R.H.Bergmans, J.H.Meulman, A.C.Kruseman, and
H.H.Brongersma

Department of Physics, Eindhoven University of Technology
P.O. Box 513, NL 5600 MB Eindhoven, The Netherlands

ABSTRACT

Low Energy Ion Scattering (LEIS) provides information on the composition and structure of surfaces. EARISS (Energy and Angular Resolved Ion Scattering Spectrometer) is a new implementation of this technique. Because of its high detection efficiency, EARISS enables measurements at extremely low ion doses. This virtually eliminates surface damage, one of the major drawbacks of conventional LEIS. The capability of EARISS to obtain a simultaneous energy and azimuthal-angle scattering image of a sample is demonstrated for the (110) surface of a GaAs sample. The Ga and As peaks can be seen directly in one single measurement. A meaningful LEIS energy spectrum is already obtained in 30 s, at a target current of only 20 pA.

INTRODUCTION

In Low Energy Ion Scattering (LEIS), ions with a primary energy of a few keV are focused onto a surface. Because of the high neutralization probability for inert gas ions of these energies an extreme surface sensitivity can be obtained. Only the ions that are scattered from the top atomic layer(s) contribute significantly to the detector signal. This is only a very small fraction of the incoming ions, because most of them are neutralized. The surface also gets sputtered by the ions, and in any LEIS facility one has to try to keep this surface damage by sputtering to a minimum. Before doing the actual measurement it is usually desirable or necessary to sputter-clean the surface during some time. Thereafter the sample should be annealed to restore the surface structure. During the final measurement any change in the surface should be avoided of course. By improving the detection sensitivity and efficiency, measurements can be done with a very low primary ion dose, virtually eliminating surface damage, during the measurement.

INSTRUMENTAL

EARISS (**E**nergy and **A**ngle **R**esolved **I**on **S**cattering **S**pectrometer) [1] is a particular implementation of LEIS. A newly developed double toroidal energy analyzer [2], that accepts all

Equilibrium Structure and Properties of Surfaces and Interfaces
Edited by A. Gonis and G.M. Stocks, Plenum Press, New York, 1992

309

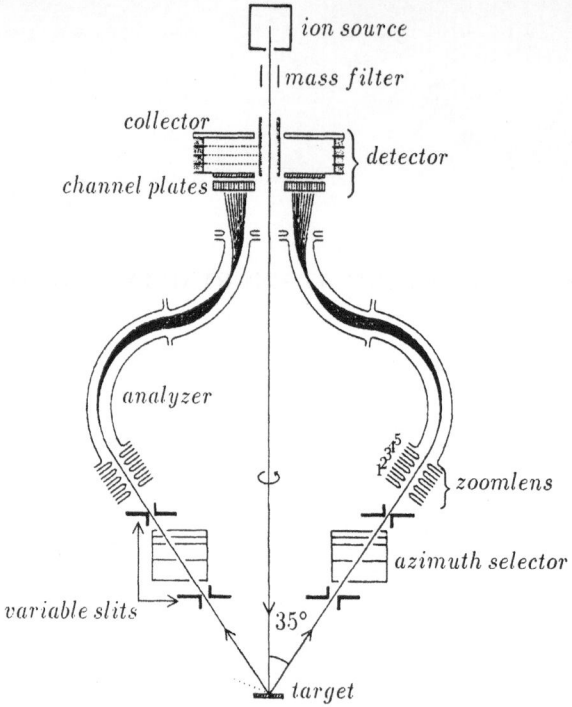

Figure 1. Schematic drawing of the EARISS instrument.

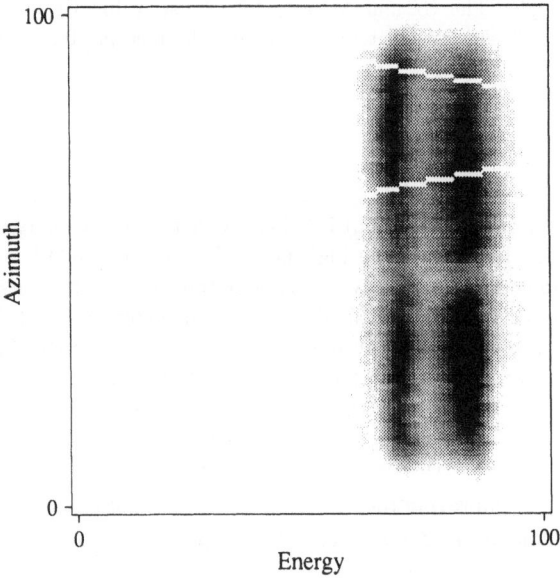

Figure 2. EARISS spectrum of the (110) surface of GaAs at an ion dose of 2 C. Ion current through the sample 11 nA, measuring time 180 s. Pass energy of the analyzer 4 keV. Primary ions 4 keV Ne$^+$. The intensity is a measure for the number of counts in a cell [5], black corresponds to about 5000 counts.

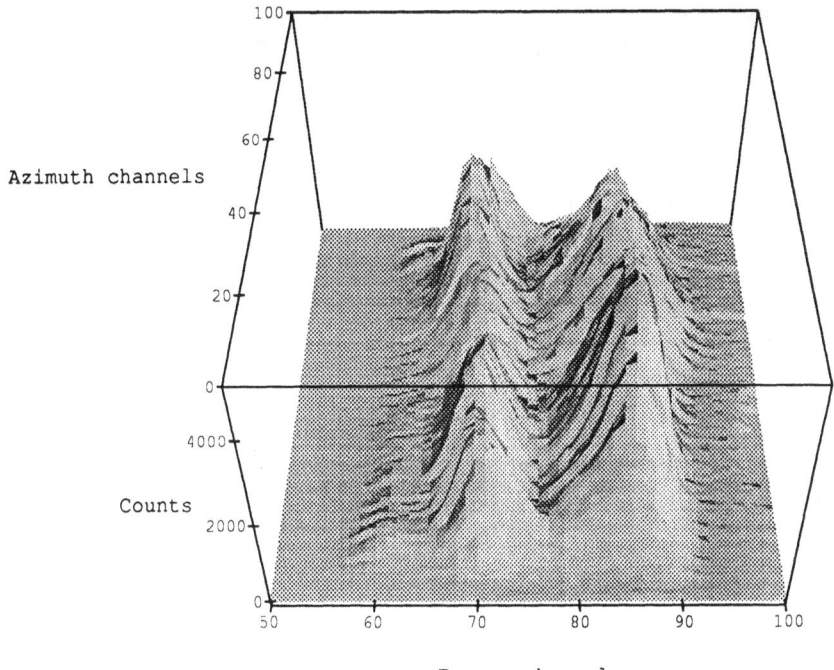

Figure 3. The same data as in figure 2 represented in a quasi 3D graph, giving better insight in the azimuthal distribution. The right hand ridge is due to Ga, the other to As. Energy increases from right to left at 8 eV/channel.

ions that are scattered over 145 ° is used. This analyzer is combined with a fast 2-dimensional detector [3], that maps energy and azimuthal scattering angle simultaneously. Figure 1 is a schematic of the EARISS analyzer and detector system. The instrument is cylindrically symmetric around the central axis, along which the incoming ion beam is directed onto the sample surface (target).

Ions that enter the analyzer at its so called pass energy will follow the central path through the analyzer. Ions with lower energy will be more strongly deflected in both sections and thus arrive at the outside edge of the detector. Higher energy ions will be projected near the central axis. Only ions within an energy window of ± 6% of the pass energy will reach the detector, the others are lost.

By setting the accelerating voltage on one of the zoom lens elements one can choose the centre of the energy range to coincide with the pass energy of the analyzer. The detector consists of two channel plates, that convert each incoming ion into an avalanche of electrons, and the actual two-dimensional collector plate consisting of sickles and rings [3]. This collector plate provides two charge pulses which are proportional to the energy and the azimuthal angle of the detected ion, respectively. These signals are amplified, normalized, counted and finally stored in a 100x100 memory.

EARISS has several advantages over conventional LEIS. The energy window of the analyzer is quite wide, i.e. 12% of the analyzer pass energy, making scanning of the energy range unnecessary. The detector and its electronics are fast, they can process up to 10^6 events/s. Moreover, the instrument has a very low background signal. EARISS is at least two orders of magnitude more efficient than conventional cylindrical mirror analyzers. Thus, it is possible to

Counts (* 10^5)

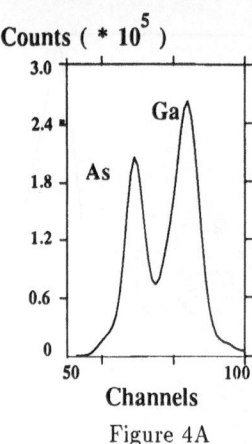

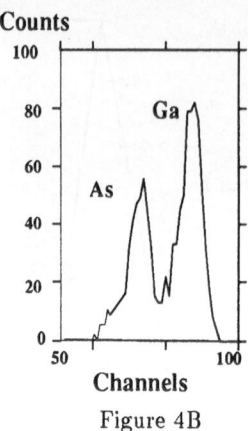

Figure 4A

Figure 4B

Energy distribution after summation over all the azimuthal channels for the (110) surface of GaAs. Energy increases from right to left, contrary to the channel numbers, at 8.5 eV/channel.
A: Left : Ion dose 2 μC. See Figure 2.
B: Right: Low ion dose 600 pC.

lower the primary ion dose, minimizing the surface damage, and still obtain meaningful results. Both the energy- and angular-distribution data are obtained in a single run, with ion doses as low as 100 pC, not requiring any energy or azimuthal-angle scanning.

The information on the azimuthal scattering angle can be used to obtain information on the surface structure [4]. The azimuth selector in front of the analyzer (as shown in figure 1) will be installed in the near future , which will strongly improve the capabilities.

For the measurements reported here only the energy resolution is of interest. Thus the azimuthal information was completely neglected by summing over all the azimuthal channels, further enhancing the energy sensitivity. The measured data were corrected for imaging errors of the detection system.

RESULTS

Figure 2 shows a typical example of the 2-D spectra that are obtained. This is almost [5] a direct representation of the 100x100 channel analyzer memory that stores the results. The two white staircase like stripes are artifacts caused by a flaw in the computer program that corrects the data for imaging errors of the detection system. These are results for GaAs (110), using 4 keV Ne^+ ions. Although, the average penetration depth of these Ne^+ ions in GaAs is quite large, i.e. $\approx$ 8 nm, only the ions that are scattered from the outermost atomic layer are of relevance as explained in the previous section. The current through the sample is 11 nA during the measuring time of 180 s. Only 50 of the energy channels could be used in this particular measurement.

Figure 3 is a different representation of the same data. Here one can clearly see large variations in the azimuthal response. These are not caused by the sample surface, but due to characteristics of the channel plates. They can be taken out by normalization [4].

By summation over several azimuthal channels, a 1-D energy distribution is obtained. Summing over all the azimuthal angles, and thus neglecting the azimuthal information completely, the full energy efficiency of EARISS can be exploited, see figure 4A. The left hand peak at about channel 70 is at 1461 eV and is due to ^{75}As (there is only one As isotope). The other peak is from Ga. The main contribution to the latter peak will be from the ^{69}Ga isotope at 1332 eV. Both peaks are asymmetrically broadened by the contribution from the ^{71}Ga isotope which is in between the two at 1376 eV. Because the instrument was set for greatest sensitivity and not for best resolution (the variable slits in figure 1 were fully opened), the three peaks were not resolved in this measurement. The ion current at the sample was 11 nA and this measurement took 180 s. This figure is comparable to a conventional LEIS spectrum that has to be accumulated from an energy scan, however.

Knowing that the background count rate, after summation over the azimuths, is of the order of 0.01 counts/energy-channel/s, it is clear from figure 4 that it should be possible to lower the primary ion dose appreciably. Indeed at a sample current of only 20 pA, measuring for 30 s, the peaks can still be detected, see figure 4B.

As the primary ion beam diameter at the sample is about 0.5 cm, we have a dose rate of 10^9 ions/cm^{2}s. Assuming a sputter efficiency of 3 atoms/incident ion, a sputter rate of the order of about 10^{-5} monolayer/s is obtained. So it is justified to state that the surface remains undamaged during the measurement.

Similar measurements on an $Au_{20}Pd_{80}$ alloy (ion dose 100 pC, 2 keV Ne^+) and a graphite sample (Highly Oriented Pyrolytic Graphite (0001), ion dose 17 nC, 3 keV $^4He^+$) were reported elsewhere [6]. The latter suggest that nondestructive measurements on polymers are feasible.

CONCLUSION

EARISS has proven capable of detecting elements at the surface of various materials at such low doses of impinging ions that surface damage due to sputtering is virtually eliminated. This opens the possibility for looking at fragile structures such as the surfaces of polymers, reconstructed semiconductor surfaces, and highly dispersed catalysts with low loading. At present the instrument is being optimized for this purpose.

REFERENCES

1. P.A.J. Ackermans, P.F.H.M. van der Meulen, H. Ottevanger, F.E. van Straten, and H.H. Brongersma
   Nucl.Instr. and Meth. **B35**, 541, (1988).
2. G.J.A. Hellings, H. Ottevanger, C.L.M.C. Knibbeler, J. van Engelshoven, and H.H. Brongersma
   Journal of Electron Spectroscopy and Related Phenomena **49**, 359, (1989).
3. C.L.C.M. Knibbeler, G.J.A. Hellings, H.J. Maaskamp, H. Ottevanger, and H.H. Brongersma
   Rev. Sci. Instr. **58**, 125, (1987).
4. R.H. Bergmans, A.C. Kruseman, and H.H. Brongersma
   to be published.
5. The original data are in color, unfortunately these could not be reproduced here.
6. R.H. Bergmans, W.J. Huppertz, R.G. van Welzenis, and H.H. Brongersma
   10th Conf. on Ion Beam Analysis, Eindhoven July 1991.
   Nucl.Instr. and Meth. **B64**, 584, (1992).

X-RAY PHOTOELECTRON DIFFRACTION STUDY OF A HIGH-TEMPERATURE

SURFACE PHASE TRANSITION ON Ge(111)

T.T. Tran, D.J. Friedman[*], Y.J. Kim[†], S. Thevuthasan[#], G.S. Herman,
and C.S. Fadley[†#]

Department of Chemistry
University of Hawaii
Honolulu, HI 96822, USA

ABSTRACT

We have studied a high-temperature surface phase transition on Ge(111) with x-ray photoelectron diffraction (XPD)[1]. A reversible disordering transition beginning at about 970 K or 240 K below the bulk melting point (T_m) and ending at about 1150 K or 60 K below T_m is observed. The Ge 3p intensity was measured as a function of both emission direction and temperature. Azimuthal XPD data at $\theta = 19°$, including nearest-neighbor forward-scattering directions and yielding high surface sensitivity, show an abrupt drop in intensity of ~35% over the interval of 970-1150 K. Data taken at $\theta = 55°$ for which second-nearest-neighbor scattering directions and more bulk sensitivity are involved also exhibit a similar abrupt decrease in intensity over the range of 970-1100K. In conjunction with model single-scattering cluster calculations, these data indicate that the disordering process involves considerably more than the surface double layer of Ge atoms, a conclusion which disagrees with certain prior studies.

INTRODUCTION

Evidence for a reversible surface phase transition on Ge(111) has been found previously by McRae et al.[2,3] in temperature-dependent low-energy electron diffraction (LEED) and by Denier van der Gon et al.[4] in medium-energy ion scattering (MEIS) experiments. This transition occurs near 1050 K or about 160 K below the bulk melting point. From the LEED data, it has been proposed[2,3] that this is not a surface melting or surface roughening transition but rather a domain disordering transition in which the domains are laterally strained to a depth of one double layer. However, beyond the suggestion based upon molecular mechanics modeling that there is loss of registry between strained domains and deeper layers[3], no precise structural conclusions have been possible concerning the type of disorder involved. From the MEIS data, it has been proposed[4] that an "incomplete melting" occurs, with the formation of a thin and uniform film of positionally disordered atoms on the Ge surface. The depth of disordering has been observed to remain constant up

Equilibrium Structure and Properties of Surfaces and Interfaces
Edited by A. Gonis and G.M. Stocks, Plenum Press, New York, 1992

315

to temperatures within 25 K of the melting point and estimated to be from 1 to 1.5 Ge(111) monolayers.

In an attempt to further understand this phase transition, we have studied this system with x-ray photoelectron diffraction (XPD), a surface structure probe that is primarily sensitive to short-range order in the first 3-5 shells of neighbors around each emitter[5]. We have examined the Ge 3p photoelectron intensity from Ge(111) as a function of both the polar and azimuthal angles of emission and temperature.

EXPERIMENT

The polar and azimuthal dependence of Ge 3p core-level intensities emitted from a Ge(111) surface at a kinetic energy of 1365 eV was studied as a function of temperature from 300 K to 1200 K using Al Kα radiation as an excitation source. In order to vary surface sensitivity, azimuthal scans were performed at two polar angles of emission relative to the surface of θ=19° and 55°, as shown in Fig. 1. For an estimated inelastic attenuation length of 22.0 Å, the mean depths from which elastically scattered electrons can be emitted are approximately 2 and 6 Ge(111) double layers, respectively.

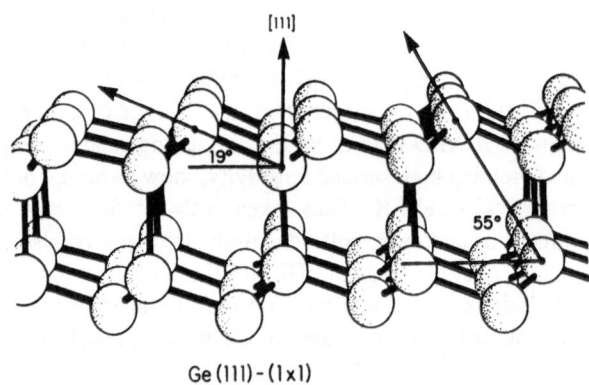

Ge (111) - (1x1)

Fig. 1. An unreconstructed Ge(111) surface showing certain near-neighbor scattering events at takeoff angles of θ=19° and 55°.

The measurements were performed on a Vacuum Generators ESCALAB5 spectrometer modified for automated angle scanning as described elsewhere[6]. Heating was done by a noninductively wound resistive button heater and temperatures were measured with a thermocouple-calibrated infrared pyrometer. The sample was a mirror-polished Ge wafer (n-type, Sb doped, 5-30 Ω-cm) oriented to within ±1.0° of (111). Surface cleaning involved sputtering (10^{-5} torr Ar$^+$, 800 eV, 850 K, 45° off normal incidence, 20 min.) and annealing (970 K, 30 min.). This treatment was found to lead to a sharp c(2x8) LEED pattern at ambient temperature. Surface cleanliness was monitored by XPS core-level peaks, and no detectable contaminant peaks were found after a continuous series of azimuthal scans at different temperatures.

RESULTS AND DISCUSSION

Fig. 2 shows three azimuthal scans taken at 300 K, 800 K, and 1110 K and at a surface sensitive polar angle of 19°. As temperature is increased, the azimuthal curves gradually lose and/or change some of their fine structure. More specifically, in the azimuthal scan taken at 1110 K (that is, above the transition point) the two peaks in the azimuths [11$\bar{2}$] (ϕ=0°) and [$\bar{1}$2$\bar{1}$] (ϕ=60°) are found to be much reduced in both absolute and relative intensities. The fine structure in the middle angles from about ϕ=14° to 52° also is reduced in intensity and slightly changes its form. Furthermore, the overall anisotropy, which we measure as $(I_{max} - I_{min})/I_{max} = \Delta I/I_{max}$, decreases significantly from 0.36 to 0.29, or by 19%, in going from 800 K to 1110 K. In Fig. 3, the intensity of the [11$\bar{2}$] forward scattering peak corresponding to nearest-neighbor scattering is plotted against temperature. The points here were obtained with both increasing and decreasing temperature, and the form of the curve was found to be identical in both cases. It is thus clear that an abrupt and reversible drop of about 35% in intensity occurs over the interval 970-1150 K.

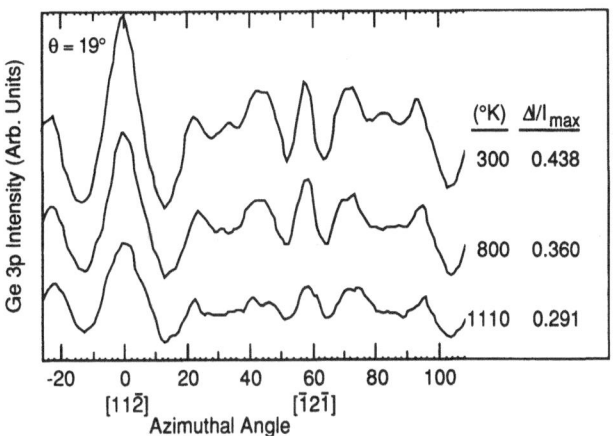

Fig. 2. Temperature-dependent azimuthal XPD data for Ge 3p emission at 1365 eV from Ge(111) at a low takeoff angle of θ=19°. This scan includes a nearest-neighbor forward scattering direction at ϕ=0° (cf. Fig. 1).

This drop cannot be explained by thermal vibrations. Debye-Waller (D-W) effects are expected to yield a smooth and linear temperature dependence in the intensity of the forward-scattering peak, unlike the step-like transition observed. Model calculations of such a D-W attenuation in a simple single scattering cluster (SSC) approximation are shown as the dashed line in the figure. Thus, Debye-Waller effects may qualitatively explain the experimental behavior below 970 K, but the transition itself must be due to a larger-scale motion of the surface atoms.

Data very similar to those shown in Figs. 2 and 3 were also obtained for the more bulk sensitive polar angle of 55°. The polar angle chosen here causes the emission direction to sweep through second-nearest-neighbor forward-scattering directions, as shown in Fig. 1. Again, with increasing temperatures, the mid-angle fine structure in the azimuthal curves gradually gets washed out and there is a corresponding decrease in the intensity of the forward-scattering peaks along the [11$\bar{2}$] and [$\bar{1}$2$\bar{1}$] azimuths. The overall anisotropy here decreases from 0.27 for the scan taken at 840 K to 0.24 for that taken at 1130 K, or by 11%, over the transition. A plot of the intensity of the [11$\bar{2}$] peak versus temperature also shows an abrupt and reversible drop of about 20% over the temperature range of 970-1150 K. These data thus suggest that the same disordering process is observed when a greater number of layers (averaging about 6 double layers) of the Ge(111) surface are probed.

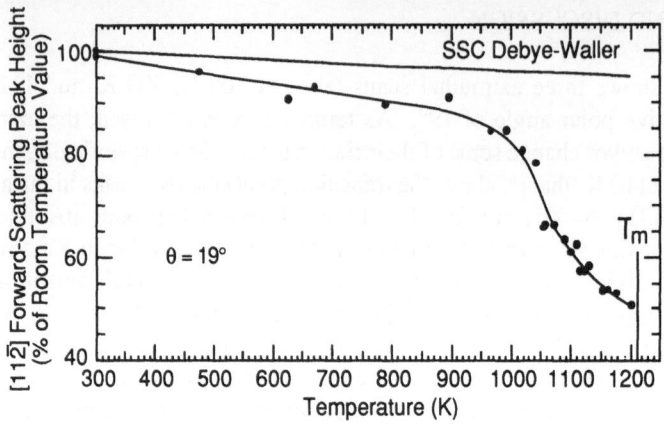

Fig. 3. The detailed temperature dependence of the height of the nearest-neighbor forward scattering peak along [11$\bar{2}$] (ϕ=0°). Also shown is the temperature dependence of intensity expected on the basis of Debye-Waller effects in a single-scattering cluster model.

Further evidence from model SSC calculations also supports this multilayer disordering mechanism. We performed SSC calculations with spherical-wave scattering for three clusters of Ge atoms with a 19° takeoff angle. Results from a first cluster in which the top double layer is compressed by 2% on the vertical axis so as to reflect the difference between the densities of the solid (5.22 g/cm^3) and the liquid (5.53 g/cm^3) at the melting point, and those from a second cluster in which the top double layer is completely flattened, thus neglecting all 19° scattering events in this double layer, are not found to differ significantly from results obtained with a third cluster of ideally terminated, non-reconstructed Ge(111).

Our experimental data will be combined with further model calculations to better understand the disordering process and its depth of penetration. However, our experimental and theoretical results thus far suggest a multilayer disordering mechanism more extensive in depth than that suggested by LEED and MEIS results for the same transition; the latter results indicate a depth of disordering that is limited to only 1-2 double layers by contrast.

REFERENCES

* Present address: S.E.R.I., Golden, Colorado.
† Present address: Lawrence Berkeley Laboratory, Berkeley, California.
\# Present address: Department of Physics, University of California-Davis.

1. A preliminary account of this work appears in: T.T. Tran, D.J. Friedman, Y.J. Kim, G.A. Rizzi, and C.S. Fadley in The Structure of Surfaces III, S.Y. Tong, M.A. van Hove, K. Takayanagi, and X. Xie, Eds. (Springer-Verlag, Berlin 1991) p. 522.
2. E.G. McRae and R.A. Malic, Phys. Rev. Lett. **58**, 1437 (1987); E.G. McRae and R.A. Malic, Phys. Rev. B**38**, 13183 (1988).
3. E.G. McRae, J.M. Landwehr, J.E. McRae, G.H. Gilmer, and M.H. Grabow, Phys. Rev. B**38**, 13178 (1988).
4. A.W. Denier van der Gon, J.M. Gay, J.W.M. Frenken, and J.F. van der Veen, Surf. Sci. **241**, 235 (1991).
5. C.S. Fadley, Physica Scripta T**17**, 39 (1987); C.S. Fadley in Synchrotron Radiation Research: Advances in Surface and Interface Science, ed. by R.Z. Bachrach, (Plenum Press, New York, 1990) in press.
6. R.C. White, C.S. Fadley, R. Trehan, J. Electron Spectrosc. Relat. Phenom. **41**, 95 (1986); J. Osterwalder, M. Sagurton, P.J. Orders, C.S. Fadley, B.D. Hermsmeier, and D.J. Friedman, J. Electron Spectrosc. Relat. Phenom. **48**, 55 (1989).

X-RAY AND LIGHT SCATTERING STUDIES OF ELECTRODE SURFACES AND INTERFACES

C. A. Melendres

Argonne National Laboratory
Materials Science and Chemical Technology Divisions
9700 South Cass Avenue
Argonne, Illinois 60439 (USA)

ABSTRACT

We review and illustrate the application of specular X-ray reflection and laser Raman spectroscopy to the study of electrode surfaces "in situ" in aqueous solution environments. As example of the former, we present results of x-ray reflectivity studies of copper in borate buffer solution. Information on the development of surface and interface roughness, the formation and reduction of surface films, as well as the density and thickness of the oxide layer have been obtained as a function of applied potential. We also show how knowledge of the composition and structure of electrochemically formed thin surface films on metals maybe gained by laser Raman scattering and demonstrate this with results of studies on the passive film on nickel.

SPECULAR X-RAY REFLECTION FROM ELECTRODE SURFACES

The advent of high intensity synchrotron radiation sources has recently led to a renaissance in the use of x-ray techniques for the study of surface and interface phenomena. X-ray reflectivity (XRR) measurements appear to have great potential for the "in situ" characterization of the electrode/solution interface which have not been generally appreciated. Inspite of the early demonstration of the utility of the technique for investigation of surfaces in ambient air by Kiessig[9], its extension to the study of the solid/liquid interface under potential control has been relatively few. To our knowledge, Bosio et. al[2),3] appear to have carried out the first reflectivity measurements on an electrode "in situ"; however, they were mainly interested in EXAFS information on the surface oxide on Ni and obtained this at grazing angle of incidence in a technique they called REFLEXAFS. Ocko et. al[4], more recently, studied the reconstruction of the Au(001) surface as a function of electrode potential using X-ray reflectivity and diffraction techniques. They showed that at -0.4 V (vs Ag/AgCl reference, in 0.01 M HClO$_4$ solution), the gold surface exhibits a hexagonal reconstructed layer with a mass density 21% greater than the underlying bulk layers. The reconstruction disappears above 0.5 V but is reversible when the potential is held again below -0.3 V. We have recently used x-ray reflectivity measurements to study the changes in surface structure of a copper electrode as a function of applied potential in borate solution[5]. We present here briefly some of the results inorder to demonstrate the utility of the technique for "in situ" surface and interface characterization.

Equilibrium Structure and Properties of Surfaces and Interfaces
Edited by A. Gonis and G.M. Stocks, Plenum Press, New York, 1992

319

Experimental

An experimental set-up for "in situ" x-ray reflectivity measurements from electrode surfaces is shown schematically in Fig. 1. X-rays from a synchrotron or a rotating anode x-ray generator are monochromatized, collimated, and focussed onto the electrode surface at an angle near the critical angle of incidence. The angle of incidence (θ) is scanned using a spectrometer that also serves to support and align the sample. The specularly reflected x-rays, with intensity I_R, are detected using a NaI scintillation detector. This signal is ratioed against the monitored intensity of the incident beam, I_O, in order to obtain the normalized reflectivity, I_R/I_O.

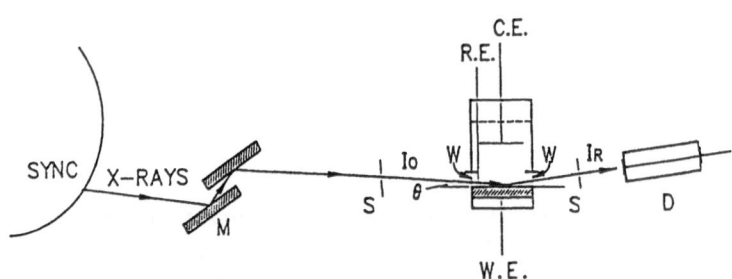

Fig. 1. Schematic of Set-up for "In Situ" X-ray Reflectivity Measurements

SYNC = Synchrotron source.
M = Si(111) double crystal monchromator.
S = Slit.
W = Teflon window
R. E. = Reference electrode.
C. E. = Counter electrode.
W. E. = Working electrode.
D = Scintillation detector.
I_o, I_R = Incident and reflected X-ray beams, respectively.
θ = Angle of incidence.

The copper electrode consisted of ~250 Å film which was vacuum evaporated onto a polished Si(111) substrate; the active electrode area was ~ 0.32 cm wide x 3.65 cm long. The electrode was mounted in an x-ray/electrochemical cell fitted with 0.025 mm-thick teflon (FEP) windows. Details of this cell have been published.[6] The 1.48 Å x-rays used suffered a 95% decrease in intensity in traversing a solution path length of ~0.32 cm. The x-ray reflectivity was measured "in situ" as a function of Q ($= \frac{4\pi sin\theta}{\lambda}$), the momentum transfer vector, and also at different potentials in borate buffer solution (pH 8.4).

Results and Discussion

The electrochemical behavior of copper in borate solution is shown in Fig. 2. Peaks I and II in the current-potential curve indicate the oxidation of the surface to form Cu(I) and Cu(II) oxides; subsequent reduction of the surface films occurs during the reverse scan of the potential (peaks III and IV).

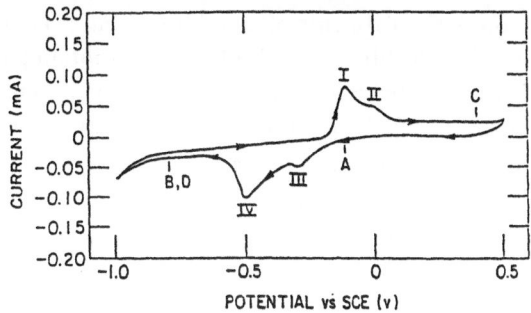

Fig. 2. Cyclic voltammogram of Cu-on-Si electrode in borate buffer solution (pH 8.4), Scan rate = 10 mV/sec.

Fig. 3 shows the reflectivity of the electrode measured while under potential control at four different points (i.e., at A, B, C and D, Fig. 2) in the current-potential curve. The reflectivity curves indicate the condition of the electrode surface at open circuit (A, -0.12 V), upon reduction of the surface film (B, -0.8 V), on anodic oxidation to form the surface oxides (C, + 0.4 V), and on subsequent cathodic reduction at -0.8 V (D).

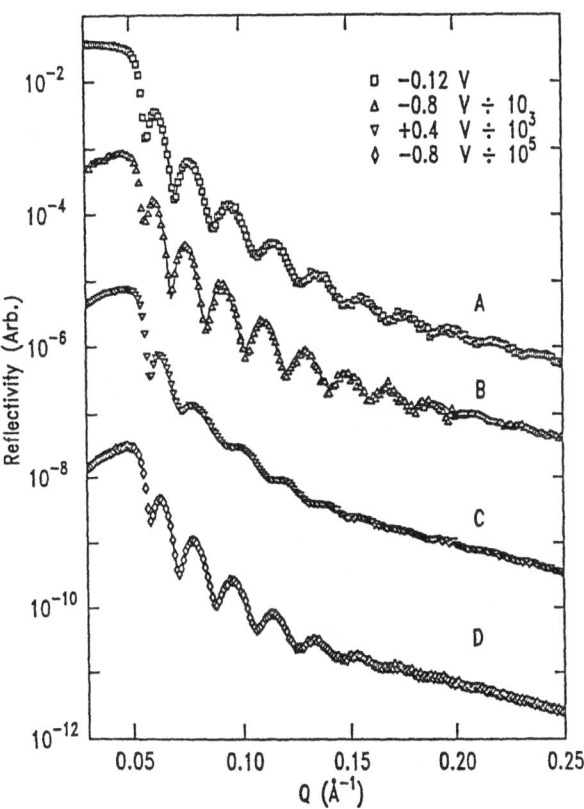

Fig. 3. Reflectivity of Cu-on-Si electrode at various potentials in borate buffer solution (pH 8.4). A, B, C, D correspond to potentials indicated in cyclic voltammogram of Fig. 2.

A number of theoretical treatments of x-ray scattering from surface and multi-layer films have appeared in the literature.[7,8,9] In order to obtain physical parameters from the experimentally measured reflectograms, application is made of Fresnel's laws of reflection for treatment of the data. In our analysis, we employed a kinematic approximation with conformally rough interfaces. This means, that in our system consisting of copper with a copper oxide film in contact with the solution, the roughness of the copper oxide/solution interface is correlated with that of the copper/copper oxide interface. In this case, the reflectivity, with appropriate corrections, has been derived by You,[10] as:

$$\frac{I_R}{I_O} = A^2 + B^2 \exp[-Q_{sol}\sigma_1] + C^2 \exp[-Q_{sol}\sigma_2]$$
$$+ 2AB\cos(Q_{sol}d_1)\exp[-Q_{sol}(\sigma_1+\sigma_c)]$$
$$+ 2AC\cos[Q_{sol}(d_1+\Delta d)]\exp[-Q_{sol}(\sigma_2+\sigma_c)]$$
$$+ 2BC\cos(Q_{sol}\Delta d)\exp[-Q_{sol}(\sigma_1+\sigma_2)]$$

where:

$$A = \left(\frac{Q_{Si}-Q_{Cu}}{Q_{Si}+Q_{Cu}}\right)\left(\frac{4Q_{sol}Q_{Cu}}{(Q_{sol}+Q_{Cu})^2}\right)$$

σ_1 = rms roughness of the Cu/Cu$_2$O interface

σ_2 = rms roughness of the Cu$_2$O/solution interface

$$B = \left(\frac{Q_{Cu}-Q_{Cu_2O}}{Q_{Cu}+Q_{Cu_2O}}\right)$$

σ_c = correlated roughness of oxide film

d_1 = thickness of the copper film

$$C = \left(\frac{Q_{Cu_2O}-Q_{sol}}{Q_{Cu_2O}+Q_{sol}}\right)$$

Δd = thickness of the Cu$_2$O film

ρ_n = the density of n

$Q_n = (Q^2 - 4.285\times10^{-4}\rho_n)^{1/2}$, for n = Cu, Si, Cu$_2$O, and the solution (sol)

Moreover, our data appear to be better fitted by assuming a Lorentzian distribution function instead of the Gaussian function commonly employed by others.[8] A non-linear least squares fitting procedure using the foregoing equations allowed the film thickness d, density ρ, and σ parameters to be obtained from the experimental data. Taking the density of pure copper to be 8.92 g/cm^3 and that of the copper oxide film to be 5.5 g/cm^3, the parameters given in Table I are derived.[10] The initial thickness of the copper film on silicon was 285 Å. This changes with potential due to the formation and reduction of the surface oxide (here taken as Cu$_2$O). The correlated roughness varies also with potential but the interface roughness seem relatively unaffected. The important point to emphasize here is that we are now beginning to obtain atomistic information on the physical structure of metals in liquid environments "in situ" under the influence of an applied potential; some of this information are not obtainable by any other means. One can expect application of the technique to other interesting problems in electrochemical surface science and technology in the near future.

322

TABLE I. Parameters derived from least squares fit to experimental data.

E/V	$d_{Cu}/\text{Å}$	$d_{Cu2O}/\text{Å}$	$\sigma_c/\text{Å}$	$\sigma_{Cu_2O/Cu}/\text{Å}$	$\sigma_{Cu_2O/sol}/\text{Å}$
−0.12	273	12	16	4.6	4.4
−0.8	285	(0)	11	3.3	4.6
+0.4	250	35	25	5.6	4.4
−0.8	276	(9)	12	5.3	4.6

LASER RAMAN SPECTROSCOPY OF SURFACE FILMS ON METALS

Another technique that is employed in our laboratory to probe the structure of electrode surfaces "in situ" is Raman spectroscopy, i.e., normal, resonance enhanced or surface enhanced Raman Spectroscopy (SERS). Significant information on the composition of corrosion films on metals have been obtained by the technique.[11] The structure of adsorbed layers has also been extensively studies, mainly by SERS.[12] We have reviewed the former application recently and refer the interested reader to this work and the references quoted therein for details of the technique and applications.[11] We will briefly illustrate here the general principles involved by describing results which we have recently obtained on the composition of the passive films on Ni in aqueous solutions.

Experimental

A schematic diagram of the experimental set-up is shown in Fig. 4. A special spectroelectrochemical cell is employed consisting of 3 separate compartments to house the electrodes. The metal to be studied is made the working electrode and is positioned very close (~1 mm) to the quartz optical window of the cell. The laser beam is directed at about 30° with respect to the window (and electrode surface) and the scattered light collected and analyzed using a SPEX Industries model 1403 double monochromator. Detection is made using a photomultiplier tube with pulse counting electronics.

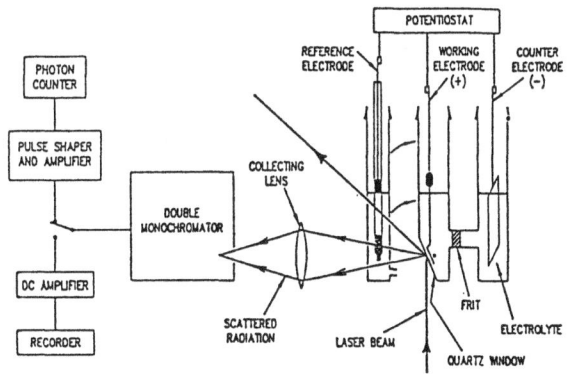

Fig. 4. Experimental set-up for laser Raman spectro-electrochemical studies.

High purity Ni foils (0.318 cm wide x 0.254 mm thick) were polished with 600 grit emery paper and then cleaned thoroughly with acetone and high purity water. Raman spectroscopic and electrochemical measurements were carried out as a function of solution pH and applied electrical potential. Because of the thinness of the passive film on Ni (< 20 Å) and its low Raman scattering cross-section, it was necessary to make use of the technique of surface enhancement. This was done by electrodeposition of a silver overlayer on the Ni surface following a procedure that has been described previously.[13] Scanning electron microscopic study of the deposited silver shows average particle size in the order of 500 to 1000 Å with only partial coverage of the Ni surface.[14]

Results and Discussion

Fig. 5 shows the current-potential behavior of Ni in 0.1 M NaOH solution (pH 12) and the effect of the Ag overlayer on the electrochemical behavior. The anodic dissolution behavior is essentially the same; the higher cathodic current at about -0.4 V is due to the catalytic activity of Ag towards the reduction of dissolved O_2. The anodic peak (1) at $\sim$0.25 V and the cathodic one (2) at $\sim$0.2 V are due to the anodic oxidation and reduction of the Ag, respectively. The peaks I and II are due to oxidation of nickel which leads to the formation of surface films which were identified by SERS and discussed below.

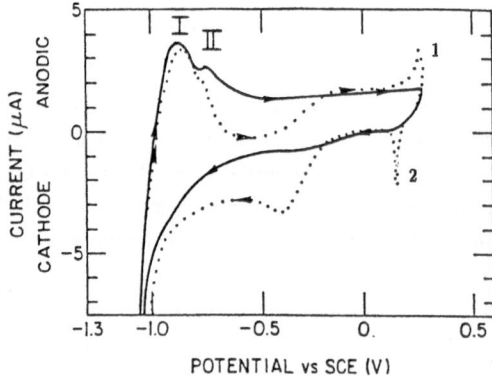

Fig. 5. Cyclic Voltammogram of Ni in 0.1 M
NaOH solution. Scan rate = 5 mV/sec

— without Ag overlayer
... with Ag overlayer

Fig. 6 shows "in situ" SER spectra of the surface film on Ni as a function of potential. Following the electrodeposition of silver and precleaning of the electrode at -1.3 V for 15 minutes, the "in situ" spectrum of the surface obtained on holding the potential at -1.0 V for 1 hour is shown in Fig. 6a. The absence of a discernible Raman band indicates that the surface is relatively free of an oxide or hydroxide film initially at this potential. On increasing the potential to -0.8 V (in the so called prepassive region, Fig. 5), a band at about 450 cm^{-1} starts to appear indicating the formation of a surface film. This band is assigned to the Ni-OH vibration of Ni(OH)$_2$ in accordance with the work of DeSilvestro et. al.[15]. They have studied extensively the SER spectra of Ni(OH)$_2$ prepared by cathodic deposition on a roughened gold electrode. Further confirmation of the correctness of this assignment is the observation

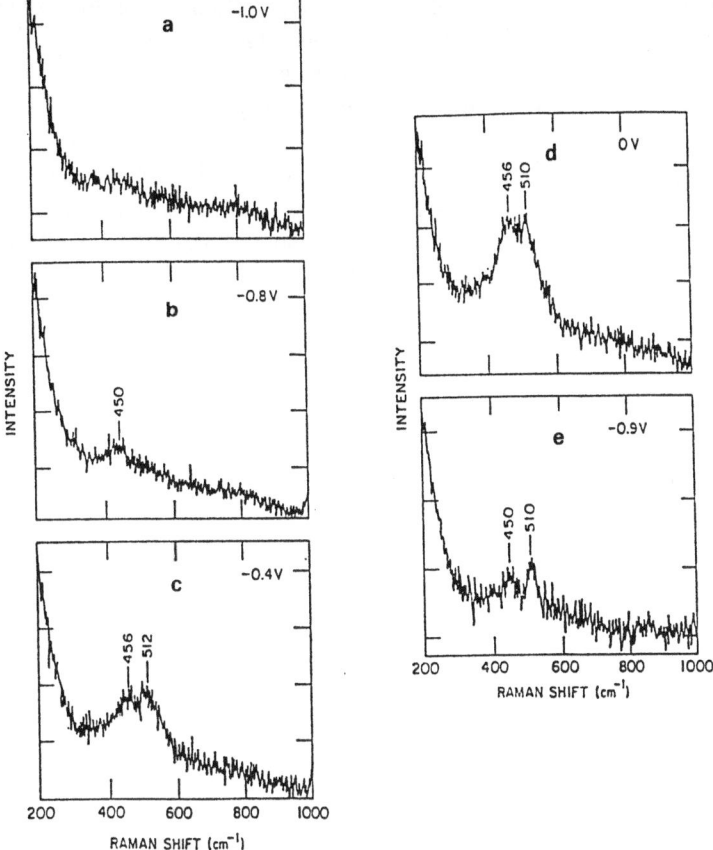

Fig. 6. Laser Raman Spectra of Nickel in 0.1 M NaOH
Solution at Various Potentials.
Ar⁺ Laser, 514.5 nm line, P∼100 mW.

of the OH stretching vibration of $Ni(OH)_2$ at about 3630 cm^{-1} as shown in Fig. 7a to 7c.

On scanning the potential further and holding at -0.4 V, we obtained the "in situ" spectrum shown in Fig. 6c. A new band is evident at about 512 cm^{-1}, along with the old one at 456 cm^{-1}. We assign this new band to NiO by comparison with the spectrum of standard sample of NiO.[14] Increasing the potential to 0 V resulted only in an increase in intensity of the two bands (Fig. 6d) presumably due to an increase in the thickness of the corresponding surface films with potential. A corresponding increase in intensity of the OH vibration is also observed (Figs. 7a to c).

On reversing the potential scan and holding at -0.9 V, a considerable reduction in the intensity of the bands at 450 and 510 cm^{-1} is observed (Fig. 6e). The bands, however, do not completely disappear even after 2 hours suggesting slow kinetics or some irreversibility in the reduction of the surface film.

Similar results are obtained in 0.15 M NaCl solution where the current-potential relation is shown in Fig. 8. The same 2 peaks (I and II) are observed as in 0.1 M NaOH. In contrast, however, breakdown of the passive film appears to occur at about 0.2 V in the presence of chloride. "In situ" Raman spectrum obtained upon potentiostating at -0.75 V is shown in Fig. 9a. The band at 450 cm^{-1} assigned to Ni(OH)$_2$ is again clearly evident. On increasing the potential to -0.4 V, new bands appear at 215 and 510 cm^{-1} (Fig. 9b). The former is assigned to Cl$^-$ adsorbed on the Ag overlayer which is commonly observed in SERS studies in chloride solutions[16]. It is interesting to note once again how the appearance of the 510 cm^{-1} band due to NiO correlates with the occurrence of wave II in the voltammogram. Spectra obtained in the region of OH vibration once more shows a band at about 3640 cm^{-1} at potentials where the 450 cm^{-1} band is also observed.

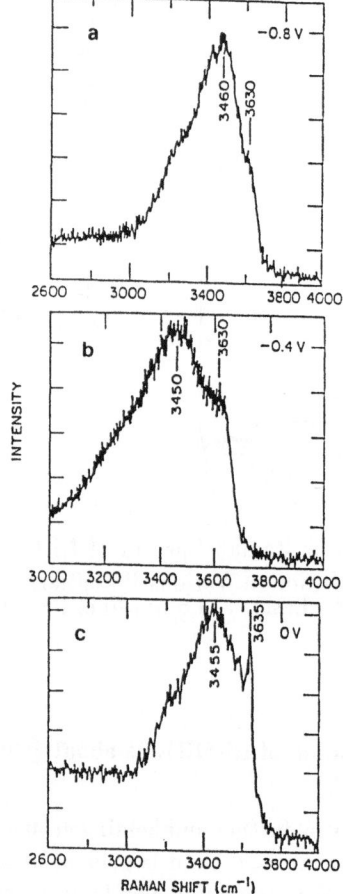

Fig. 7. Laser Raman Spectra of Ni in the OH Stretching Region.

Essentially the same spectroscopic results were obtained for Ni in acidified 0.15 M NaCl solution (pH 2.7), although the intensity of the 450 and 510 cm^{-1} bands are considerably lower due presumably to the thinness of the film.

The principal results that emanate from this work are:

1) The observation of the presence of both Ni(OH)$_2$ and NiO in the surface film(s) on nickel in the passive region of potential for all the solution studied from pH 2.7 to 12.

2) The correlation between the formation of Ni(OH)$_2$ with the first anodic wave (I) and that of NiO with the second wave (II) in the current-potential curve.

There has been no consensus on the composition of the passive film on nickel despite over 50 years of research investigations. Some workers[17,18] claim that the film is $Ni(OH)_2$ while others believe that it is NiO.[19,20] Our results show unequivocally that both species are present in the surface film on Ni in the passive potential region. We also showed that $Ni(OH)_2$ and NiO are formed separately, consistent with thermodynamic calculations,[14] and contrary to theories that NiO is produced by dehydration[17] of $Ni(OH)_2$.

The foregoing is just one of many examples that can be cited on the use of laser Raman scattering for the characterization of electrode surfaces "in situ". The purpose of this exposition is to give the participants of the NATO-ASI an overview of the complex but interesting problems associated with the structure of surfaces and interfaces of metals under applied electrical potential in aqueous solution environments. Hopefully, this would encourage the participants to take an active interest in these problems.

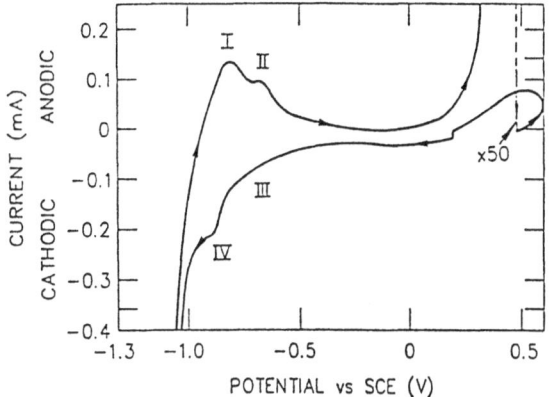

Fig. 8.   Cyclic voltammogram of nickel in 0.15 M NaCl solution (pH 8.5). Scan rate = 10mV/sec.

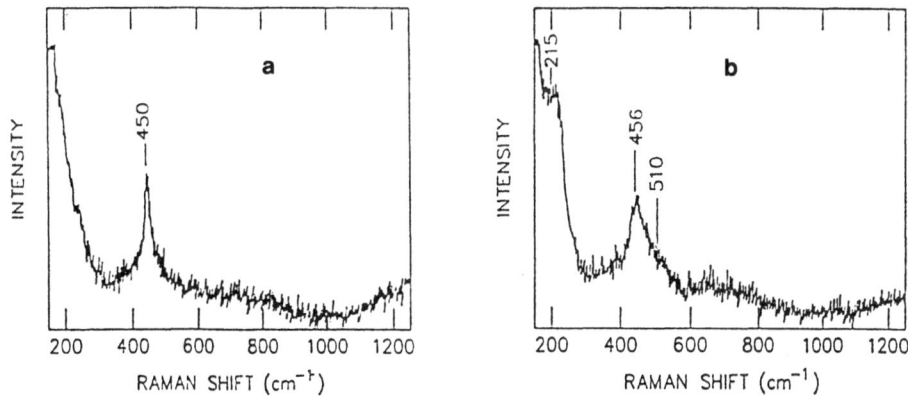

Fig. 9.   Surface enhanced Raman spectrum of corrosion film on nickel in 0.15 M NaCl at various potentials. a) E = -0.75 V, b) E= -0.40 V. Ar^+ laser, 514.5 nm line, 100 mW.

ACKNOWLEDGEMENT

The work on x-ray reflectivity was carried out in collaboration with H. You, Z. Nagy, V. A. Maroni, and W. Yun, all of Argonne National Laboratory. M. Pankuch carried out most of the laser Raman scattering measurements. The author gratefully acknowledges the contribution of these colleagues. Support for these researches was provided by the Division of Materials Science, U.S. Department of Energy.

REFERENCES

1) H. Kiessig, Ann. Phys. 10, 714 (1931).

2) L. Bosio, R. Cortes, A. Defrain, M. Froment, and A. M. Lebrun, in Passivity of Metals and Semiconductors, M. Froment (editor), Elsevier, Amsterdam (1983), p. 131.

3) L. Bosio, R. Cortes, A. Defrain and M. Froment, J. Electroanal. Chem. 180, 265 (1984).

4) B. M. Ocko, J. Wang, A. Davenport and H. Isaacs, Phys. Rev. Lett. 65, 1466 (1990).

5) C. A. Melendres, H. You, V. A. Maroni, Z. Nagy, and W. Yun, J. Electroanal. Chem. 297, 549 (1991).

6) Z. Nagy, H. You, R. Yonco, C. A. Melendres, W. Yun and V. A. Maroni, Electrochim Acta 36, 209 (1991).

7) L. G. Parratt, Phys. Rev. 95, 359 (1954).

8) R. Cowley and T. W. Ryan, J. Phys. D: App. Phys. 20, 61 (1987).

9) S. K. Sinha, E. B. Sirota, and S. Garoff, Phys. Rev. B. 38, 2297 (1988).

10) H. You, C. A. Melendres, Z. Nagy, V. A. Maroni, W. Yun and R. M. Yonco, paper submitted to Phys. Rev. B.

11) C. A. Melendres in Spectroscopic and Diffraction Techniques in Interfacial Electrochemistry, C. Gutierrez and C. Melendres (editors), Kluwer Academic Publishers, Dordrecht, (1990), p. 181.

12) R. K. Chang and T. E. Furtak (editors), Surface Enhanced Raman Scattering, Plenum Press, NY (1982).

13) C. A. Melendres, J. Acho, and R. L. Knight, J. Electrochem. Soc. 138, 877 (1991).

14) C. A. Melendres and M. Pankuch, paper submitted for publication in the J. of Electroanal. Chem.

15) J. Desilvestro, D. A. Corrigan, and M. J. Weaver, J. Electrochem. Soc. 135, 885 (1988).

16) R. K. Chang and B. L. Laube, in CRC Critical Rev. Solid State Materials Sci. 12, (1984).

17) W. Paik and Z. Szklarska-Smialowska, Surface Science 96, 401 (1980).

18) B. Beden and A. Bewick, Electrochimica Acta 33, 1695 (1988).

19) W. Visscher and E. Barendrecht, Surface Science 135, 436 (1983).

20) B. MacDougall and M. Cohen, J. Electrochem. Soc. 123, 191 (1976).

POINT-TO-POINT RESOLUTION IN SCANNING AUGER ELECTRON SPECTROSCOPY AT HIGH PRIMARY BEAM ENERGIES FOR SURFACE AND INTERFACE ANALYSIS

A. G. Nassiopoulos and N. M. Glezos

Institute of Microelectronics
NCSR Demokritos, P.O. Box 60228
15310 Aghia Paraskevi, Athens, Greece

Abstract

The point-to-point resolution of Scanning Auger Electron Spectroscopy has been determined by Monte-Carlo calculations. The primary beam energy used is in the range 10-100keV and the primary beam spot size in the range of 20-80 Å, which is nominal to the highest performance Scanning Auger Spectrometers. Three kinds of samples are used: bulk materials, thin unsupported films and thin films on a high-Z material substrate. The angle of incidence is also changed from 0 to 85° with respect to the normal to the surface. The radial distribution of the primary beam induced AES signal and of the signal due to backscattering, is calculated separately from the Monte-Carlo programme. A point-spread-function is then obtained from which the point-to-point resolution is calculated.

For the energy range 20-100 keV an analytical expression is also given for the lateral extent of the primary beam induced signal. Its validity is verified by the Monte-Carlo programme.

I.INTRODUCTION

Auger Electron Spectroscopy is now combined with Transmission Electron Microscopes, using high energy electron beams. So the analysis is made at high primary beam energies and at minimum spot size[1,2] . It is now well established that high lateral resolution may, in this case, be achieved[3-6] . So it is important to know what are the exact resolution limits for the actual minimum spot sizes used, situated in the range of 20-80 Å. Three factors are expected to influence the lateral resolution in AES: a) the incident beam spot size and the incident beam broadening within the sample, b) backscattering and c) X-rays (characteristic or continuous) induced within the sample by the incident or by backscattered electrons. Signals from b) and c) create Auger electrons in their way out of the sample. These two factors may have, in some cases, an important contribution to the total Auger signal. This is for example the case of thin films of low-Z material deposited on a high-Z material substrate. When the analysis is done at energies greater than 20keV the enhancement signal from back-scattering and X-rays induced from the substrate is sometimes greater than the primary beam induced signal[6,7].

Nevertheless this enhancement signal is laterally extended over an area many times greater than that of the primary beam induced signal (radial extent of the order of few microns, compared to some tens of Å). So it constitutes a low background over which the primary beam induced signal is superimposed. The point-to-point resolution, defined as the minimum distance between two points for which the two points are resolved, and calculated by using the Rayleigh criterion, is found to be independed of the enhancement signal.

Previous studies on bulk specimens have shown this general behavior[3-5]. Experimental evidence is also demonstrated in a recent paper for a 100 keV primary beam electrons and an incident beam spot size of the order of 80 Å[11].

In this paper, Monte-Carlo simulations have been used in order to calculate systematically the lateral extent of the Auger signal due to the primary beam and to backscattered electrons in the whole energy range 5-100 keV and for spot sizes ranging from 20 to 80 Å. Three different cases are considered: thin unsupported films, bulk materials and thin films on a bulk substrate. No significant difference is found between these three cases as it concerns the lateral resolution and this is because in all cases the only factor which governs the lateral resolution is the primary beam spot size and the beam broadening within the surface of the film (the few tens of Å). This is found to be valid for different angles of incidence, ranging from 0 to 90° with respect to the normal to the surface. So for the calculation of the resolving power the problem may be simplified by considering only the radial extent of the primary beam induced Auger signal. It is, thus, simple to write an analytical expression for this radial distribution. From this expression the resolving power is easily deduced for different primary beam spot sizes , energies and angles of incidence.

II. GENERAL FORMALISM AND MONTE-CARLO RESULTS.

The Monte-Carlo programme has been described in detail elsewhere[11]. A single scattering model is used. The step length for elastic scattering is variable and equal, at each step, to the mean free path for elastic scattering (single scattering model). Elastic scattering is described by a screened Rutherford cross section with a Thomas-Fermi potential. The inelastic scattering cross section is described by Gryzinski's formula.

The Auger signal per electron at a given step is given by:

$$I_{Auger} = (\rho \cdot N/A) \cdot \sigma(E_o, E_i) \cdot step \cdot \alpha_{ijk} \cdot \exp(-x/\lambda \cos\theta) \quad \textbf{(1)}$$

where ρ and A are the density and atomic weight respectively, N is Avogadro's number, $\sigma(E_o, E_i)$ is the ionization cross section given by Gryzinski's formula, α_{ijk} is the Auger yield, x is the position of creation of the Auger electron, λ the Auger electron attenuation length and θ the escape angle.

The geometry used is indicated in fig. 1. In this figure, θ_o is the angle of incidence with respect to the normal to the sample and n is the escape angle. The angle between the incident and detected beam is maintained constant ($\xi=110^\circ$). A(x',y') is the point of incidence and d is the film thickness. B(x,y) is the exit point of an Auger electron produced at the point C of the electron trajectory. When the angle of incidence is changed, the detection angle is changed consequently. The above configuration has been chosen to coincide with that used in the VG Scanning Auger Microscope (VG HB 501 A STEM) in order to compare the obtained results with experiments by Cazaux et al[8-10]. The Auger signal $I_{Auger}(r)$ from a sample is written as:

$$I_{Auger}(r) = I_p(r) + I_b(r) \qquad (2)$$

where $I_p(r)$ is the radial distribution function of the Auger signal induced by the primary beam and $I_b(r)$ that of the signal induced by backscattering (it has been demonstrated previously[6] that the signal due to X-rays from the substrate is always negligible as it concerns the resolving power).

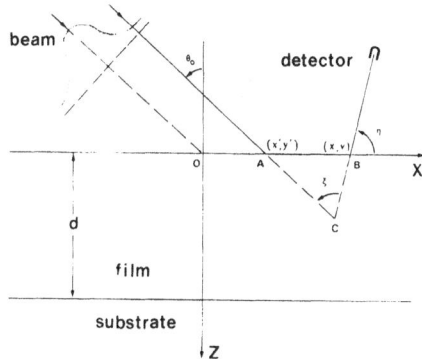

Fig.1   Configuration of the setup used in the analysis.
        The beam is incident at the point O and it has
        a gaussian distribution around this point. θ_o is
        the angle of incidence and n is the exit angle.

Fig. 2 indicates the functions $I_p(r)$ and $I_b(r)$ for the case of a thin film of low-Z material (Al, 400 Å thick) on a high-Z material substrate (Au). In this case the enhancement of the Auger signal due to both backscattering and X-rays from the substrate is found to be greater than the primary beam induced signal, the more important contribution being that of backscattered electrons. So it is interesting to investigate the influence of this important enhancement signal to the resolving power of the technique. Two primary beam energies are considered in fig. 2, $E_o=20keV$ and $E_o=100keV$. The incident beam is normally distributed (gaussian curve) around the point of incidence, with a standard deviation $\sigma_o=40Å$. The angle of incidence is $\theta_o=40^\circ$. The obtained functions $I_p(r)$ and $I_b(r)$ may easily be approached by gaussian curves, shifted with respect to the point of incidence by an amount r_o. This shift is due to the asymmetry of incidence, illustrated in fig. 1.

The function $I_p(r)$ is concentrated around the point of

incidence (σ_o=60Å for both E_o=20keV and E_o=100keV). On the other hand $I_b(r)$ is radially distributed over a large area (σ_R=383Å for E_o=20keV, σ_R=1747Å for E_o=100keV). Its maximum intensity at r=r_o is small compared to the maximum intensity of the primary beam induced signal ((I_b/I_p)$_{max}$=0.110 for E_o=20keV and 0.030 for E_o=100keV). So it constitutes a low background over which the primary beam induced signal is superimposed.

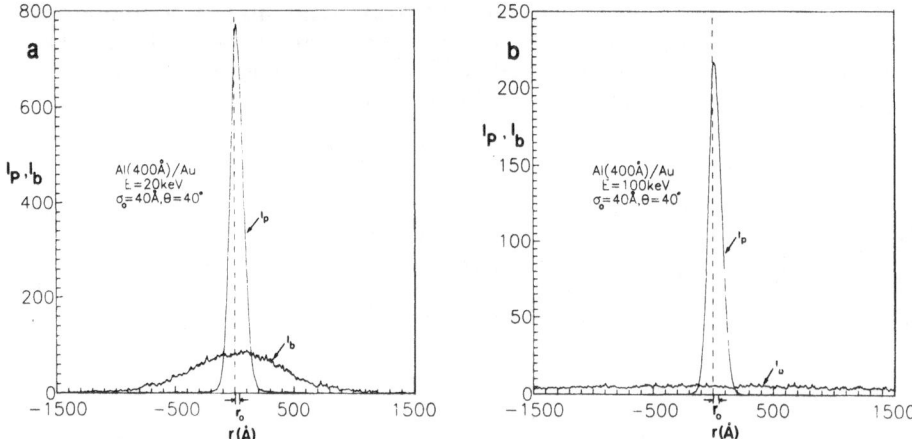

Fig.2 Radial distribution of the signals I_p and I_b respectively
for two primary beam energies,(a)E_o=20keV and (b) 100keV.
σ_o=40Å and θ_o=40°.

The resolving power is calculated by considering the Rayleigh's criterion, according to which two points are resolved if there is a minimum of 26.4% dip in intensity at the midpoint between them. Let us consider the obtained Auger signal expressed as the sum of two gaussian curves:

$$I_{Auger}(r) = \frac{J_p}{\pi\sigma_p{}^2} e^{-\frac{r^2}{\sigma_p{}^2}} + \frac{J_b}{\pi\sigma_R{}^2} e^{-\frac{r^2}{\sigma_R{}^2}} \qquad (3)$$

where J_p and J_b denote the primary and backscattered signal obtained by integrating each gaussian curve separately.
    If we consider the zero of the r-axis at the mid-point between two adjacent points 1 and 2, then the Point-Spread-Function given by equation 3 above may be written as:

$$I_{1,2} = \frac{J_p}{\pi\sigma_p{}^2}[e^{-\frac{r_+{}^2}{\sigma_p{}^2}} + e^{-\frac{r_-{}^2}{\sigma_p{}^2}}] + \frac{J_b}{\pi\sigma_R{}^2}[e^{-\frac{r_+{}^2}{\sigma_R{}^2}} + e^{-\frac{r_-{}^2}{\sigma_R{}^2}}] \qquad (4)$$

where $r_\pm = r \pm d/2$.
    The resolving power d_{min} is the solution of the equation:
        $I_{1,2}(0)$=73.6% of $I_{1,2}(r_{min})$ where $r_{min}=d_{min}/2$.
    For the two cases considered in fig. 2, d_{min} is found to be close to $2\sigma_p$ (d_{min}=120Å for E_o=20keV and d_{min}=119Å for E_o= 100-keV). So backscattering does not influence the resolving power, defined by the Rayleigh's criterion.

The same result is found to be valid in the whole energy range from 5 to 100keV and for different angles of incidence. Indeed, when the angle of incidence is increased, the resolving power is increased too, but this increase is only due to the primary beam induced signal broadening. This is illustrated in fig. 3, where the functions $I_p(r)$ and $I_b(r)$ are drawn for different angles of incidence. The signal $I_p(r)$ varies significantly with the angle of incidence, but this is not the case for the signal $I_b(r)$, which stays always negligible for the lateral resolution with respect to $I_p(r)$. Fig. 4 shows the variation of d_{min} as a function of θ_o. It follows the variation of $2\sigma_p$ as a function of the same angle.

The above result is valid for both bulk materials, thin unsupported films and thin films on a bulk substrate. In all cases the resolution is governed by the primary beam broadening within the few surface layers of the sample.

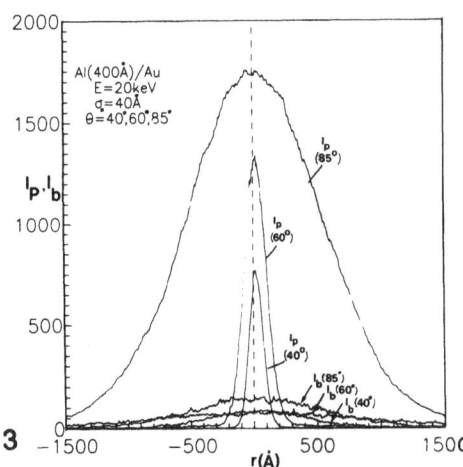
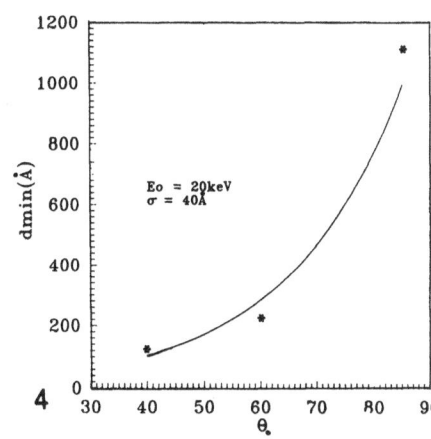

Fig.3 Radial distribution of the signals I_p and I_b for different angles of incidence ($\theta_o=40°$, $60°$ and $85°$).

Fig.4 Resolving power d_{min}, obtained by Monte-Carlo calculations, as a function of θ_o for $E_o=20keV$, $\sigma_o=40Å$.

III. ANALYTICAL EXPRESSION FOR THE RADIAL DISTRIBUTION OF THE PRIMARY BEAM INDUCED AUGER SIGNAL.

It has been demonstrated above that for the beam conditions considered the only signal that influences the resolving power in the whole energy range 5-100keV is that induced by the primary electron beam. So the radial distribution of this signal is sufficient for the calculation of the lateral resolution. This signal can be easily approached by an analytical expression.

Indeed, by considering an exponential decrease of the number of both primary and Auger electrons along their trajectory, the following expression is obtained for the primary beam induced Auger intensity, for the case of a thin film (thickness d) on a bulk substrate[10]:

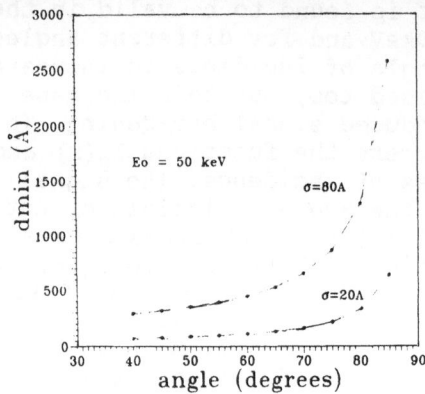

Fig.5 Resolving power d_{min}, as a function of θ_o for E_o=50keV and for σ_o=20 and 80Å.

$$I_A(\chi,\psi) = \frac{J_o}{2\pi\sigma^2}N.\sigma(E_o,E_i)\frac{\cos\theta_o.\cos(\theta_o-\xi)}{\sin\xi}e^{[-\alpha(\chi-\frac{\alpha}{2})-\psi^2]} \times$$

$$\times [erf(\chi-\frac{\alpha}{2})-erf(\chi-\frac{\alpha}{2}-D)] \qquad (5)$$

In the above expression $I_A(\chi,\psi)$ (in e-/s.cm^2) is the Auger intensity at each point χ,ψ where χ,ψ are reduced coordinates equal to:

$$\chi = x\cos\theta_o/2\sigma_o \quad \text{and} \quad \psi = y/2\sigma_o$$

σ_o is the gaussian dispersion of the incident beam and J_o(e-/s) is the primary beam intensity. N (atoms/cm^3) is the atomic density of the sample, $\sigma_o(E_o,Ei)$ the inelastic cross section for ionization of the energy level i, E_o the primary beam energy and erf the usual error function. θ_o and ξ are illustrated in fig. 1. The dimensionless quantities D and μ are given by the expressions:

$$D=\frac{d}{\sqrt{2}\sigma_o}\cdot\frac{\sin\xi}{\cos(\xi-\theta_o)} \ , \ \alpha=\frac{\sqrt{2}\sigma_o}{\cos\theta_o}\cdot\mu(E_o) \qquad (6a)$$

$$\mu(E_o)=\frac{1}{\sin\xi}\{\frac{\cos\theta_o}{\lambda_A}+\frac{\cos(\xi-\theta_o)}{\lambda_P(E_o)}\} \qquad (6b)$$

where λ_A and $\lambda_P(E_o)$ are the Auger and primary electron mean free path respectively and d is the film thickness.

The validity of expression (5) above has been tested by comparing with Monte-Carlo results. It may easily be used for the calculation of the resolving power d_{min} as a function of different parameters of the incident beam.

Fig. 5 shows the variation of d_{min} as a function of the angle of incidence θ_o, obtained by using expression (5) for two different standard deviations σ_o of the incident beam (20Å and 80Å). For E_o=50keV and σ_o=80Å the resolving power d_{min} varies from 300Å at 40° to 2600Å at 85°. A smaller variation is observed for σ_o=20Å.

IV. CONCLUSION

Monte-Carlo calculations have been used in order to simulate the radial extent of Auger electrons in the energy range 5-100keV. It is found that the Auger signal due to backscattering is laterally distributed over an area many times greater than that of the primary beam induced signal. So it constitutes a low background over which the primary beam induced signal is superimposed. The lateral resolution of the technique is found to be only affected by the broadening of the primary beam within the few tens of Å from the surface. This broadening has been calculated analytically as a function of the angle of incidence for two gaussian dispersions σ_o of the beam, $\sigma_o=20$Å and $\sigma_o=80$Å. It is of the order of a few tens to a few hundreds of Å at small angles of incidence. At $\theta_o=85°$ it may reach some thousants of Å when $\sigma_o=80$Å.

REFERENCES

[1] J. Chazelas, J. Cazaux, G. Gillmann, J. Lynch and R. Szymanski, Surf. Interface Anal., 12 (1988) 45.

[2] P. Kruit, A. J. Bleeker and F. J. Pijper, Inst. Phys. Conf. Ser.93 (1988) 249.

[3] M. M. El Gomati and M. Prutton, Surf. Science, 72 (1978) 485.

[4] M. M. El Gomati, M. Prutton, B. Lamb and C. G. Tuppen, Surf. Interface Anal., 11 (1988) 251.

[5] M. Tholomier, E. Vicario and N. Dogmane, J. Microsc. Spectr. Electron., 12 (1987) 449.

[6] E. Valamontes, A. G. Nassiopoulos and N. Glezos, Surf. Interface Anal., 16 (1990) 203.

[7] A.G.Nassiopoulos and E.Valamontes, Surf.Interface.Anal., 15 (1990) 405.

[8] J. Cazaux, J. Chazelas, M. N. Charasse and J. P. Hirtz, Ultramicroscopy, 25 (1988) 31.

[9] J. Cazaux, Surf. Interface Anal., 14 (1989) 354.

[10] J.Cazaux, J. Microsc. Spectr. Electr.,13(1988) 315.

[11] N.M.Glezos and A.G.Nassiopoulos,Surf. Sc., 254 (1991)309

SURFACE RELATED ELECTRONIC STATES ON CLEAVED HgSe AND CdSe SURFACES AS

OBSERVED BY MEANS OF PHOTOEMISSION YIELD SPECTROSCOPY

B. J. Kowalski, A. Sarem[1], B. A. Orłowski

Institute of Physics,
Polish Academy of Sciences
Al. Lotników 32/46, 02-668 Warsaw, Poland

HgSe and CdSe belong to the II–VI compounds family. Until now, the surface properties of these crystals have not been investigated as intensively as those of Si and III–V compounds. Lately, however, growing of the epilayers of II–VI materials on III–V substrates and applications in the technology of quantum wells and superlattices turned out to be possible and useful. This makes the investigation of the surface structure of II–VI crystals increasingly necessary and important.

The purpose of this paper is to contribute to the experimental investigation of the surface electronic structure as well as of the adsorption processes on the cleaved surfaces of HgSe and CdSe.

Both materials investigated are sp^3-bonded crystals with tetrahedral coordination. However, there are important differences between them:

- HgSe is a zero gap semiconductor with a zinc blende (cubic) structure,

- CdSe is a wide gap (1.74 eV) semiconductor with a wurtzite (hexagonal) structure.

These characteristics may influence markedly the surface properties (e.g. the band bending cannot be observed for the crystals with high carrier concentration). Thus, the comparison of the results obtained should be interesting, even though the compounds are of similar chemical character and the surfaces investigated are related (apart from the different crystal structure).

Cleavage under UHV conditions (10^{-9} $Torr$) was used to obtain the clean HgSe(110) and CdSe(11$\bar{2}$0) surfaces. However, the cleavage generates a lot of defects (e.g. steps) on the surface. Furthermore, it is known that there is a second non-polar plane for the wurtzite-type crystals: (10$\bar{1}$0). This makes the quality of the cleavage surface worse than that of the cubic materials (in spite of the alignment of the cleavage knives with the chosen cleavage planes).

Photoemission yield spectroscopy (PYS) was applied to investigate the electronic structure of the crystals. The high sensitivity and good energy resolution are the main advantages of the method. The usefulness of PYS in surface physics was shown clearly during investigations of Si surfaces [1,2,3]. However, we pay for this with a complicated analysis of the spectra. Since PYS consists in measuring the total photoemission current as a function of photon energy, the number of electrons recorded at a particular energy of light integrates the contribution of all transitions from the occupied states from the Fermi energy down to $E_F - h\nu - \varphi$ (φ is the work function of the sample). If we assume that the transition matrix elements and the escape function of electrons do not depend on the energy of light, it is possible to consider the first derivative of the photoemission yield versus energy ($dY/d(h\nu)$) as an effective density of states. If we consider a small energy range, this approach usually turns out to be reasonable [2].

In the experiments reported, photoemission yield was measured using a Seya— Namioka type

---

[1]Permanent address: Department of Physics, Faculty of Science, Tishreen University, Lattakia, Syria.

Equilibrium Structure and Properties of Surfaces and Interfaces
Edited by A. Gonis and G.M. Stocks, Plenum Press, New York, 1992

vacuum-ultraviolet monochromator ($4 < h\nu < 12\ eV$) with a flowing hydrogen low-pressure discharge lamp ($p = 10^{-2}\ Torr$) as a light source. A LiF window separates the monochromator from the stainless steel UHV chamber. A photomultiplier with a luminophor (sodium salicylate) deposited on a glass plate was used as a light detector. The photoemission current was measured by an ECCO vibrating reed electrometer (sensitivity $10^{-15}\ A$). The experimental set-up described enables us to measure the photoemission yield for a wide energy range. However, since our work reported in this paper was devoted to the investigation of the electronic states in the ionization energy edge region, the described procedure of differentiation was applied to the relatively small part of the spectrum experimentally observed (about 1.5 eV just above the valence band edge). The results are shown in Figs. 1 and 2 for HgSe and CdSe, respectively.

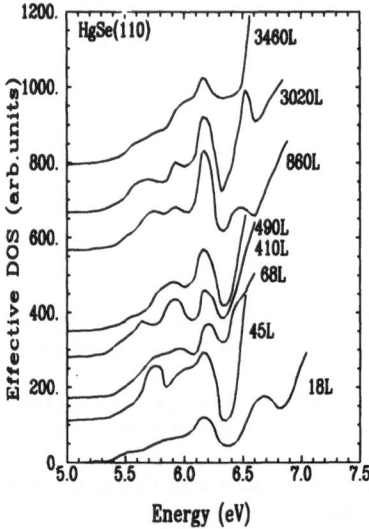

Figure 1. Linear plot of the effective density of states (the derivative of the photoemission yield spectra) in the ionization energy edge region for the HgSe(110) surface cleaved in UHV and annealed at room temperature at $5 \cdot 10^{-9}$ Torr.

Interaction with the ambient atmosphere ($p = 5 \cdot 10^{-9}\ Torr$) was the phenomenon that enabled us to distinguish the surface related features in the spectra. It has already been shown that H_2O, CO_2, CO, N_2 and O_2 molecules occur on the real surface of II-VI compounds [4]. This set of compounds corresponds well to the low-pressure atmosphere in a stainless steel UHV chamber. Thus we can expect that the adsorption of these gases occurs during the annealing of the cleaved sample at room temperature at $5 \cdot 10^{-9}$ Torr. This, together with the surface electronic structure of the crystals, determines the shape of the spectra shown in Figs.1 and 2.

Fig. 1 illustrates the evolution of the distribution of the effective density of states observed for the HgSe(110) surface as a result of the influence of the ambient atmosphere. The numbers correspond to the total exposures measured from the cleavage of the sample ($10^{-6} Torr \cdot s = 1L$). The lowest curve was obtained just after the cleavage. The changes in the shape of the curves observed during the experiment enable us to distinguish two regions with different behaviour caused by the gas adsorption. The first one consists of four maxima (5.5–5.6, 5.7–5.8, 5.95, and 6.15 eV). Their height is sensitive to the state of the surface but the position is almost constant. In the second region (6.4–7.0 eV) the valence band edge can be observed. Its position changes according to the growth of the maxima revealed just over the edge. Three stages of this evolution can be distinguished: 18–45 L, 68–490 L, and 860–3460 L.

According to these observations, the suggestion can be put forward that the first group of states is connected with the surface of the crystal. Apart from their sensitivity to the annealing, the similarity with states observed in PYS experiments for a cleaved Si surface [1] and the correspondence with the results of the ARUPS experiments [5] support this prop sal. The rich structure of this band can be connected with the presence of defects on the cleaved surface. Steps and other imperfections lead to the formation of the electronic states with different binding energies. This has already been observed for other crystals [3,6]. The relative heights of the maxima depend on the state of the surface, which can be changed by the adsorption of gases [2,3,7].

The changes observed in the region of the valence band edge should probably be interpreted as a result of the change in the near-surface layer rather than the surface states. The intensity of the emission observed in this region is comparable with that coming from the bulk states of the valence band.

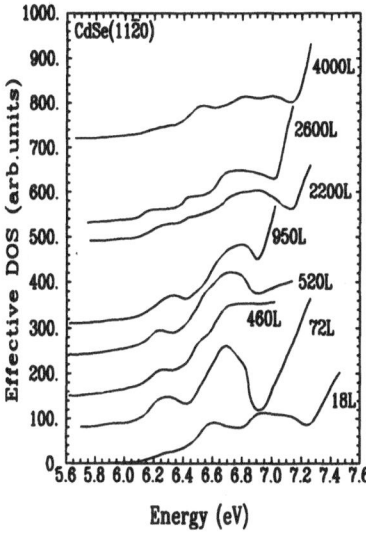

Figure 2. Same as Figure 1 for the CdSe(11$\bar{2}$0) surface.

The structure of the curves shown in Fig. 2 is poorer than that of the HgSe(110) surface spectra. For the freshly cleaved CdSe crystal (the lowest curve in Fig. 2) the valence band edge with an ionization energy of about 7 eV and a band of states sensitive to the gas adsorption (see the other curves in Fig. 2) are observed. The CdSe(11$\bar{2}$0) has been investigated by both experimental (UPS) and theoretical methods [8]. No occupied surface states were revealed in the gap. Thus, the broad band in the distribution of the density of states observed in the energy region from 6.1 to 7.2 eV (in the lowest curve in Fig. 2) must be connected mainly with the defect induced surface states. The influence of the ambient atmosphere leads easily (for 18-72 L) to a shift of the spectrum towards the lower energies. A change of the relative heights of the maxima is observed, too. This shift is characteristic of dipole layer formation rather than band bending. The latter cannot lead to a change of the binding energy of the surface states. Further adsorption causes independent changes in the position of the valence band edge with respect to the other states. In a wide gap material the band bending can be induced by a change of the charge bounded on the surface. This leads to the observed shift of the valence band edge, while the positions of the surface states remain constant. The influence of the adsorption on the shape of the band of the surface related states confirms the complex character of the band. It is composed of contributions connected probably with different defects created on the surface. The prolonged interaction of the ambient atmosphere decreases the number of states observed in the experiment (see the last curve in Fig. 2) and neutralizes in part the dipole layer created in the first stage of the adsorption (compare the last two curves in Fig. 2).

In summary, the photoemission yield measurements enabled us to observe the evolution of the HgSe(110) and CdSe(11$\bar{2}$0) surfaces cleaved in UHV and then annealed at room temperature at 5 · 10^{-9} Torr. The occupied electronic states were revealed on both HgSe and CdSe. The states are sensitive to adsorption of gases on the surfaces and originate probably from the defects created on the cleaved surfaces. The changes of the valence band edge position on the surface were observed during the adsorption, too.

We expect that further experiments using other techniques will help us to learn more about different characteristics of the surfaces. Then, our argumentation concerning the origin of the surface related states observed on the cleaved surfaces of the II–VI crystals will be much more comprehensive.

References

[1] G. M. Guichar, C. A. Sébenne, G. A. Garry, M. Balkanski, *Surface Science* **58**, 374 (1976)

[2] C. A. Sébenne, *Il Nuovo Cimento* **39B**, 768 (1977)

[3] I. Andriamanantenasoa, J. P. Lacharme, C. A. Sébenne, F. Proix, *Semicond. Sci. Technol.* **2**, 145 (1987)

[4] I. A. Kirovskaya, *Inorg. Mater.* **25**, 1246 (1989)

[5] The surface state was observed at about 0.5 eV above the valence band edge — B. A. Orłowski, J. Bonnet, C. Hricovini, R. Pinchaux, *Proc. ECOSS-12, Stockholm, Sweden, 1991* — to be published

[6] H. Ibach, K. Horn, R. Dorn, H. Lüth, *Surf. Science* **38**, 433 (1973)

[7] E. G. Seebauer, *J. Vac. Sci. Technol.* **A7**, 3279 (1989)

[8] Y. R. Wang, C. B. Duke, K. Stevens, A. Kahn, K. O. Magnusson, S. A. Flodström, *Surf. Science* **206**, L817 (1988)

VOLUME AND INTERFACIAL PROPERTIES

OF METAL/RARE EARTH OXIDE/METAL STRUCTURES

Tadeusz Wiktorczyk

Institute of Physics, Technical University
of Wrocław
Wybrzeże Wyspiańskiego 27, 50-370 Wrocław, Poland

INTRODUCTION

Thin films of rare earth metal oxides (TFREMO) as well as films of the other rare-earth metal compounds have been the subject of extensive examinations. TFREMO exhibit interesting optical properties and a good chemical and mechanical stability. Their excellent dielectric and insulating properties are the reason that they are considered as dielectric materials for thin-film (TF) capacitors, TF transistors (MOSFET) and TF electroluminescent devices. Different preparation methods have been employed to produce TFREMO[1-3].

We have applied a method of electron-gun assisted thermal evaporation for deposition of high-quality TFREMO[3-5]. In this paper the preparation technique and results of the examination of dielectric properties for some TFREMO are presented, namely for dysprosium, ytterbium and holmium oxides (Dy_2O_3, Yb_2O_3 and Ho_2O_3). These oxides are representatives of sesquioxides of heavy rare-earth metals. Electrical properties are discussed taking into account the volume of TFREMO and metal-insulator (M-I) interfaces. It has been shown on the basis of different experimental tests, that near-electrode regions a few nanometers thick cannot be neglected in metal-TFREMO-metal structures.

PREPARATION OF TFREMO AND MIM STRUCTURES

Re_2O_3 thin films (where Re=Dy, Yb or Ho) were produced by thermal evaporation method in vacuum apparatus. Deposition of

Equilibrium Structure and Properties of Surfaces and Interfaces
Edited by A. Gonis and G.M. Stocks, Plenum Press, New York, 1992

341

TFREMO was carried out by the reactive evaporation, with rate from 0.008 nm/s to 0.16 nm/s, at a constant oxygen pressure from 0.8 mPa to 10.7 mPa with the help of 1.5 kW electron gun[3,4]. Re_2O_3 of 99.9% purity was used for evaporation. Quartz or glass plates were used as substrates to study optical and electrical properties. Films for structural examinations were deposited onto NaCl monocrystals. For electrical measurements TF Al-Re_2O_3-Al sandwich structures were prepared. The presented results refer to the films of different thickness (from 20 nm to 1.5 μm).

STRUCTURE OF TFREMO

Electron diffraction patterns and electron micrographs show that TFREMO prepared at room temperature and at oxygen pressure from p = 0.8 mPa to 10.7 mPa are amorphous or poly-crystalline. The electron micrographs for these film reveal one or a few diffusive rings ascribed to rare-earth sesquioxide. Usually, the films deposited at low oxygen pressure were com-pletly amorphous. Recrystallization occurred in films irradi-ated with electron beam. As a result of this pulse heating 1 μm grains were formed. The identification of the structure was performed on the basis of diffraction patterns for recrystal-lized films. All observed diffraction rings for films of ytter-bium, dysprosium and holmium oxides (up to 22 diffraction fringes) belong to C-type Re_2O_3 (see also[3] and[4]). all films deposited onto heated substrates (from 500 K to 700 K) were polycrystalline.

DIELECTRIC PROPERTIES OF TFREMO IN A WIDE FREQUENCY RANGE

Electrical measurements from very low frequencies (from 10^{-5} Hz) up to radio frequencies were carried out for Al-Re_2O_3-Al thin film capacitors with different experimental techniques, schematically shown in Fig. 1. Dispersion characteristics of the complex capacitance, dielectric permittivity or susceptibil-ity were evaluated from these measurements (see also[5-9] for further details). In Fig. 2 a frequency dependence of the real and imaginary part of the relative dielectric permittivity is shown for ytterbium oxide films, in the frequency spectrum from 10^{-5} Hz up to 10^{15} Hz. Results for films of dysprosium oxide and holmium oxide are very similar[4,9]. The dielectric permittivity increases from $\varepsilon_r = 3.42$ for Dy_2O_3, 3.62 for Yb_2O_3 and 3.56 for Ho_2O_3, in the optical region, to $\varepsilon_r = 10.2\text{-}16$ for radio frequencies and up to $\varepsilon_r = 10^2\text{-}10^4$ for very low frequencies[4-10].

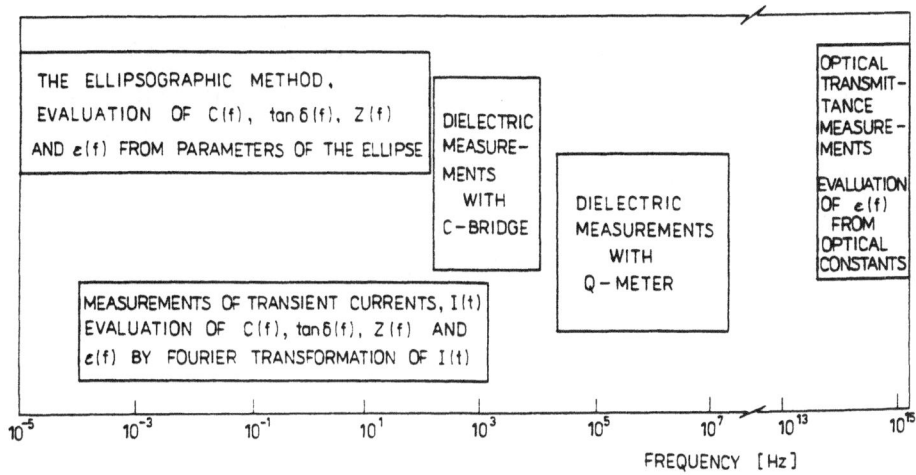

Fig. 1. Schematic representation of the dielectric measurement for Re_2O_3 thin films in the frequency region from 10^{-5} Hz to 10^{15} Hz.

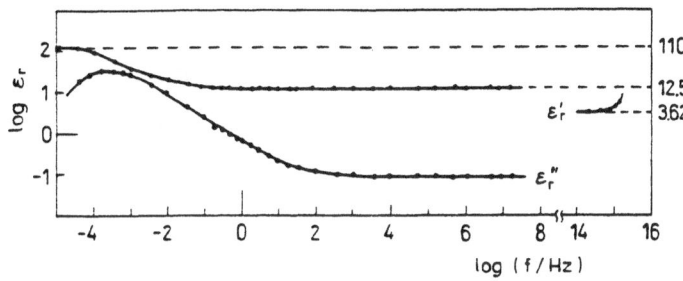

Fig. 2. Frequency dependence of ε_r' and ε_r'' for Yb_2O_3 thin films.

Different mechanisms of dielectric polarization can be distinguished in this very wide frequency region: electron polarization, ionic polarization, polarization due to hopping of the carriers and interfacial polarization connected with near-electrode regions (Schottky barriers at M-I boundaries)[4-8].

THE INFLUENCE NEAR-ELECTRODE REGIONS ON PROPERTIES OF Al-Re_2O_3-Al STRUCTURES

The interfacial polarization mechanism is connected with near-electrode regions existing at each Al-Re_2O_3 boundary.

These regions are responsible for specific low-frequency properties of Al-Re$_2$O$_3$-Al TF structures, namely:

(a) - high values of the low-frequency (or high-temperature) capacitance[5,6],

(b) - anomalously high values of the low-frequency permittivity (see Fig. 2 and 3). Values of ε_r for the low-frequency region (denoted in Fig. 3 as "region II") depend on specimen thickness and have to be taken as "apparent dielectric permittivity". Values of the high-frequency ε_r determine volume properties of Re$_2$O$_3$ ("region I" in Fig. 3),

(c) - pronounced dielectric dissipation peaks[5-7],

(d) - strong voltage-dependence of the capacitance[10] and

(e) - dependence of the dielectric properties on metal used for electrodes[10].

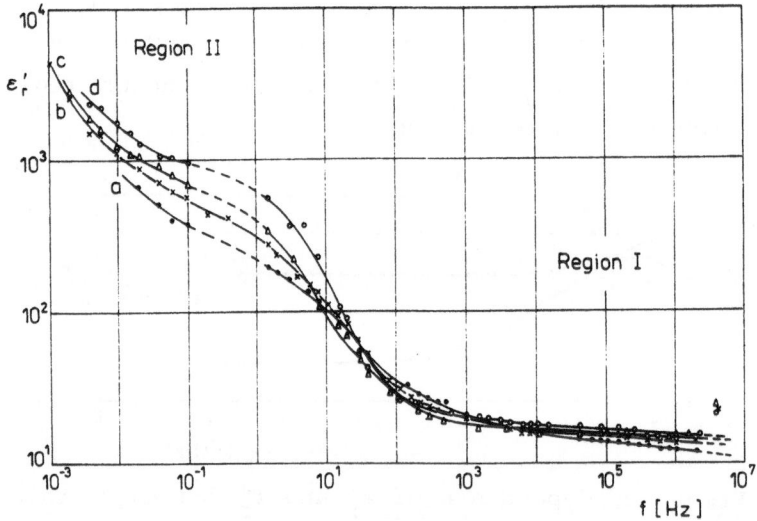

Fig. 3. Curves of the dielectric permittivity versus frequency for Al/Dy$_2$O$_3$/Al TF capacitors measured at the temperature T=543K. Each curve shows results for capacitors with different insulator thickness: (a)-96 nm, (b)-167 nm, (c)-226 nm, (d)-273 nm and (e)-348 nm.

THE IMPEDANCE DIAGNOSTICS OF Al-Re$_2$O$_3$-Al THIN-FILM SANDWICHES

For inhomogeneous materials (e.g. for specimens in which electrode-ascribed processes, internal barriers or grain boundaries play an important role) the complex impedance characteristics are very useful in presentation and interpretation of electrical properties[11]. Results of dielectric measurements for Al-Re$_2$O$_3$-Al sandwiches were presented in the complex impedance plots ($Z''=F(Z')$). The examples of these characteristics are shown in Fig. 4 where Z'-Z'' plots are presented for Al-Dy$_2$O$_3$-Al structures with different insulator thickness. Two separate regions can be distinguished in each plot: high frequency region (region I) and low-frequency region (region II). These regions strictly correspond to the volume and interface effects in the examined MIM structures and they make it possible to determine electrodes resistance (R_e), Dy$_2$O$_3$ film resistance (R_v) and sometimes resistances of the near-electrodes regions (R_b) (see Fig. 4). Our results suggest that volume resistance depends on insu-

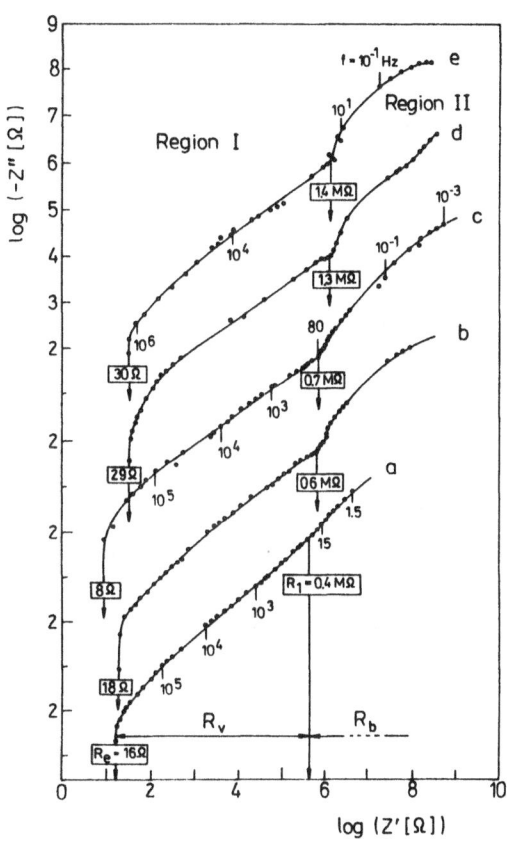

Fig. 4. Z'-Z'' plots for Al-Dy$_2$O$_3$-Al thin film sandwiches with different insulator thickness: (a)-96 nm, (b)-167 nm, (c)-226 nm, (d)-273 nm and (e)-348 nm, measured at the temperature of 543 K. A description of the imaginary axis deals with the thickest specimen. Z'-Z'' plots for the other specimens are vertically shifted.

lator thickness and on a temperature ($R = R_b \exp(E/kT$, where $E = 1$ eV)) whereas the barrier resistance is extremely high and strongly depends on voltage. This behaviour is typical for Schottky barriers.

CONCLUSIONS

Our studies lead to conclusion that $Al-Re_2O_3-Al$ TF sandwich structures can be represented by three regions, namely: the region of the volume of TFREMO and two near-electrode regions (Schottky barrier regions), which are connected with aluminium electrodes (see the model in Fig. 5). It was shown, that 4-8 nm thick Schottky barriers are formed at each $Al-Re_2O_3$ boundary. These barriers are responsible for M-I interfacial polarization

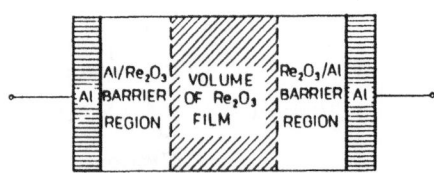

Fig. 5. Schematic illustration of $Al-Re_2O_3-Al$ TF structure

mechanism and for low-frequency properties of $Al-Re_2O_3-Al$ TF structures. The high-frequency properties of $Al-Re_2O_3-Al$ TF structures are connected with properties of the volume Re_2O_3 itself.

REFERENCES

1. M. Gasgnier, Phys. Stat. Sol. A, 14:14 (1989).
2. K. Takagi, T. Fukuzawa and K. Susa, J. Appl. Phys., 61:1030 (1987).
3. T. Wiktorczyk and C. Wesołowska, Thin Solid Films, 91:9 (1982).
4. T. Wiktorczyk, Eur. J. Sol. State Inorg. Chem., 28:581(1991).
5. T. Wiktorczyk and C. Wesołowska, Vacuum, 37:107 (1987).
6. T. Wiktorczyk, K. Nitsch and Z. Bober, Thin Solid Films, 157:13 (1988).
7. T. Wiktorczyk, Solid State Commun., 67:143 (1988).
8. T. Wiktorczyk, Materials Science, 14:31 (1988).
9. T. Wiktorczyk and K. Nitsch, to be published.
10. T. Wiktorczyk, Proc. of the 6-th Int. Symposium on Electrets, p. 607-611, 1-3 Sept. 1988, Oxford, England, Ed. by D. K. Das Gupta and A. W. Pattullo, Univ. of Wales, Bangor, U.K. (Cat. No. 88CH2593-2).
11. J. R. Macdonald, Space Charge Polarization, in: "Solid State Ionic", M. Kleitz and J. Dupuy, eds., D. Reidel Publishing Co., Dordrecht-Holland (1976).

EXCITATION SPECTRA OF ADSORBATES ON DIELECTRIC SURFACES

Constantine Mavroyannis

National Research Council of Canada
Ottawa, Ontario,
Canada K1A 0R6

INTRODUCTION

There had been extensive experimental studies[1], where the optical excitation spectra of adsorbed rare-gas atoms on various metal surfaces were measured as a function of coverage. At high coverage, the adsorbed spectra were found to be similar to comparable excitations in the gas phase and condensed rare gases. On the other hand, striking changes in the optical spectra have been observed[1] in the limit of low coverage, where the adsorption spectra disappear on some metal surfaces and persist on others. The spectra of Xe adsorbed on Ti, Al and Au and Kr on Au and Mg are virtually eliminated while those of Xe and Ar on Mg, Kr and Ar on Al and Ar on Au persist[1].

A microscopic theory had been developed[2] concerning the optical spectra of neutral atoms physisorbed on metal surfaces. At low coverage and when the damping of the surface plasmons is much greater than the effective radiative damping, the disappearance of the peaks of the spectral lines is due to the existence of the symmetric and antisymmetric states, whose relative intensities take positive and negative values, respectively. Hence, they may have a chance to cancel each other out whenever the frequency profiles of the peaks in question coincide. Numerical calculations have been found to be compatible[3] with the experimental observations[1]. The theory has been recently extended[4], hereafter referred to as I, to consider the optical spectra of neutral atoms physisorbed on dielectric surfaces. In I a simple model has been used, where the neutral atom A and its image B are a distance 2R apart and interact through the dipole-dipole interaction[5]. The dielectric function describing the dielectric surface, which separates the atom A and its image B, is taken as

$$\varepsilon_d(\omega) = 1 - \frac{(\varepsilon_d - 1)E_g^2}{\omega^2 - E_g^2 + i\omega\gamma_{eh}} \tag{1}$$

Equilibrium Structure and Properties of Surfaces and Interfaces
Edited by A. Gonis and G.M. Stocks, Plenum Press, New York, 1992

347

where $\varepsilon_d \equiv \varepsilon_d(0)$ is the static dielectric constant and E_g is the frequency gap of the surface electron-hole pairs, respectively, while γ_{eh} ($\gamma_{eh} \ll \omega$) is the intrinsic damping of the electron-hole pairs due to the decay processes with phonons, impurities and other surface excitations. For the sake of convenience, we have considered in eq. (1) only one excited state and $\varepsilon_\infty = 1$; in general, such simplifications are not necessary. Each electron of the atom A with charge e will have an image with a fictitious charge $-e\varepsilon_1$, where the screening factor $\varepsilon_1 \equiv \varepsilon_1(\omega)$, defined as

$$\varepsilon_1(\omega) = \frac{\varepsilon_d(\omega) - 1}{\varepsilon_d(\omega) + 1} \tag{2}$$

is due to boundary conditions at the plane separating the empty space and the homogeneous isotropic dielectric medium, provided that retardation effects have been neglected. Using Eq. (1), the expression (2) for $\varepsilon_1(\omega)$ becomes

$$\varepsilon_1(\omega) = 1 - \frac{(\varepsilon_d - 1) E_g^2}{\omega^2 - \omega_g^2 + i\omega\gamma_{eh}}, \quad \text{for} \quad \omega_g = E_g[(\varepsilon_d - 1)/2]^{1/2} \tag{3}$$

The interaction between the adsorbed atom and its image is taken to be of the dipole-dipole type[5], which is screened by ε_1, namely, by the dielectric function of the surface $\varepsilon_d(\omega)$. Using a microscopic Hamiltonian, which describes the aforementioned model, the optical excitation spectra due to radiative and nonradiative processes have been considered in I in great detail. We shall review here the excitation spectra due to the nonradiative processes.

NONRADIATIVE EXCITATION SPECTRA

In the limit when the frequency mode due to the surface electron-hole pair is heavily damped, namely, when the nonradiative damping γ_{eh} is much greater than the radiative damping γ_0 ($\gamma_{eh} \gg \gamma_0$), the radiative contributions to the damping can be neglected. In this limit the spectral function per atom takes the form[4]

$$\gamma_{eh} I_j(\omega) = \frac{N_j(\omega_{j+})\,\lambda_j(\omega_{j+}) - X_{j+} M_j \gamma_{eh}}{X_{j+}^2 + \lambda_j^2(\omega_{j+})} + \frac{N_j(\omega_{j-})\,\lambda_j(\omega_{j-}) - X_{j-} M_j \gamma_{eh}}{X_{j-}^2 + \lambda_j^2(\omega_{j-})}, \tag{4}$$

where $X_{j\pm} = (\omega - \omega_{j\pm})/\gamma_{eh}$ defines the reduced frequency, and the energies of excitation ω_{j+} and ω_{j-} are determined by

$$\omega_{j\pm}^2 = \frac{1}{2}\left\{ E_{vo}^2 + \omega_g^2 \pm \left[\left(E_{vo}^2 - \omega_g^2 \right)^2 \pm 4 E_{vo}^2 \omega_g^2 u \right]^{1/2} \right\}. \tag{5}$$

The functions $N_j(\omega_{j\pm}), \lambda_j(\omega_{j\pm})$ and M_j are defined as

$$N_j(\omega_{j\pm}) = \pm \left(\frac{2E_{v0}}{\omega_{j\pm}}\right) \frac{\left[\omega_{j\pm}^2 - \omega_g^2(1\mp u)\right]}{\left(\omega_{j+}^2 - \omega_{j-}^2\right)}, \qquad M_j = 2E_{v0}/\left(\omega_{j+}^2 - \omega_{j-}^2\right), \qquad (6)$$

$$\gamma_{eh}(\omega_{j\pm}) = \gamma_{eh}\lambda_j(\omega_{j\pm}), \qquad \lambda_j(\omega_{j\pm}) = \pm\frac{1}{2}\frac{\left(\omega_{j\pm}^2 - E_{v0}^2\right)}{\left(\omega_{j+}^2 - \omega_{j-}^2\right)}, \qquad (7)$$

$$u = \alpha(\varepsilon_d - 1)/(\varepsilon_d + 1), \qquad \alpha \equiv \alpha(\vec{R}) = \alpha_s/2R^3, \qquad (8)$$

with α_s being the static polarizability of the atom. The relative maximum intensity per atom at the frequency $\omega = \omega_{j\pm}$ in units of γ_{eh} is determined by

$$\gamma_{eh}I_j(\omega_{j\pm}) = N_j(\omega_{j\pm})/\lambda_j(\omega_{j\pm}), \qquad (9)$$

which defines the maximum height of the peak in question. The spectral function $\gamma_{eh}I_j(\omega)$ describes asymmetric Lorentzian lines, which are peaked at the frequencies $\omega = \omega_{j\pm}$ and having spectral widths of the order of $\gamma_{eh}\lambda_j(\omega_{j\pm})$. Inspection of Eq. (5) reveals that one of the energies is close to the value of E_{v0} while the other is close to the value of ω_g. Thus, when E_{v0} is greater than ω_g, $E_{v0} > \omega_g$, then the excitations at $\omega = \omega_{j+}(E_{v0} < \omega_{j+})$ and at $\omega = \omega_{j-}(\omega_g > \omega_{j-})$ may be referred to as the atomic-like and the surface electron-hole-like excitations, respectively; the reverse is true when $\omega_g > E_{v0}$. As an application, we shall make use of Eq. (4) to compute the excitation spectra of rare-gas atoms, He, Ne, Ar and Kr which are physisorbed on the surface of graphite.

Using the values of α_s, ε_d and E_g given in ref. 6, the computed spectra from Eq. (4) of the atomic-like excitations are illustrated in Figs. 1 and 2, where the relative intensity per atom $I_j(\omega)$ in units of γ_{eh}, $\gamma_{eh}I_j(\omega)$, is plotted versus the relative frequency $X_{j-} = (\omega - \omega_{j-})/\gamma_{eh}$ in units of eV/γ_{eh} for the symmetric j=s and antisymmetric j=a modes, respectively. The graphs in Figs. 1 and 2 illustrate the net lineshapes, which result from the combination of the red shifted peaks of the symmetric (positive intensities) modes and the blue shifted peaks of the antisymmetric (negative intensities) modes of the two atomic excitations under investigation. Thus, each line in Figs. 1 and 2 represents the net result of the cancellation process, which corresponds to an excited configuration for a given value of R for the physisorbed atom in question.

The spectra of the atomic-like excitations of He and Ne on graphite are illustrated in Fig. 1, where the net lineshapes arising from the cancellation processes for the four excited configurations corresponding to different distance R are denoted by (a)-(d), respectively. Inspection of Fig. 1 reveals that the frequency profile of the (b) configuration for He and that of the (d) configuration for Ne describe absorption-amplification type spectra, which are going from positive to

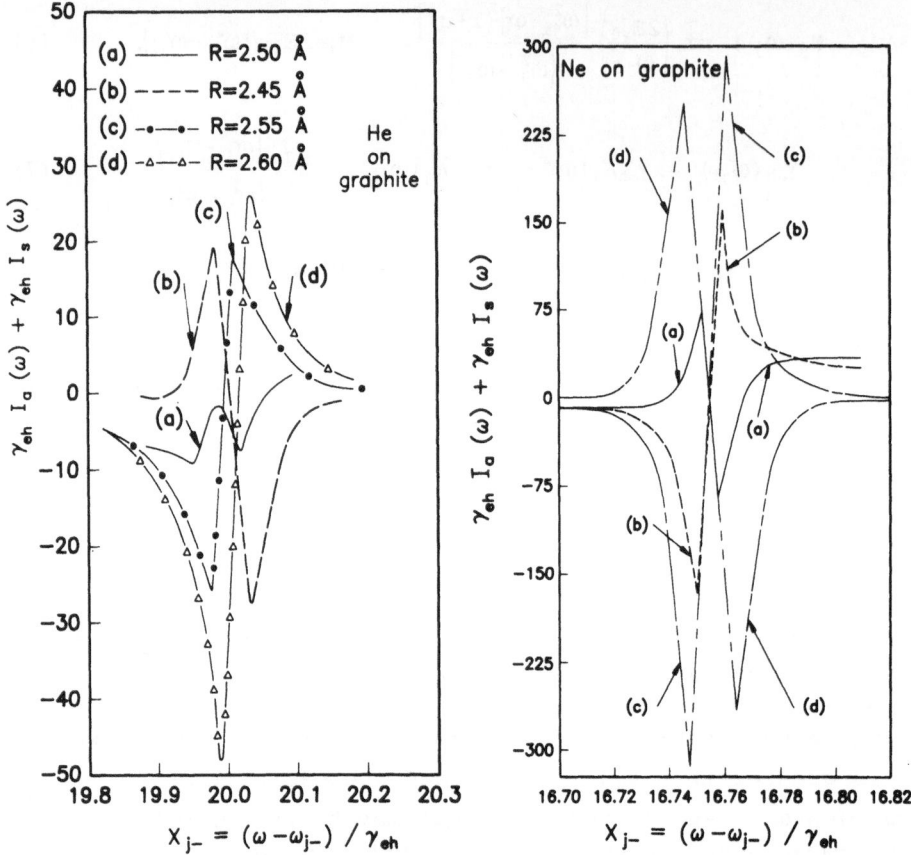

Figure 1. Spectra of the atomic-like excitations of excited He and Ne at the frequencies E_{v0}=19.81, 20.21 eV and E_{v0}=16.67, 16.84 eV, respectively, which are physisorbed on graphite at various distances R from the surface of the dielectric. For Ne on graphite: (a) R=3.10Å, (b) R=3.15Å, (c) R=3.20Å and (d) R=3.0Å.

negative intensities while those of the (c) and (d) configurations for He and those of (b) and (c) configurations for Ne are of the opposite type, namely, of the amplification-absorption type spectra. On the other hand, the (a) configuration (solid line) for He consists of two peaks separated by a shoulder, which have very small negative intensities and appear on each side of the peak near $\omega = \omega_{j-}$. Therefore, the (a) configuration for He is unstable so that the vanishing configuration for the excited state of the adsorbed atom in question, if it exists, it will be for values of R near 2.50Å. Here, by an unstable configuration, it is meant that an infinitesimal perturbation or a very small disturbance in the proper direction may cause the disappearance of the unstable configuration and, consequently, the vanishing of the shape function in question. The (a) configuration (for Ne) represents an absorption-amplification type spectra, where the absorption and the amplification frequencies are nearby, and being located between the (c) and (d) configurations as having the appropriate value of R equal to R=3.10Å; therefore, it seems that the (a) configuration for Ne is the most likely one to be closed to the vanishing one. Wide range of values R have

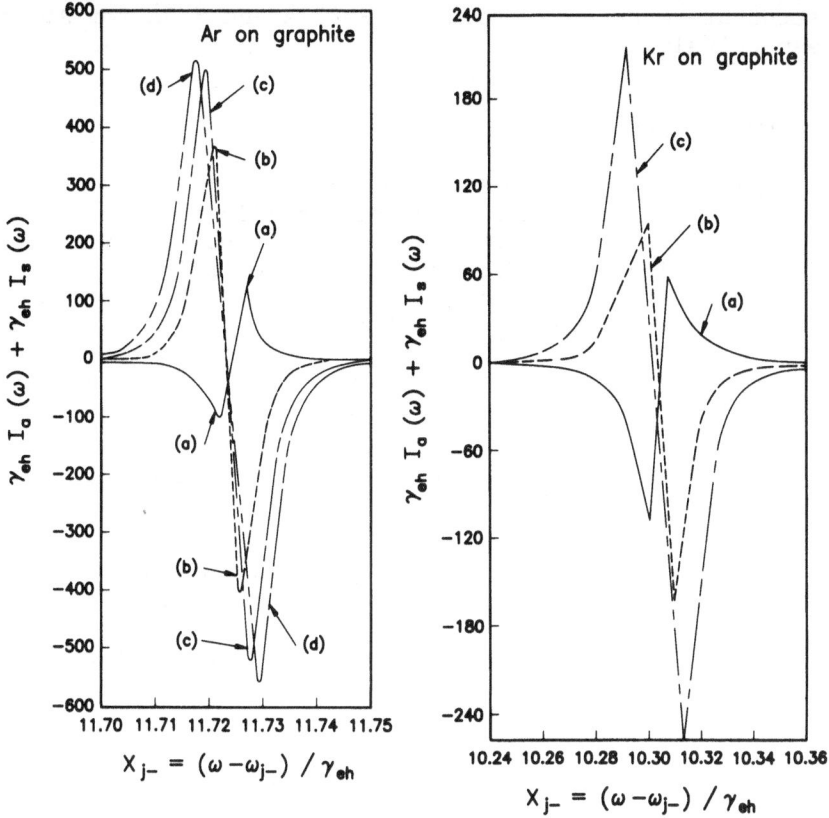

Figure 2. As in Fig. 1 but for the excited Ar and Kr at the frequencies E_{v0}=11.62, 11.83 eV and E_{v0}=10.03, 10.60 eV, respectively, which are physisorbed on graphite at various distances R from the surface of the dielectric. For Ar on graphite: (a) R=3.50Å, (b) R=3.47Å, (c) R=3.45Å and (d) R=3.43Å. For Kr on graphite: (a) R=2.75Å, (b) R=2.70Å and (c) R=2.68Å.

been used in the literature[7] for He on graphite. For instance, Carlos and Cole[7a] have obtained the values of R=2.8Å and 2.4Å for the 6-12 potential and for the Yukawa-6 potential, respectively, while neutron scattering experiments[7b] give R=2.85Å. An estimate of R=3.0Å has been also given by Toigo and Cole[7c]. The values of R=2.98Å, 2.8Å and 2.85Å have been reported by Steele[8a], Vidali, Cole and Klein[8b] and Toigo and Cole[7c], respectively.

Figure 2 illustrates the net lineshapes of the symmetric and antisymmetric modes and describes the excited configurations of Ar and Kr adsorbed on graphite, respectively. The (b), (c) and (d) configurations for Ar and those of (b) and (c) for Kr describe absorption-amplification type spectra while those of the (a) configurations for both Ar and Kr represent spectra of the amplification-absorption type ones. Therefore, the most probable vanishing configurations, if there are any, should be those near the (a) configurations, which have the appropriate values of R equal to R=3.50Å and R=2.75Å for Ar and Kr, respectively. Steele[8a] has reported the values of R=3.28Å and 3.48Å for Ar and Kr on graphite, respectively. Low-energy electron diffraction and surface extended X-ray absorption fine-structure data have produced[9] R=3.2Å and 3.3Å for Ar and Kr, respectively.

CONCLUDING REMARKS

We have computed here the excited geometrical configurations, which arise from the cancellation processes between the lineshapes of the peaks describing the symmetric and antisymmetric modes, respectively. The excited configurations of He, Ne, Ar and Kr which are physisorbed on graphite have been considered as a function of R from the atom to the dielectric surface. Figures 1 and 2 illustrate the excited configurations of He, Ne, Ar and Kr at different distances R, respectively. We may conclude from Figs. 1 and 2 that for given values of E_{vo}, E_g, α_s and ε_d, the vanishing of the peak in question determines the value of the coupling function α, which in turn defines the distance R from the atom to the surface of the dielectric. Therefore, the vanishing of a particular peak, which represents an excited configuration of the adsorbed atom, may determine the distance R from the atom to the dielectric surface. Unfortunately, to our knowledge, there are no optical experimental data to compare with our computed spectra.

REFERENCES

1.   D. Gibbs, J.E. Cunningham, and C.P. Flynn, Phys. Rev. B29, 5292 (1984); J.E. Cunningham, D. Gibbes, and C.P. Flynn, Phys. Rev. B29, 5304 (1984); Y.C. Chen, J.E. Cunningham, and C.P. Flynn, Phys. Rev. B30, 7317 (1984); C.P. Flynn, surface Sci. 158, 84 (1985) and references therein.
2.   C. Mavroyannis, Mol. Phys. 64, 457 (1988); Can. J. Chem. 66, 741 (1988).
3.   C. Mavroyannis, Appl. Phys. A47, 157 (1988); Mol. Phys. 65, 977 (1988).
4.   C. Mavroyannis, J. Chem. Phys. 91, 1294 (1989).
5.   J.E. Lennard-Jones, Trans. Farad. Soc. 28, 334 (1932); C. Mavroyannis, Mol. Phys. 6, 593 (1963); C. Mavroyannis and D.A. Hutchinson, Solid State Commun. 23, 463 (1977); C. Mavroyannis, Mol. Phys. 36, 1565 (1978).
6.   C. Mavroyannis, Chem. Phys. Lett. 56, 263 (1978).
7.(a)   W.E. Carlos and M.W. Cole, Surf. Sci. 91, 339 (1980); (b) K. Carneiro, L.P. Passel, W. Thomlinson and H. Taub, Phys. Rev. B24, 1170 (1981); (c) F. Toigo and M.W. Cole, Phys. Rev. B32, 6989 (1985).
8.(a)   W.A. Steele, J. Phys. Chem. 82, 817 (1978); (b) G. Vidali, M.W. Cole and J.R. Klein, Phys. Rev. B28, 3064 (1983).
9.   C.G. Shaw, S.C. Fain, M.D. Chinn and M.F. Toney, Surf. Sci. 97, 128 (1980); C. Bouldin and E.A. Stern, Phys. Rev. B25, 3462 (1982).

CORONA POLARIZATION EFFECTS

ON POLYMER SURFACES

J.N. Marat-Mendes, Maria Raposo and Paulo-António Ribeiro

Universidade Nova de Lisboa[#]
Faculdade Ciências e Tecnologia
Torre, 2825 Monte de Caparica - Portugal

ABSTRACT

Corona polarization in air is an usual method for depositing and injecting charges in polymers for production of electrets. The charges remain either on the surface or near it We have studied the effect of the polarization atmosphere, the polarization current and the time of polarization on the stability of the electrical charge of the electrets. For these studies we have used the technique of Thermally Stimulated Currents (TSC) that gives depolarization currents due to the jumping of the charges from traps, when the sample is heated up at a constant rate.

For polymers (Teflon-FEP) polarization in air for a very short time and small corona currents the TSC measurements present a single peak, denoting the existence of one level of energy traps. For longer times and/or higher corona currents new peaks appear indicating the formation of new traps. Polarizations in O_2, CO_2 and dry air atmosphere, show also modifications on the initial trap distribution, while polarization in N_2 atmosphere does not modify the initial single level energy trap.

We interpreted the results considering that the corona discharge can produce charged particles as well as neutral excited molecules. The interaction of these excited molecules with the polymer surface produce new traps, that can be filled up by the incoming charges, as happens for the preexisting traps of the polymer surface.

INTRODUCTION

Some materials like organic waxes, ceramics, glass-ceramics and polymers are able to store considerable amount of electrical charges for long period of time. A material with this capability is called an electret. The stored charges may be real charges, polarization charges or both [1,2]. Fluorocarbon polymers such as the fluoretlylenepropylene (Teflon-FEP) were found to possess excellent charge storing properties and owing to their mechanical properties find very useful applications in transducers like electret microphones. One of most widely used methods of

---

[#]and Centro de Fisica Molecular das Universidades de Lisboa (INIC)

Equilibrium Structure and Properties of Surfaces and Interfaces
Edited by A. Gonis and G.M. Stocks, Plenum Press, New York, 1992

353

polarizing these polymers is by corona charging in air. With this method the ions produced by a soft electrical discharge are accelerated against the polymer, the deposited charge remaining in traps at the sample surface or near it [3].

The charged material can be characterized with respect to the total charges as well as the charge distribution and charge stability, by measuring the external potential and discharge currents under isothermal or nonisothermal conditions. The thermal stimulated discharge currents technique (TSC) is one of the methods that can be used to investigate the traps distribution and the charges activation energies and relaxation times of the polymer electrets.

On the other hand, corona discharge provides also a good means of treating polymer surfaces, changing their electrical properties [4]. The surface modifications have been explained in therms of the action of the activated neutral molecular species generated by the corona discharge and transported to the sample surface by the corona wind [5]. The neutral activated species (atoms, radicals,...) were identified as vibrational or electronic excited states [6]. The surface modifications have also been associated with bond breaking and the creation of polar groups [7] studies with aluminium samples [8] showed the formation of a layer of a glassy liquid substance on the sample surface, that seems to consist of some strong reactive chemical compounds resembling liquid electrolytes, that eventually solidifies and cracks.Our previous works of negative corona polarization of Teflon FEP in air showed also the formation of a liquid phase on the polymer surface [5].

THE CHARGE INJECTION IN POLYMERS

Electrical charges produced by a soft corona discharge are accelerated against the polymer surface. The charge is transfered to the polymer remaining in traps at surface or near it. Figure 1 shows a schematic energy diagram for a polymer.

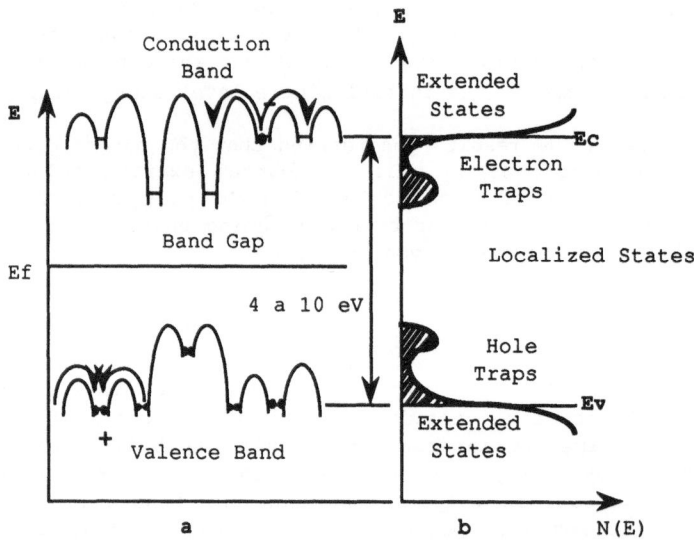

Figure 1 - a) Energy diagram for a polymer. b) Density of states N(E) for a polymer.

Samples of Teflon FEP $12,7\mu m$ thick were used. The polarization experimental set up, refered as corona triode, is shown in figure 2. In order to have a good electrical contact of the polymer with the back electrode the sample was metallized on that surface with a 500Å vacuum evaporated aluminium.

The corona triode is formed by a corona tip (needle) and a flat back electrode, 11mm apart. Between these two electrodes a metallic grid is placed. This grid is a mesh of 750μm spaced brass wires 160μm of diameter. The corona tip is connected to a negative regulated current source. A positive voltage is applied to back electrode in order to controle the electret potential. The grid is connected to ground and it has the effect of introducing a uniform charge distribution over the electret. The corona chamber is connected to a vacuum pump and gas inlet in order to make polarizations in controled atmospheres.

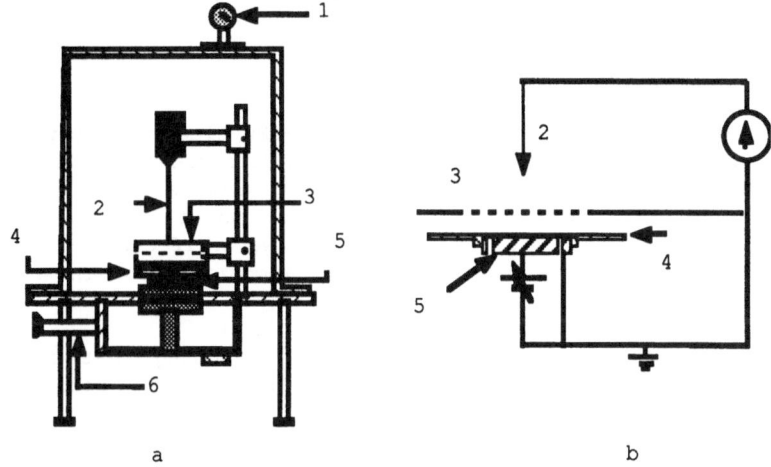

a   b

Figure 2 - a) The Corona Chamber :1-Manometer; 2-Corona tip (sewing needle); 3-Metallic Grid; 4- Electret on a holder; 5-Metallic sample holder (Back electrode); 6- to vacuum pump and gas inlet. b) The Corona triode.

ELECTRET CHARACTERIZATION

The charge uniformity on the electret surface was studied with an electrostatic capacitive probe, scanning the surface radially.

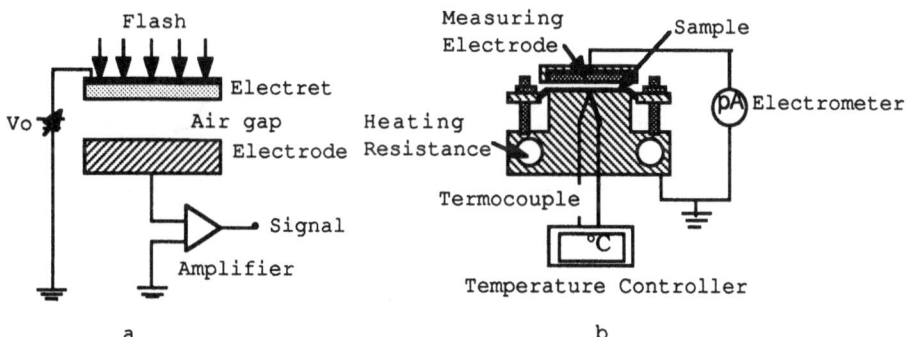

a   b

Figure 3 - a) The heat pulse technique experimental setup. b) The TSC experimental setup.

The charge distribution in the volume of the polymer was studied by applying an heat pulse [9] to the metallized side of the sample which introduces a temperature rise, leading to an expansion profile. Due to that heat diffusion this profile changes in time giving an electrical signal that allows the determination of a charge centroide inside the surface. The set up for the centroide determination is shown in Fig. 3.

The charge stability of the polymer electret was studied using an open circuit thermal stimulated discharge currents technique (TSC) [5]. In this technique the sample, placed in a gap between two flat short-circuits electrodes, is heated at a constant rate ($b = (3,5°C/min)$, figure 4. An electrometer measures the induced current in the upper electrode.

Infrared FT measurements of the electret were also taken in order to study the surface modifications of the polymer after corona treatment.

RESULTS AND DISCUSSION

Experimental results of negative Teflon-FEP electrets show that the time and the corona current of polarization have a great influence on the surface charge distribution. Figure 4 shows a surface potential profile (A) of a Teflon-FEP sample corona charged for 1 minute with a current of 1mA. A good charge uniformity was obtained. However if the sample is left for a long time under the corona discharge a modification in the charge uniformity appears (B).

This modification has been interpreted assuming that during the corona discharge process, besides the formation of ions, neutral activated species are also formed. The ions transfer charge to the polymer. The neutral species are assumed to modify the surface traps. Experiments have been carried on in air in order to study the effect of these neutral activated species. It has been observed that the effect of the surface treatment by the activated species is qualitatively the same before polarization, during polarization of after polarization. Also if during the polarization process the activated species are deviated with an air blower no changes will occur on the charge distribution.

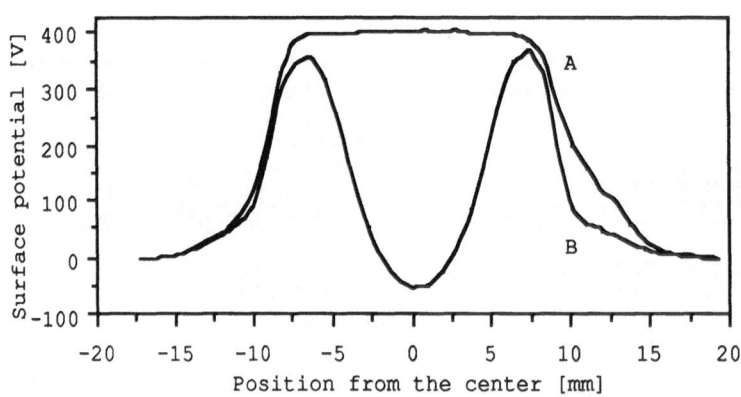

Figure 4 - Surface potential profile of a sample immediately after charging (A) and after 30 min of treatment time (B). Polarization: I_C =1μA; t_C = 1min. Treatment: I_C = 50μA; V_p = 1000V.

Measurements of the charge centroide show however that it is always localised near the polymer surface and does not change with corona treatment. This is an indication that no injection of charge in the bulk of the polymer takes place.

Results of the charge stability obtained with the TSC technique for corona polarization in different atmospheres during 10 seconds and 30 minutes are shown in figures 5a) and 5b) respectively.

It can be seen that for short polarization times (10s), independently of the corona atmosphere, the current curves present only a single peak centered at 210°C. This peak is interpreted assuming the release of charges from the surface traps during the heating process. The existence of a single peak means that a single type of traps are involved.Calculations of the activation energy of the trap level gives a value of 1,9eV and a natural relaxation frequency of $10^{17}s^{-1}$.

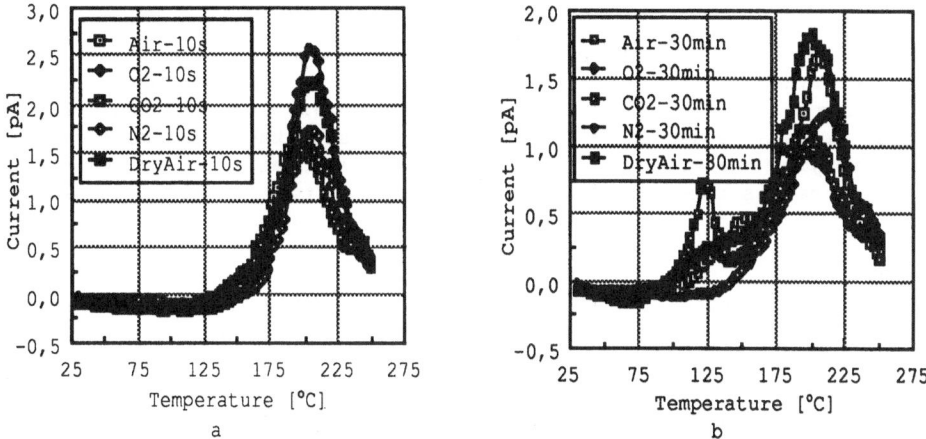

Figure 5 - TSC results of Teflon-FEP samples charged in various atmospheres: a) during 10 seconds. b) during 30 minutes

For longer polarizations times the TSC curves start growing at lower temperatures, what is the indication that charges are being released from new shallow traps. The formation of these traps is associated with the neutral activated species as has been observed with the surface charge measurements, and is even more clear by the results shown in figure 6a), for a sample polarized in CO_2 during 30 minutes (Curve I) and blowing away the excited molecules during the polarization process (Curve II).

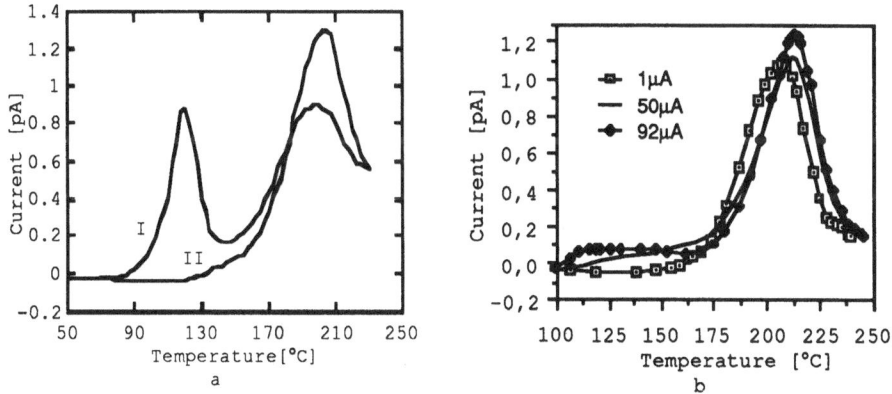

Figure 6 - a) - TSC results of the samples charged during 30 minutes in CO_2 . Curve I - usual polarization. Curve II - blowing a gas flux during the polarization process. b)- TSC measurements for different corona polarizing currents (t_C=3min, V_p=+400).

As has been said for the surface charge measurements, also the TSC curves show that the effect of the neutral activated species is qualitatively the same if these impinge the polymer surface before, during or after polarization.It should be pointed out that no change in the TSC results were observed for the corona polarization in the N_2 atmosphere what is an indication that no activated species are produced in this gas.

As in different gas atmospheres the corona discharge produce different kinds of neutral activated species, the effect of these species in the polymer surface will depend on the corona atmosphere and therefore different kind of shallow traps must appear as can be seen in figure 5b).

Results of polarization for different corona currents were also obtained. Figure 6b) presents some thermograms for samples polarized in air during 3 minutes for different corona polarizing currents. These results are also consistent with the hypothesis of the trap modification of the polymer surface by the neutral actived species. For higher corona currents one has the formation of a higher density of activated species and therefore an increase in the rate of polymer surface trap modification.

The shallow traps produced by the neutral activated species can be removed by cleaning the polymer surface with alcohol or eliminated by heating the sample. Samples polarized during 30 minutes in a CO_2 atmosphere, heated up to 160°C, cooled down to R.T., then polarized for 10 seconds and heated again in order to obtain the thermogram, no charge released from shallow traps appears. Futhermore, it is interesting to mention, that for a long time polarization of the polymer samples in humid air the surface becomes whitish, covered with a fine film of liquid.

Infrared F.T. spectroscopy measurements were carried on this liquid and the results are shown in figure 7. In these results we identified the presence of groups HO^-, CO_3^-, NO_3^-, and NO_2^-.No liquid was observed on the polymer surface for polarizations in other gas atmospheres, besides the humid air.

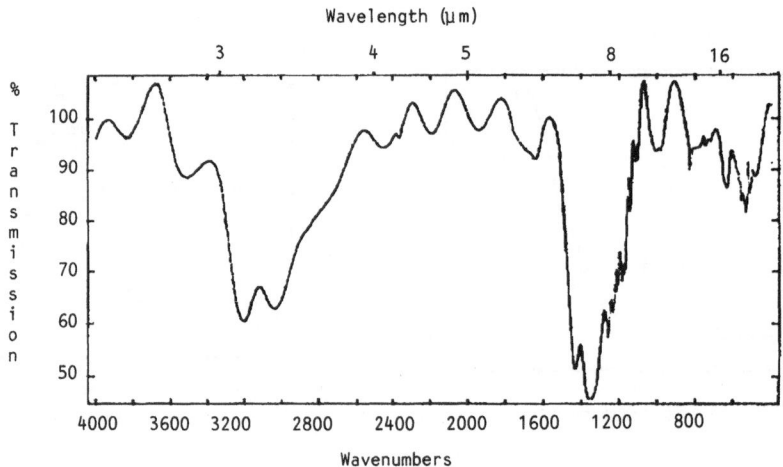

Figure 7 - FT-IR spectrum of the liquid formed on the polymer surface during corona polarization in air.

CONCLUSIONS

Corona discharge in gas atmospheres besides charged particles produces also neutral activated species (atoms, molecules, radicals,...). When the charged particles impinge the polymer surface they transfer the electrical charges to existing trapping sites.

The nature of the activated species depends evidently on the atmosphere. When some of these activated species interact with the polymer surface they create new trapping levels or modify the preexisting ones. The new traps have a much lower activation energy than the preexisting ones. In this way the electrical properties of the polymer are changed by the corona discharge. The result is the production of electrets of poor charge stability and therefore with a shorter life time. Therefore in addition of the scientific interest this is a problem of great technological importance since corona discharge is a promising method for electret production. Further research is needed in order to understand the interaction of the excited molecules with the polymer surface.

ACKNOWLEDGMENTS

The authors are grateful to Prof. C. Maycock for his help in the FT-IR spectroscopy measurements, to the Junta Nacional de Investigação Científica e Tecnológica (JNICT) and Instituto Nacional de Investigação Científica (INIC) for the financial support.

REFERENCES

[1] Sessler, G.M.; Electrets (Topics in Applied Physics: Vol 33). Springer, Berlin, 1980.
[2] Hilczer B. and Malecki J.; Electrets (Studies in Electrical and Electronic Engineering 14); Elsevier, Amesterdan,1986.
[3] von Seggern, H. J. Appl. Phys. 50, 2817 (1979).
[4] Bartnikas, R. Engineering Dielectrics, Vol IIB, ASTM (STP926) Philadelphia, 1987.
[5] Dias C.J.; Marat-Mendes, J.N. and Giacometti J.A. J.Phys. D: Appl. Phys. 22, 663 (1989).
[6] Baum,E.A. Lewis T.J. and Toomar R. J. Phys D, Appl. Phys 11,963 (1977).
[7] Goldman A. and Amouroux J. Electrical Breakdown and Discharge in Gases, Part B, ed E.E. and L.H. Leussen ,New York, Plenum, (1983).
[8] Goldman A. and Sigmond R.S. J. Electrochem. Soc. 132, 2842 (1988).
[9] Collins R.E., J. Appl. Physics, 51 (6), 2973 (1980).

THE REAL-SPACE MULTIPLE-SCATTERING THEORY: A FIRST-PRINCIPLES METHOD FOR THE COMPUTATION OF THE ELECTRONIC STRUCTURE OF EXTENDED DEFECTS

Erik C. Sowa,[*] A. Gonis[*] and X. –G. Zhang[**]

[*] Lawrence Livermore National Laboratory, L356, Livermore, CA 94551
[**] University of Kentucky, Lexington, KY 40506

ABSTRACT

We describe the real-space multiple-scattering theory. This theory is designed explicitly for the treatment of systems with isolated extended defects, such as surfaces and interfaces. It avoids the use of artificial slab or supercell boundary conditions by taking advantage of the principle of removal invariance, which is possessed by systems that have the property of semi-infinite periodicity. We illustrate the method with calculations on grain boundaries in Cu.

The theoretical treatment of pure, periodic crystals is made practical by the use of Bloch's theorem. By introducing reciprocal space and using Fourier transforms, one can reduce the size of the problem from the order of Avogadro's number of atoms to the number of atoms in the unit cell of the crystal in question. Unfortunately, the required translational invariance is broken by the presence of an interface. This issue can be side-stepped, but not completely avoided, by approximating the isolated interface as a periodic array of repeating slabs of finite thickness, i.e. a supercell. However, the results should be checked for dependence on slab thickness on a case-by-case basis. Large repeat distances in the interface plane combined with slabs thick enough to avoid interference can result in a unit cell with too many atoms for current techniques on current computers to handle.

Another way to approximate the interface system is to forgo the use of Bloch's theorem and consider a free cluster of atoms with the geometry of the interface at the center of the cluster. As before, the results must be checked for dependence on cluster size; typically, a cluster large enough to eliminate the effect of the cluster surfaces on the interface at the center of the cluster is too large to fit into current computers.

Many of the techniques developed for perfect crystals make the assumption of translational invariance from the start. The Green-function method as developed by Korringa [1] and Kohn and Rostoker [2] (KKR) is, however, more flexible when it comes to systems with reduced symmetry. When cast in the language of multiple-scattering theory (MST), the Green-function method does not require translational invariance explicitly, although Bloch's theorem may still be used in calculations when appropriate. However, one is still faced with the challenge of reducing the size of the problem without making unwarranted approximations such as artificial periodic boundary conditions.

The answer to that challenge lies in replacing the concept of translational invariance with that of semi-infinite periodicity (SIP), which is defined by the regular repetition along a given direction of a scattering unit, e.g. an atom or plane of atoms, starting from a given point. The property of SIP is possessed, for example, by a bicrystal, since far away from the interface itself, the system becomes bulk-like. Clearly, SIP is a property of a more general class of systems than full translational invariance. The symmetry principle which we shall apply to such systems is *removal invariance*, which states simply that the scattering properties of a system with SIP will not be changed by the addition to, or removal from, its free end of an integral number of scattering units. Using this principle in conjunction with MST, one can solve the Schrödinger equation to obtain the Green function, and hence the charge density and

Equilibrium Structure and Properties of Surfaces and Interfaces
Edited by A. Gonis and G.M. Stocks, Plenum Press, New York, 1992

361

the local electronic density of states (DOS), directly in real space. The key idea is to treat the interface system as a finite cluster in real space, but use removal invariance to renormalize the boundary sites of the cluster to represent the scattering properties of the infinite medium surrounding the cluster. The RSMST [3, 4] converges much more rapidly with respect to cluster size than a free cluster. Unlike Bloch's theorem, one cannot use removal invariance to directly construct the wave function of the extended system. Replacing reciprocal-space methods with real-space methods represents a trading of formal power, in a restricted domain, for wider applicability.

These basic principles may be applied to solving the equations of MST, as is shown in Figure 1. One begins with a semi-infinite half-line, Figure 1(a), and the problem of finding its "renormalized" scattering matrix $T^{(1)}$. The half-line is formed by atoms, each with a "bare" scattering matrix t, which begin at the point 0 and are repeated to the right with a spacing of a between them. Applying the removal invariance principle, one may replace the infinite arrangement of scatterers in Figure 1(a) by the finite arrangement of Figure 1(b). Figure 1(b) represents a self-consistent condition for $T^{(1)}$ which can be solved by iteration or other, more efficient means. The "renormalization" procedure described here may be repeated for more complex arrangements of atoms, such as half-planes and half-spaces, until one has all of the elements necessary to dress the edges of a three-dimensional cluster with the appropriate interface geometry. This is shown schematically in Figure 1(c), where the central atom has been deliberately shown with a different scattering matrix to represent the fact that the RSMST can treat interfaces with relaxations and/or impurities, as long as the system recovers SIP away from the interface region. Note that "dressing" of the cluster is done independently for each part of a system that is characterized by its own SIP, allowing the treatment of interfaces between essentially arbitrary crystal structures.

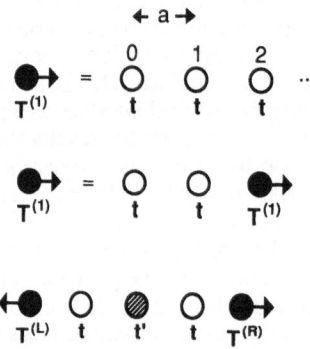

Figure 1. Pictorial representation of the RSMST method, as discussed in the text.

The RSMST has been implemented as a computer code which solves the Schrödinger equation and determines the local DOS at any atom of interest in a cluster. At its present stage of development, our code can be applied to known atomistic configurations with known electronic one-particle potentials. We use the self-consistent potentials for bulk Cu given by Moruzzi, Janak and Williams (MJW) [5]. This is only one part of a complete electronic-structure code – a Poisson equation solver and a means of evaluating electrostatic energies are currently being developed. These issues will be discussed in the next section; here we will show the DOS results. In all cases, we will show the DOS at a site in the interface region for both relaxed and unrelaxed interfaces, where for the purposes of illustration we have taken relaxed coordinates from the results of simulations using the embedded atom method. [6, 7, 8] (The atomic coordinates of the unrelaxed grain boundaries are easily found through an appropriate twisting or tilting of one half of the underlying lattice.) In particular, we have used the code on Cu twist and tilt grain boundaries, and a Nb twist grain boundary. These DOS are compared to the DOS at bulk-like sites far from the interface.

Figure 2 is a guide to the interpretation of the figures which follow; the $\Sigma 5$ (310) tilt grain boundary structure in Cu is shown, and the DOS for the atom at the interface which moves the most upon relaxation is compared to the bulk Cu DOS computed using the same method. The clusters used for all the calculations shown here contain approximately 120 atoms; clusters of this size require the use of a supercomputer with large memory such as a Cray-2. Note that the integral of the DOS yields the charge density, and the integral of the energy-weighted DOS yields the band energy, an important contribution to the total energy of the system.

We first consider the case of the $\Sigma 5$(310) tilt grain boundary in Cu, formed by cutting a Cu crystal to expose its (310) faces, rotating one half with respect to the other by 180° about the normal to the interface plane, and joining them. (The choice of their relative alignment, or the three microscopic degrees of freedom in the interface, is somewhat arbitrary for the unrelaxed configuration.) The unrelaxed configuration chosen is not mirror-symmetric. The primary relaxation predicted by the embedded atom method is that the atom in Figure 2 with the arrow moves to a position on the grain boundary, forming a mirror plane at the interface.

The local DOS at the black atom in the interface plane, which happens to be a coincidence site (a site which is common to the lattices on both sides of the interface), is shown with solid lines in Figure 3. In the unrelaxed configuration, Figure 3(a), there is considerable smearing and loss of structure of the DOS compared to the bulk (dotted lines). This is associated with the lowered symmetry and consequent destruction of the associated Van Hove singularities. One peak in particular, the second from the right, has essentially disappeared – it is associated with the periodicity of the (100) plane and is present in the twist boundary to be shown later. Upon relaxation, some of the peaks recover some amplitude (Figure 3(b)). The local DOS at the moving (white) atom exhibits similar features; the DOS for the relaxed configuration is shown in Figure 2.

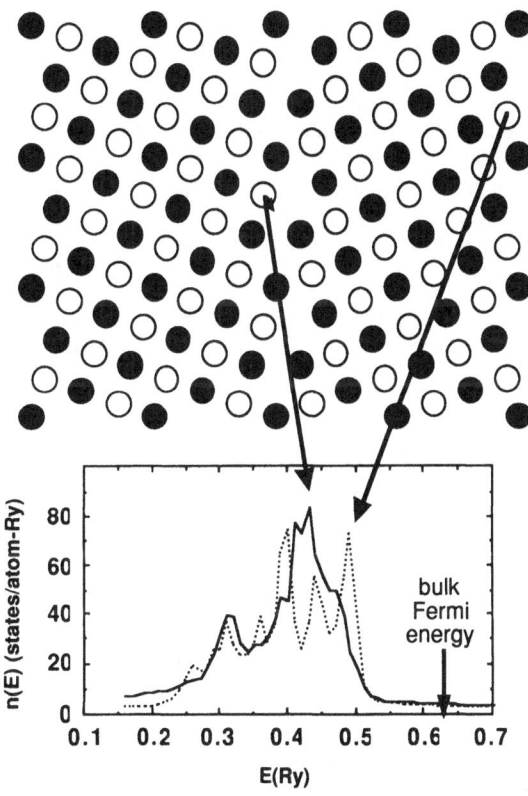

Figure 2. This figure shows the geometry of the $\Sigma 5$ (310) tilt grain boundary in Cu. The shading of the atoms represents location perpendicular to the plane of the figure. The small arrow on the atom at the center of the interface region indicates the direction in which it moves upon relaxation.Two local DOS curves·are shown for this structure – the solid line is the DOS at a site in the interface region, while the dotted line is the DOS at a site far enough away from the interface to be bulk-like. Both curves are computed with the RSMST method.

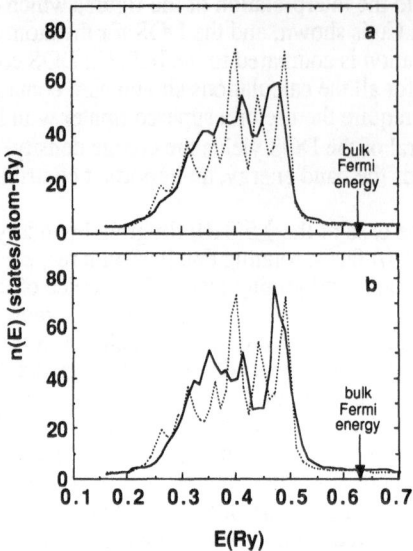

Figure 3. Electronic density of states associated with a coincidence site in a Cu $\Sigma 5$ (310) tilt grain boundary, solid lines, compared with bulk Cu (dotted lines). Panels (a) and (b) are for the unrelaxed and relaxed configurations, respectively.

We now turn to the case of the $\Sigma 5$ (100) 36.9° twist grain boundary. This is formed by cutting a crystal (either fcc or bcc) along the (100) plane and twisting one half 36.9° with respect to the other. Embedded atom method calculations for fcc Cu indicate that the primary effect upon relaxation is for the grains to split apart by approximately 20% of the bulk interplanar spacing. Figure 4 shows the results: we see the same sort of smearing and loss of structure as in the tilt boundary. One primary difference is that the second peak from the right remains present – since this is associated with the (100) plane, and these planes form the interface, this is to be expected. We also note that, after the twist is applied, the atoms are too close together, giving the grain boundary a slightly broader band than the bulk as seen in panel (a). Upon relaxation, this condition is relieved, and in fact panel (b) shows a band narrower than the bulk band. We have also performed this calculation for bcc Nb.

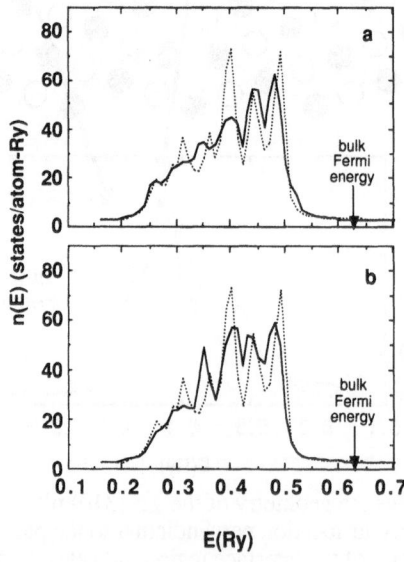

Figure 4. Results analogous to those shown in Figure 3,
but for a coincidence site in a Cu $\Sigma 5$ (100) 36.9° twist grain boundary.

As the calculations just presented illustrate, the RSMST method allows one to solve the Schrödinger equation for materials with extended defects, such as surfaces and interfaces. The solution of the Schrödinger equation is only part of a complete electronic-structure calculation. One of the primary goals of such a complete calculation is to be able to calculate energy differences, which are not only useful for finding the ground state of the system, but can also be used to predict properties such as elastic constants. The standard procedure for calculating total energies is to solve the Hohenberg-Kohn-Sham equations of density-functional theory self-consistently in the local-density approximation. This is a very computationally intensive procedure, especially since at a grain boundary there are many inequivalent atoms. For this reason, it is important that the "innermost loop," i.e. the implementation of removal invariance, be as fast and numerically accurate as possible. Since we performed the calculations shown in this paper, we have rewritten the code with this in mind, and achieved promising results.

Even with the more efficient code, the self-consistency cycle may demand more computer resources than are available. Recently, there have been a number of calculations [9, 10, 11, 12, 13] with an alternative energy functional, the Harris-Foulkes energy, which does not require charge self-consistency.The Harris-Foulkes scheme is also a density-functional method. It assumes that the extended system is composed of weakly interacting charge fragments, which when superimposed are not very different from the self-consistent charge density. In the original treatment by Harris [9], frozen atomic-like charge-density fragments are used to compute the energy of the interacting system, but bulk-like fragments such as those from the MJW [5] calculations which we use in the RSMST may also be used. We have implemented such a scheme for computing total-energy differences in this manner which is currently being tested and refined.

We are also working with a "hybrid" code that uses Bloch waves parallel to, and the RSMST perpendicular to, the interface. This code is based on MacLaren's Layer KKR code[14], but uses removal invariance instead of layer doubling It is charge-self-consistent and can compute total energies, but its applicability is limited to interfaces with two-dimensional periodicity. When this work is completed, we will have a tool which is useful both by itself and for benchmarking the more general RSMST code.

ACKNOWLEDGMENT

This work was performed under the auspices of the Division of Materials Science of the Office of Basic Energy Sciences, U. S. Department of Energy, and the Lawrence Livermore National Laboratory under contract No. W-7405-ENG-48. Computations were performed at the National Energy Research Supercomputing Center.

REFERENCES

1. J. Korringa, Physica **13**, 392 (1947).
2. W. Kohn and N. Rostoker, Phys. Rev. **94**, 1111 (1954).
3. X.–G. Zhang, A. Gonis and J. M. MacLaren, Phys. Rev. B. **40**, 3694 (1989).
4. X.–G. Zhang and A. Gonis, Phys. Rev. Lett. **62**, 1161 (1989).
5. V. L. Moruzzi, J. F. Janak and A. R. Williams, *Calculated Electronic Properties of Metals* (Pergamon Press, New York, NY, 1978).
6. M. S. Daw and M. I. Baskes, Phys. Rev. B **29**, 6443 (1984).
7. S. M. Foiles, M. I. Baskes and M. S. Daw, Phys. Rev. B **33**, 7983 (1986).
8. Erik C. Sowa, A. Gonis, X. –G. Zhang and S. M. Foiles, Phys. Rev. B **40**, 9993 (1989).
9. J. Harris, Phys. Rev. B **31**, 1770 (1985).
10. W. Matthew C. Foulkes and Roger Haydock, Phys. Rev. B **39**, 12520 (1989).
11. F. W. Averill and G. S. Painter, Phys. Rev. B **41**, 10344 (1990).
12. A. T. Paxton, M. Methfessel and H. M. Polatoglou, Phys. Rev. B **41**, 8127 (1990).
13. H. M. Polatoglou and M. Methfessel, Phys. Rev. B **41**, 5898 (1990).
14. J. M. MacLaren, S. Crampin, D. D. Vvedensky, R. C. Albers and J. B. Pendry, Comp. Phys. Comm. **60**, 365 (1990).

INDEX